Tara
Tara Ho
32 Lowe
Dublin 2
Ph: (01) 762121
Fax: (01) 763684

G000147507

PHP   USA   MAG

Tara Consultants Ltd.
Tara House
32 Lower Baggot St.
Dublin 2
Ph (01) 76 21 21
Fax (01) 76 34 34

PHILANTHROPIC GIVING

YALE STUDIES ON NONPROFIT ORGANIZATIONS

Program on Non-Profit Organizations
Institution for Social and Policy Studies
Yale University

JOHN G. SIMON AND PAUL DiMAGGIO, CO-CHAIRMEN
BRADFORD GRAY, DIRECTOR

Philanthropic Giving

Studies in Varieties and Goals

Edited by

RICHARD MAGAT

New York · Oxford
OXFORD UNIVERSITY PRESS
1989

Oxford University Press

Oxford   New York   Toronto
Delhi   Bombay   Calcutta   Madras   Karachi
Petaling Jaya   Singapore   Hong Kong   Tokyo
Nairobi   Dar es Salaam   Cape Town
Melbourne   Auckland

and associated companies in
Berlin   Ibadan

Copyright © 1989 by Yale University

Published by Oxford University Press, Inc.,
200 Madison Avenue, New York, New York 10016

Oxford is a registered trademark of Oxford University Press

All rights reserved. No part of this publication may be reproduced,
stored in a retrieval system, or transmitted, in any form or by any means,
electronic, mechanical, photocopying, recording, or otherwise,
without the prior permission of Oxford University Press.

Library of Congress Cataloging-in-Publication Data

Philanthropic giving : studies in varieties and goals / edited by
  Richard Magat.
      p.   cm. — (Yale studies on nonprofit organizations)
    Bibliography: p.
    Includes index.
    ISBN 0-19-505050-9
    1. Charities—United States—Finance.   2. Corporations, Non-
profit—United States—Finance.   3. Endowments—United States.
4. Philanthropists—United States.   5. Corporations—United States—
Charitable contributions.   6. Humanitarianism—United States—
History.   I. Magat, Richard.   II. Series.
HV91.P55   1989
361.7'0973—dc19   89-2957
                                                            CIP

9 8 7 6 5 4 3 2 1
Printed in the United States of America
on acid-free paper

Series Foreword

This volume and its siblings, comprising the Yale Studies on Nonprofit Organizations, were produced by an interdisciplinary research enterprise, the Program on Non-Profit Organizations, located within Yale University's Institution for Social and Policy Studies.[1] The Program had its origins in a series of discussions initiated by the present author in the mid-1970s while serving as president of Yale. These discussions began with a number of Yale colleagues, especially Professor Charles E. Lindblom, Director of the Institution, and Professor John G. Simon of the Law School faculty. We later enlisted a number of other helpful counselors in and out of academic life.

These conversations reflected widespread agreement that there was a serious and somewhat surprising gap in American scholarship. The United States relies more heavily than any other country on the voluntary nonprofit sector to conduct the nation's social, cultural, and economic business—to bring us into the world, to educate and entertain us, even to bury us. Indeed, the United States can be distinguished from all other societies by virtue of the work load it assigns to its "third sector," as compared to business firms or government agencies. Yet this nonprofit universe had been the least well studied, the least well understood aspect of our national life. And the nonprofit institutions themselves were lacking any connective theory of their governance and function. As just one result, public and private bodies were forced to make policy and management decisions, large and small, affecting the nonprofit sector from a position of relative ignorance.

To redress this startling imbalance, and with the initial assistance of the

[1] The sharp-eyed editors at Oxford University Press requested that we explain the presence of an intrusive hyphen in the word "Non-Profit" in the Program's title, and suggested that the explanation might be of interest to this volume's readers. The explanation is simple: At the Program's inception, it adopted the convention, in wider currency than it is today but even at that time incorrect, of hyphenating "non-profit." Since then the Program has mended its way wherever the term "nonprofit" is not used as part of the Program's title. But in the Program's title, for reasons both sentimental and pragmatic, the hyphen remains, as a kind of trademark.

late John D. Rockefeller III (soon joined by a few foundation donors), the Program on Non-Profit Organizations was launched in 1977. It seeks to achieve three principal goals:

1. to build a subtantial body of information, analysis, and theory relating to non-profit organizations;
2. to enlist the energies and enthusiasms of the scholarly community in research and teaching related to the world of nonprofit organizations; and
3. to assist decision makers, in and out of the voluntary sector, to address major policy and management dilemmas confronting the sector.

Toward the first and second of these goals the Program has employed a range of strategies: research grants to senior and junior scholars at Yale and at forty-one other institutions; provision of space and amenities to visiting scholars pursuing their research in the Program's offices; supervision of graduate and professional students working on topics germane to the Program's mission; and a summer graduate fellowship program for students from universities around the country.

The Program's participants represent a wide spectrum of academic disciplines—the social sciences, the humanities, law, medicine, and management. Moreover, they have used a variety of research strategies, ranging from theoretical economic modeling to field studies in African villages. These efforts, supported by fifty foundation, corporate, government, and individual donors to the Program, have gradually generated a mountain of research on virtually every nonprofit species—for example, day-care centers and private foundations, symphony orchestras and wildlife advocacy groups—and on voluntary institutions in twenty other countries. At this writing the Program has published 100 working papers and has sponsored, in whole or in part, research resulting in no fewer than 175 journal articles and book chapters. Thirty-two books have been either published or accepted for publication. Moreover, as the work has progressed and as Program-affiliated scholars (of whom, by now, there have been approximately 150) establish links to one another and to students of the nonprofit sector not associated with the Program, previously isolated researchers are forging themselves into an impressive and lively international network.

The Program has approached the third goal, that of assisting those who confront policy and management dilemmas, in many ways. Researchers have tried to design their projects in a way that would bring these dilemmas to the fore. Program participants have met with literally hundreds of nonprofit organizations, either individually or at conferences, to present and discuss the implications of research being conducted by the Program. Data and analyses have been presented to federal, state, and local legislative and executive branch officials and to journalists from print and electronic media throughout the United States to assist them in their efforts to learn more about the third sector and the problems it faces.

Crucial to the accomplishment of all three goals is the wide sharing of the Program's intellectual output not only with academicians but also with nonprofit practitioners and policy makers. This dissemination task has been an increasing preoccupation of the Program in recent years. More vigorous promotion of its working paper series, cooperation with a variety of non-academic organizations, the publication of a handbook of research on nonprofit organizations, and the establishment of a newsletter (published with increasing regularity for a broad and predominantly nonacademic list of subscribers) have all helped to disseminate the Program's research results.

These efforts, however, needed supplementation. Thus, the Program's working papers, although circulated relatively widely, have been for the most part drafts rather than finished papers, produced in a humble format that renders them unsuitable for the relative immortality of library shelves. Moreover, many of the publications resulting from the Program's work have never found their way into working paper form. Indeed, the multi-disciplinary products of Program-sponsored research have displayed a disconcerting tendency upon publication to fly off to separate disciplinary corners of the scholarly globe, unlikely to be reassembled by any but the most dogged, diligent denizens of the most comprehensive of university libraries.

Sensitive to these problems, the Lilly Endowment made a generous grant to the Program to enable it to overcome this tendency toward centrifugality. The Yale Studies on Nonprofit Organizations represent a particularly important part of this endeavor. Each book features the work of scholars from several disciplines. Each contains a variety of papers, many unpublished, others available only in small-circulation specialized periodicals, on a theme of general interest to readers in many regions of the nonprofit universe. Most of these papers are products of Program-sponsored research, although each volume contains a few other contributions selected in the interest of thematic consistency and breadth.

Thus, the present volume, edited by Richard Magat, Visiting Fellow at the Foundation Center and at the Program on Non-Profit Organizations, addresses the indispensable prerequisite to nearly all of the efforts of non-profit institutions: philanthropic giving, from individuals, corporations, and private foundations. The papers assembled here explore the historical and individual origins of giving, the various avenues by which contributions reach their intended beneficiaries, and the consequence of the philanthropic tradition, so strong in the United States and gaining force in much of the rest of the world.

As the reader will already have observed, I do not write this foreword as a stranger. I am very much a member of the family, someone who was present at the creation of the Program on Non-Profit Organizations and continues to chair its Advisory Committee, and who also serves Oxford as Master of University College. What this extended family is doing to advance knowledge about the third sector is a source of considerable satisfaction. From its birth at a luncheon chat more than a decade ago, the Program on Non-Profit Organizations has occupied an increasingly important role as the leading academic center for research on voluntary institutions both in America and abroad. And now the publication by Oxford University Press of this volume and the other Yale Studies on Nonprofit Organizations enlarges the reach of the Yale Program by making its research more widely available within the scholarly community and to the larger world beyond.

London   Kingman Brewster[2]

---

[2] This foreword was written before Mr. Brewster's death late in 1988. His continuing interest in the Program on Non-Profit Organizations, of which he was an early advocate, was evident in the foreword the other volumes in this series have carried.—R.M.

Preface

Although not all nonprofit institutions are the object of giving as commonly understood, the vast array of educational, cultural, scientific, and community organizations rely altogether, or in great part, on donations from a variety of private sources. Furthermore, many nonprofit organizations arose from the concern of individual donors, although many others were associations of the type that so impressed Tocqueville, where money was not a critical element, at least not at the outset.

The purpose of this book, reflected in the subtitle, is to illuminate the variety of methods of, and influences on, the main sources of private giving (both individual and instutitional) and to illustrate the venture entailed in many acts of giving, though much giving is traditional and routine. Thus, *Philanthropic Giving: Studies in Varieties and Goals* connects with the volumes in this series that deal with the objects of giving—the arts, private education, and community organizations—and with those that discuss the economics and management of nonprofit organizations—subjects in which funding plays a large, if not dominant, role. Its ties to the volume on international perspectives on nonprofit organizations are weaker, although private giving affects such groups more than is widely supposed.

Variety also characterizes the types of papers in this book and the interdisciplinary composition of the authors. Some of the contributions are technical, others historical analyses. Half of them are direct products of research sponsored by the Yale Program on Non-Profit Organizations. Several others represent research presented in preliminary form at the Program's seminars. Three chapters were commissioned especially for this book. The remainder have been adopted by the authors from a book and from journals or conferences of quite limited circulation or attendance.

In addition to scholars, the book should, I hope, be valuable to the growing number of undergraduate and graduate students who are taking an interest in the nonprofit sector, to professional fund raisers, trustees of nonprofit organizations, legislators and public officials whose policies affect private giving, and to those who give—especially wealthy women and men and their advisors and the staffs of corporate and foundation sources of funds. And since the largest number of philanthropic givers are individuals who are not wealthy, the book concerns the public at large.

ACKNOWLEDGMENTS

My first and greatest debt is to John Simon and Paul DiMaggio for the privilege of editing this book. At various stages in its planning and execution they were patient, encouraging, and generous with time from daunting schedules. To DiMaggio, an extra measure of gratitude for a close and helpful editing of my Introduction.

It hardly needs saying that I am indebted to the contributors, but special mention is due Richard Steinberg, who made suggestions for several papers early in the planning of the book. For administrative backup, I extend warm thanks to Ella Sandor and Melissa Middleton of the Yale staff.

My wife Gloria undertook the editing of references and notes for each chapter. For that, and for the intangible, unspoken support that ensures a nurturing environment for prolonged work, I am deeply grateful. My appreciation goes to my daughter, Claudia Keenan, for taking on crisis typing at just the wrong time for her, and to her husband Jeffrey, for mundane but well-timed logistical support.

A long, solitary weekend at the Blue Mountain Center enabled me to concentrate on the final stage of this work. I am thankful to Kate Chieco and Harriet Barlow for arranging the stay, and to Kaye Burnett and the rest of the staff. The finishing touches to the book were placed at the Foundation Center, where I am presently a visitor. All there, from Tom Buckman, the president, to the gracious and helpful library staff, have been most hospitable, and it hardly needs saying that the resources of the center about grantmakers and grantmaking are matchless.

The Lilly Endowment deserves recognition for its grant to the Program on Non-Profit Organizations in support of the publication of this series. It is one of a handful of foundations that work actively to advance understanding of philanthropy. Another is the Edward W. Hazen Foundation, to whose Trustees I am most appreciative for encouraging me to accept this and other extracurricular assignments dealing with philanthropy along with my executive duties there. And although she did not work directly on the book, I owe much to Alyson Tufts, my former assistant at Hazen, whose dedication and endless supply of tolerance and cheerfulness made her a good friend as well as a valued colleague.

New York   R.M.
November 1988

Contents

Contributors

Gerald E. Auten is a professor of economics at Bowling Green State University. He has been a visiting economist in the Office of Tax Analysis, U.S. Department of the Treasury. He has written on various aspects of the economic effects of federal tax policy, focusing primarily on charitable giving and capital gains.

Elizabeth T. Boris is vice-president for research and planning of the Council on Foundations. Author of many articles and reports and coauthor of a book on foundation careers, Dr. Boris is currently an associate professor adjunct at the University of Colorado Graduate School of Public Affairs in the nonprofit management program. She also teaches a course on foundations at George Washington University. Earlier she directed a program on Business Ethics and Social Responsibility for the Council of Better Business Bureaus.

Robert H. Bremner is professor emeritus of history at Ohio State University. He has written on American attitudes toward poverty, philanthropy, and social welfare, and public policy toward children and youth. A new edition of his book, *American Philanthropy*, appeared in 1988 (University of Chicago Press).

Emmett D. Carson is a research associate at the Joint Center for Political Studies, where he directs a national study on black philanthropy. Dr. Carson is also a lecturer at the University of Maryland.

Charles T. Clotfelter is professor of public policy studies and economics at Duke University. His research interests include tax policy and charitable behavior. He is currently director of Duke's Center for the Study of Philanthropy and Voluntarism.

Joseph Galaskiewicz is professor of sociology at the University of Minnesota. He is the author of *The Social Organization of an Urban Grants Economy: A Study of Business Philanthropy and Nonprofit Organizations*. He is currently doing research on the strategic response of nonprofit organizations to changes in their funding environment, with Wolfgang Bielefeld.

Peter Dobkin Hall is a historian, writer and consultant. Since 1979, he has been affiliated with Yale's Program on Non-Profit Organizations. He is currently a lecturer on nonprofit organizations at the School of Organization and Management, Yale University.

J. Craig Jenkins, associate professor of sociology at Ohio State University and affiliated with the Yale Program on Non-Profit Organizations, is author of *The Politics of Insurgency: The Farm Worker Movement of the 1960s* and numerous articles on social movement theory and the politics of philanthropy. He is currently completing a book *The Patrons of Social Reform,* extending the analysis included here.

Edwin B. Knauft is executive vice-president of Independent Sector, a national nonprofit organization that encourages giving, volunteering, and not-for-profit initiatives. He was formerly vice-president of corporate social responsibility at Aetna Life and Casualty. The corporate giving research study was completed while Dr. Knauft was a visiting fellow at Yale's Program on Non-Profit Organizations.

Kathleen D. McCarthy, director of the Center for the Study of Philanthropy at the Graduate Center of the City University of New York, has worked for an array of private, corporate and government donors, including the Rockefeller Foundation, the National Endowment for the Humanities and the Metropolitan Life Foundation. She is the author of *Noblesse Oblige: Charity and Cultural Philanthropy in Chicago, 1849–1929* and many articles on local, national, and international philanthropy.

Richard Magat, visiting fellow at the Foundation Center and at the Yale Program on Non-Profit Organizations, formerly was president of the Edward W. Hazen Foundation and director of communications for the Ford Foundation. He is the author of *The Ford Foundation At Work: Philanthropic Choices, Methods, and Styles,* book editor of *Foundation News* and editor of the Transaction Publishers series, Classics in Philanthropy and Society.

Teresa Odendahl is an anthropologist and writer. She was project manager of the Foundation Project, sponsored by the Yale Program on Non-Profit Organizations and the Council on Foundations. She has written widely on foundations and the charitable attitudes of the wealthy.

Francie Ostrower is a doctoral candidate in sociology at Yale University. She received a B.A. with honors from Swarthmore College, with a major in philosophy, and an M.A. and M.Phil. in sociology from Yale. She is currently working on her dissertation on the philanthropic attitudes and practices of wealthy New Yorkers.

Robert L. Payton, former president of Exxon Education Foundation, is director of the Center on Philanthropy at Indiana University.

Gabriel Rudney is an economist and research affiliate at Yale University's Institution for Social and Policy Studies (Program on Non-Profit Organizations). He was research director of the Commission on Private Philanthropy and Public Needs (Filer Commission) and served in the U.S. Treasury Department on tax policy research.

Jerald Schiff is assistant professor of economics at Tulane University and has been a visiting economist at the Department of the Treasury. He is the author of a number of articles as well as a forthcoming book on charitable giving and the behavior of private nonprofit organizations.

John H. Stanfield II is the Ethel and Edwin Cummings Professor of American Studies and Sociology, College of William and Mary. He is author of *Philanthropy and Jim Crow in American Social Science* and editor of Charles S. Johnson's *Bitter Canaan* and *Research in Social Policy: Historical and Contemporary Perspectives, Volume I.* Dr. Stanfield specializes in the organizational history of race philanthropy and race public policy.

Richard Steinberg is an assistant professor and director of undergraduate studies for the Department of Economics, Virginia Polytechnic Institute and State University. He serves as associate editor of *Evaluation Review,* consulting editor for *Journal of Voluntary Action Research,* an affiliate of the Center for Volunteer Development, and a research consultant for Yale's Program on Non-Profit Organizations. He specializes in the economics of the nonprofit sector.

Arthur H. White is president and CEO of the WSY Consulting Group, Inc., a management consulting firm that he heads along with Daniel Yankelovich and Florence Skelly. Currently, he serves as national treasurer of the Reading Is Fundamental program of the Smithsonian Institution, chairman of the Connecticut Housing Finance Authority, and vice-chairman of the Institute for Educational Leadership.

PHILANTHROPIC GIVING

Introduction

RICHARD MAGAT

Nearly forty years ago, the Russell Sage Foundation published a volume with the same title as this one. Unlike this volume, F. Emerson Andrews's *Philanthropic Giving* (1950) was not a compendium of scholarly research. It was an informed overview by one man, who for a generation was one of the few students of philanthropy. Scholarly research on philanthropic giving was then virtually nonexistent. Andrews's volume carries no bibliography, and Donald Young, director of the Russell Sage Foundation, observed in his foreword,

Current information about giving is fragmentary and biased. Few objective students have made philanthropy their central concern, although economists, social historians, specialists in the social behavior discipline, and others should have found it an exciting field for study. Actually, the bulk of what has been written on the subject consists of essays explaining personal points of view about giving, exhortations to generosity . . . and overly sympathetic accounts of . . . philanthropic individuals and agencies. Critical analyses of the relative needs and opportunities for philanthropy, of the relative advantages of the various ways of giving, and of the relative effectiveness of the many preferred means for accomplishing any philanthropic purpose are indeed rare.[1]

Research on philanthropic giving has surged in the last decade or so, although there is still plenty to catch up with—a plenty that stretches back, as some of the authors in this volume remind us, almost literally to the organization of human societies. The rich and future lode includes not only more insight into the philosophical and anthropological roots of altruism and other motivations for sharing with one's fellow man, but also the development of modern forms we take for granted but know relatively little about and understand even less.

Although several chapters touch on current matters, the reader will not find here a detailed report on the state of the art. Like many other subjects, philanthropic giving is a moving target, but one may take comfort in the fact that there are antecedents to current novelties; for example, such highly dramatic, literally theatrical wrinkles as the massive fund-raising concerts devoted to starvation abroad, poverty, and the plight of dispossessed American farmers. Thus, the star-studded rallies of two world wars, the March of Dimes, and special telethons and walkathons prefigured Live Aid, with its celebrity performers.

On a less theatrical level, note has been taken (see Teresa Odendahl's chapter, for example) of the tendency of the second and third generations of heirs to great fortunes to take new philanthropic paths, often in support of ultraliberal, out-of-the-mainstream causes. A movement in the 1970s has led to the organization of some thirty "alternative" or "progressive" foundations set up by young heirs and heiresses, and Inherited Wealth Conferences are held to draw more wealthy young people into alternative philanthropy.[2] Here, too, history provides precedents. Thus, in 1922, a twenty-one-year old Harvard undergraduate, Charles Garland, who had declined to use his share of the estate of his millionaire father for himself, gave $800,000 to start an unusual foundation. Known later as the American Fund for Public Service, it supported the National Association for the Advancement of Colored People, the Rand School of Social Science, the League for Industrial Democracy, the American Birth Control League, the magazine New Masses, the Sacco-Vanzetti Defense League, and so forth.[3]

Precedent or no, however, bold new philanthropic gestures surface now and then to capture the public imagination—and to furrow some professional brows. Witness Eugene Lang's "I Have a Dream" program to guarantee a college education for ghetto youngsters who keep their shoulders to the academic wheel. Such colorful one-person ventures brighten a seemingly gray landscape of institutionalized, professionalized philanthropy.[4]

Individual philanthropy is full of examples of idiosyncratic giving, and sometimes the same individual embodies both sophisticated giving to vexing social problems and great dollops of funds to traditional institutions. The wealthy Chicago business executive Irving B. Harris, who helped create the Ounce of Prevention Fund, which seeks to prevent family problems that can lead to child abuse and neglect, and Family Focus, which sets up model neighborhood centers that provide family support for drop-in clients, is also capable of giving $7 million in one gift to establish a graduate school for public policy studies at the University of Chicago.

If this volume is not mainly an analysis of the very newest philanthropic events and developments, it does encompass important changes in old and middle-aged forms, most of which are by no means immune to change, voluntary or otherwise. Thus, in its second half-century, the United Way must reckon with company it chose at one time not to invite to dinner and with which it is still not quite comfortable: special funds for the arts, women's funds, combined funds devoted to minority interests, and so forth. In some cases, such as the attempts of minority and environmental activist groups to gain entry into United Way coffers, open warfare has broken out.

More placid but nonetheless significant are developments in another philanthropic innovation of World War I vintage: the community foundation.

Peter D. Hall provides an interesting history and an analysis of why they are an increasingly favored vehicle for wealthy donors. Their assets have tripled since 1975.

Finally, perhaps, the most ambitious organized effort to produce a quantum rise in the level of philanthropic giving was launched as this volume was being assembled. That is the campaign of Independent Sector, itself a relative newcomer, to double charitable giving and increase volunteer activity 50 percent by 1991.

Even if successful, why all the fuss, when all philanthropic giving is just 2 percent of national income, compared to 79 percent for the commercial sector? Despite the vast disparity in scale, private philanthropy is powerful because it "provides the absolutely vital elements of flexibility, innovation, creativity, and the capacity for criticism and reform."[5] Of course, not all philanthropic giving meets these vaunted characteristics; much of it is directed toward ordinary, if important, services, and some of it is humdrum, even to the point of sustaining organizations that in any objective sense may have outlived their social usefulness (except to the few who enjoy working for them as volunteers or those who depend on them for jobs). Significant or not, however, the activities supported by philanthropic giving spring (or mutate) from a venerable, continuing tradition. That tradition is addressed in the opening chapters of this volume. Succeeding chapters go on to analyze established varieties of philanthropic experience, and several point to unfinished research tasks.

The beginning of the sharp rise in research on philanthropic giving in the mid-1960s coincides roughly with a burgeoning of nonprofit activity. Yet research has continued apace even as growth in the nonprofit sector has tapered off in the last several years. But as other volumes in this series have noted, nonprofit activity is not entirely a product of private charitable giving, especially since the 1960s. The reason, of course, is the extent to which vast sums of government money for social welfare and artistic endeavors flowed through nonprofit organizations and then on (if not entirely) to the intended beneficiaries.

As it turns out, charitable giving has risen steadily despite wide swings in the government's social and cultural role. Thus, philanthropic giving in 1964 (on the eve of President Johnson's War on Poverty) was $13.7 billion. A decade later, after adjusting for inflation, it had risen 20 percent; in 1981, the first full year of the Reagan Administration, it had risen another 19 percent in inflation-adjusted dollars. And in 1987, Americans donated an estimated $94 billion, which reflected an increase of 91 percent over the amount given in 1964, after taking inflation into account. Patterns among the main giving sectors are given in Table 1. Between 80 and 90 percent of all giving comes from individuals, and about half from families with incomes under $30,000. Twenty-three million Americans contribute 5 percent or more of their income to charitable activities.

The rising figures for individual philanthropy appear to belie the conventional wisdom that the post-Vietnam era is marked by an individual as

Table 1. Current and Constant Dollars (figures in $ billions)[a]

	Individuals		Foundations		Corporations		Bequests		Totals		
	Curr $	Const $[a]	Curr $	Const $[a]	Curr $	Const $[a]	Curr $	Const $[a]	Curr $	Const $[a]	% Change (Constant $)
1964	11.19	34.0	0.83	2.52	0.73	2.22	0.95	2.89	13.70	41.64	
1974	21.60	40.0	2.11	3.11	1.2	2.22	2.07	3.83	26.98	49.96	20
1981	46.42	47.38	3.07	3.27	2.51	2.67	3.49	3.71	55.94	59.51	19
1987	76.82	65.26	6.38	5.42	4.50	3.82	5.98	5.08	93.68	79.59	34

SOURCE: *Giving USA* (annual). New York: AAFRC TRUST for Philanthropy.
[a]Deflated by GNP Implicit Price Deflator, 1982 = 100

well as institutional retreat from social concern, a kind of privatization of the spirit. Robert Payton, in the opening chapter, espouses his belief that philanthropy lost its mooring to values when it detached from religion.

The reception given to critiques of this period of presumed consuming self-interest varies. President Carter learned that sounding a trumpet against self-indulgence is not universally welcome; but similar analyses, even at the academic level, win wide attention. The thesis of the widely noted *Habits of the Heart*[6] is that American individualism which, more so than equality, has "marched inexorably through our history," may have grown cancerous, "that it may be destroying those social integuments that Tocqueville saw as moderating its more destructive potentialities, that it may be threatening the survival of freedom itself."

On the other hand, as Barbara Ehrenreich reminds us,[7] Bellah et al are carrying forward a tradition of this sort of critique stretching back at least to the Eisenhower years (William Whyte, David Reisman) and including, more recently, Daniel Bell and Christopher Lasch. Because it is impossible to arrive at any consensual judgment of the American character, homely data such as dollars contributed and time donated may be our best indexes.

Which raises a point about omissions in this volume. The volume's coverage reflects a literal reading of its title—it is confined to financial resources involved in giving. Since two related, important activities are not covered, a few words about them are in order. One is the donation of time to the vast array of American nonprofit organizations: from service on the boards of museums and colleges to a helping hand in hospitals and schools, to running Boy and Girl Scout troops. The dollar value of volunteer time has been calculated—$100 billion in 1985—of which $19 billion is said to represent time spent by teenage volunteers.[8] Not all beneficiaries are nonprofit organizations. An estimated $20 billion in volunteer time was spent in public institutions (schools, prisons, service agencies).

Unmeasurable, of course, is the value received by the beneficiaries of volunteer activity and the social, psychic, and moral benefits to the volunteers themselves. Nearly half of the American population is said to have engaged in volunteer activity in 1985, contributing an average of 3.5 hours a week, or a total of 16.1 billion hours. The volunteering habit cuts across all age and income groups. Fifty-one percent of women were volunteers, compared to 45 percent of men. Volunteering persists with age: 44 percent between 50 and 64 years old and 43 percent of persons between 65 and 74 years old are giving volunteer time.[9]

But demographic analysis poses a concern for the future of the philanthropic impulse. A 1985 survey (by Independent Sector) showed a decline since 1980 in donation of volunteer time among persons 18 to 24 years of age (11 percent) and single persons (19 percent).

The donation of time is intertwined with cash giving. A higher proportion of volunteers donate funds to charitable organizations than persons who do not volunteer: 92 percent compared to 72 percent. The correlation is stronger in particular areas of nonprofit activities. Nine out of ten persons who volunteered for religious institutions also contributed money to them, compared to 42 percent of nonvolunteers. At least one out of two persons who reported volunteering in such other areas as arts and culture, civic, social, and fraternal groups; education; and recreation also contributed in those areas.[10]

The second important and related activity that lies outside the scope of this volume is fund raising, though a few chapters touch on it, e.g., in chapters on the effect on philanthropic giving of government tax policy. Tax policy has played a role since the first national income tax was passed in 1913. At first the effect was negligible. Today, arcane changes in Ways and Means Committee markups of pending legislation bring chills or sighs of relief not only to trade lobbyists but also to fund raisers in such major sectors of the nonprofit world as universities and museums.

Of course, not everyone bemoans the reductions of tax incentives for charitable giving. Commenting on the 1986 Tax Reform Act disincentives, Gary Becker of the University of Chicago argues that general taxpayers should not be required to subsidize groups (e.g., colleges and universities, museums, symphonies, and other artistic and civic groups) that mainly benefit the wealthy. "Little of so-called charitable contributions goes to poor people," he notes. "Nor should subsidies through the tax system help support churches, synagogues, and other religious bodies in the face of the American tradition of separating church and state."[11] Given a recent survey that indicates that 46 percent of contributed funds ($19.1 billion) are used by congregations for community services, not worship or religious education, the latter notion is arguable.[12]

Connections between tax policy and giving (at least by the wealthy) are quantifiable, as several chapters clearly suggest, but proof of other causal relations is notoriously fragile and sparse. Take the great increase in giving by individuals. Did they give more because the tin cup rattled more alluringly, or did fund raisers escalate their appeals because more philanthropic dollars were available? Was it a reflection of postwar prosperity, a reaction to the growth of government, or a result of increasingly elaborate and sophisticated fund raising, especially that aimed at large givers? In the 1970s and 1980s, for example, colleges and universities have mounted capital compaigns in the several hundred million dollar range—Stanford's is over $1 billion. Public institutions of higher education have also entered the derby for substantial private funds: e.g., Ohio State, $750 million. Also, any number of nonprofit organizations with budgets under $1 million have audaciously, and in some cases successfully, tried to build endowments.

In any event, fund raising itself has come to be a major subindustry of the nonprofit world. Curiously, it consists to a great extent of for-profit firms and individual entrepreneur-consultants.[13] A few measures of the advance of the profession will suffice. On-staff fund raisers' titles, "development officer" or "vice-president for institutional resources," indicate their enhanced status as part of the policy-making circle.

The membership of the American Association of Fund-Raising Counsel, the elite circle of large firms, has grown from thirteen when it was established in 1935 to thirty in 1987. In some universities the fund-raising departments have staffs over 250.

The expansion of professional fund raising reflects not only the growth of the economy but also the pressures in an era of government cutbacks. Government had taken on a dominant funding role in areas once ruled by the private sector and also relied on (and therefore subsidized) nonprofit organizations to carry out its programs. The cutbacks brought pressure on nonprofit organizations to fill cultural and social gaps at the very time public subsidies of these organizations were being curtailed. Although those speaking for the nonprofit organizations were reasonably successful in demonstrating the implausibility of President Reagan's suggestion that private initiatives could compensate for lost government funds, the sector became aware of the need to alter traditional habits. As John Gardner, one of the grand old men of philanthropy and the voluntary sector, advised:

Society at large and donors in particular are going to expect institutions to be more tough-minded, less indulgent, in facing the question of whether they are effective users of the resources they have. [Nongovernment institutions] are going to have to put more imagination into diversifying their sources of support. . . . Most fundraisers see that they are going to have to be experts on the possibilities of direct mail, membership, user's fees, unrelated business income, and the obtaining of funds from government—federal, state, and local—including the still-to-be developed arts of getting one's share of block grants.[14]

Although no chapter is devoted to the motivations for giving, it is touched upon in several: the historical and philosophical chapters, especially Payton, McCarthy, and Hall; those on the relation between federal tax policy and giving, Auten and Rudney, Clotfelter, Steinberg, Schiff, and White; and those on the formation of foundations, Odendahl and Hall. There is an extensive literature on motivation, although a good deal of the research deals with the altruistic behavior of children and college undergraduates in laboratory settings. But some grand empirical work has been done, such as Richard M. Titmus's *The Gift Relationship*. And Dennis R. Young's *If Not for Profit, for What?*, based on research conducted under the auspices of the Yale Program on Non-Profit Organizations, deals with factors that motivate enterprising men and women to pursue ventures in nonprofit organizations. On the basis of his case studies and the literature, Young extrapolates from entrepreneurial motivations to the behavior of nonprofit organizations and the sector as a whole.

An extensive guide to research on philanthropic motivation and altruism will be available with the forthcoming publication by Independent Sector of a major bibliography. It will also deal with voluntarism, ethics, and charity. The bibliography will also identify gaps and opportunities in research, such as longitudinal studies, studies conducted in realistic rather than laboratory contexts, and cross-disciplinary approaches. This work will be a welcome companion to the annotated bibliography, *Philanthropy and Voluntarism,* by Daphne Niobe Layton (The Foundation Center, 1987), which references 1614 books and articles, of which 244 works of scholarly quality or other exceptional value are annotated. That work was commissioned by the Trust for Philanthropy, established by the American Association of Fund-Raising Counsel and the Association of American Colleges.

Insight into philanthropic giving can be gleaned from biography as well as the type of research covered in this volume. The range is very wide: from Allan Nevins' *John D. Rockefeller* (1959) and William Greenleaf's *From These Beginnings: The Early Philanthropies of Henry and Edsel Frod, 1911–1936* (1964) to popular works like Joseph C. Goulden's *The Money Givers* (1971) and Ben Whitaker's *The Philanthropoids: Foundations and Society* (1974).

A most interesting biographical undertaking, completed and in the process of publication, is James A. Joseph's study of ten men and women, in eight foreign countries, who used their private wealth for public purposes. The philanthropists range from Lord Nuffield to Calouste Gulbenkian, Adnan Khashoggi, and others even less well-known to most Americans. Joseph, president of the Council on Foundations, is less interested in the causes to which these philanthropists gave than to their "motives, methods, and message." Their motives range from noble to self-serving, says Joseph, and he points to four stages of philanthropic consciousness: (1) the development of an altruistic personality, especially nurtured by family culture; (2) the reinforcement of compassionate values, by religion, literature, or indigenous mythology; (3) the experience of community, a social responsibility growing out of a sense of interdependence, even of strangers, and a move from passive altruism to active engagement; and (4) the dynamics of public life, a recognition that individual beneficence must be supplemented by sound government and just and humane laws and institutions.

This volume ranges from the historical and philosophical context of philanthropic giving to specific current aspects. Among the authors, economists are the largest contingent; others are historians, political scientists, and anthropologists. More than in other volumes in this series, the authors include practitioners: a notable figure in survey research (White), a former corporate contributions manager for one of the country's largest insurance companies (Knauft), and a thoughtful analyst who has spent a lifetime on both sides of the table, as a college president and as head of one of our largest corporate foundations (Payton). Some of the scholars themselves are no strangers to practice, e.g., Boris is an officer of the Council on Foundations.

Part I deals with the roots of philanthropic giving, ancient and modern. Part II covers patterns of individual giving in general and among the wealthy and the black community in particular. Part III deals with the role of government policy, taxation in particular. Parts IV and V discuss historical, managerial, and other aspects of giving by private foundations and corporations, respectively. Part VI first offers an analysis of the controversial issue of donor control of perpetual trusts, then discusses particular illustrative areas of giving—for children and youth, black education, and social-change organizations; as such, they complement the more detailed discussions, in earlier volumes in this series, of philanthropy in the arts (DiMaggio, *Nonprofit Enterprise in the Arts: Studies in Mission and Constraint,* 1986) and private education (Levy, *Private Education: Studies in Choice and Public Policy,* 1986).

The opening chapter takes up philanthropic values and brings us back to a venerable tradition that many commentators (to say nothing of boosters) say reached full flowering on the American continent. But Robert Payton questions whether the connection between ancient moral and religious antecedents and today's diverse, sometimes frenzied philanthropic giving is a strong bond, a tenuous thread, or mainly symbolic. He believes that in the institutionalization of charity and philanthropy, underlying values have become obscured. As much as organizational evolution (after all, the charity of the medieval church was organized), the distance from religion accounts for the distance between philanthropy and the love of fellow man suggested by altruism. Well into the nineteenth century, social reform movements—abolition of slavery, improvement of conditions in asylums and hospitals, and reform of prison and judicial practices—were largely inspired by religious values. But now, as private philanthropy is professionalized and as government has taken on a vast proportion of what was once charitable activity, the secular practice of philanthropy is based on utilitarian values. Payton—fund raiser, grant maker, philanthropist, philosopher—leaves us with a dozen vexing questions about the purpose, sources, and direction of modern private philanthropy; abuses by philanthropic organizations; the role of private charity in public policy, and character and values in philanthropic giving. The questions provide a provocative agenda for scholars and practitioners alike.

The religious and civic bases of philanthropy are the departure points for Kathleen McCarthy's tracing of modes of giving throughout American history. "The doctrine of social stewardship that echoed from English pulpits in the Tudor-Stuart period" and John Winthrop's "Modell of Christian Charity" carried well into the nineteenth century, to the "Christian gentleman" and his feminine counterpart, the "Benevolent lady." McCarthy's account also reminds us that public criticism of highly visible philanthropists and bitter legal struggles over their fortunes went on long before the Buck Trust case of the 1980s, in which the San Francisco Foundation failed to broaden geographical restrictions of a nearly half-billion dollar trust it was administering. Witness John D. Rockefeller's treatment at the hands of muckrakers. The effect of social cataclysms on philanthropic patterns is graphically illustrated in McCarthy's analysis of the reinterpretation of civic stewardship that began in the post-Civil War era. Did activist industrialists, preeminently Andrew Carnegie, usher in scientific philanthropy, and yield to the professionalization of the enterprise as McCarthy and Payton observe? A dissent of sorts is put forward in Part V by another historian, Peter D. Hall, though his business leaders, such as Gerard Swope and Walter S. Gifford, hardly occupied Carnegie's Olympian height.

McCarthy aptly shows how the Depression led to an accommodation between private philanthropy and the welfare state, a relationship that expanded in the post-World War II era to a synergistic, even cozy, entente. (Philanthropic responses to the Depression, incidentally, remain a subject barely touched by research.) The last quarter of McCarthy's chapter is a fine backdrop to subsequent chapters that deal with private foundations, such as Boris's and Odendahl's, including their recurrent bouts with Congress.

The transition from philosophical and historical contemplation of philanthropic giving to the data-filled realm of modern survey research is less jarring than one might expect, at least in this volume. Arthur White, a founder of one of the country's leading survey research firms, begins with a biblical imperative of charity. But if White, like Payton and McCarthy, draws ties between the religious tradition and the practice of philanthropy, he also poses a paradox. Given the "very powerful relationship between religious observance and giving, why . . . do more than half of all Americans give less than 1 percent of their incomes to charities?" His question reflects two strong currents in the study of philanthropic giving. One is the level of individual and corporate giving, which at least those with a professional interest in philanthropy believe is too low. The other is motivation. White's chapter contains speculation, but it is reinforced by an interesting construct that grows out of survey research results—the "mushiness" index of Americans' thinking on various subjects, charitable behavior included.

He then sets forth the major conclusions of the study commissioned by the Rockefeller Brothers Fund, *Charitable Behavior of Americans* (1985). Many of them square with conventional understanding of giving behavior; for example, the dominant position of churches or religious organizations as recipients of individual giving. But White's data raise qualifying flags about other traditional assumptions, such as the correlation between giving and such factors as income, age, and educational level. One aspect taken up later by Odendahl, Auten, and Rudney, is the giving patterns of the wealthy.

White also raises an interesting parallel between the generally increased competitive environment in the Reagan 1980s and efforts to promote "consumer choice" in giving. Given White's long association with the voluntary sector, the auspices of a foundation clearly interested in advancing philanthropic giving, and his own observation that the *Charitable Behavior* study was a direct contributor to Independent Sector's Daring Goals for a Caring Society campaign, it is not surprising that, even though we are neither tithing nor coming close to meeting biblical injunctions, he cites several reasons for hoping that giving will rise.

For students of philanthropy, be they historians or economists, the preoccupation with the wealthy goes beyond the public's perennial curiosity, if not envy or even hatred. Scholars of philanthropic giving seek to know and understand the extent to which the rich share their wealth in the public interest, what motivates their giving, the patterns in which they give, and to whom. Several chapters touch on the wealthy, either as particular historical figures (Bremner, McCarthy, Hall) or as a class of givers. Gerald Auten and Gabriel Rudney, economists with the U.S. Treasury, have taken previous studies of high-income giving behavior a giant step forward. Their methodology enables them to examine the variability of giving by wealthy individuals over a five-year period and of generosity among such individuals. Although the very wealthy are prime targets for fund raisers, the vast majority of them are, in the Auten-Rudney definition, not exceptionally generous. Thirty percent of those studied contributed less than 1 percent of income during the period, or half of what the typical (median) taxpayer donates annually. Just as the relatively few wealthy account for a disproportionate share of total giving, a small number of the wealthy—5 percent—account for nearly half (43 percent) of all high-income giving. Discretionary giving does seem to loom high among the generous wealthy, and more research into committed and discretionary giving by the rich would be fruitful. (In fact, Auten and Rudney's finding of how much giving comes from such a tiny segment of the population may account for what may be a record in the speed with which research on nonprofit organizations reached the public. After they presented their findings at a scholarly meeting, a front-page headline in the *Non-Profit Times,* a monthly, blared, "Yale Study Discovers Pools of Untapped Funds," and the news also was carried by wire services.)[15]

The volatility in giving from year to year is the product of many factors, including changes in discretionary income, holdings of wealth, the tax incentive to give, savings and financial investment alternatives, family obligations, and so forth. Prestige can also come into play—a wealthy person may choose to give a single very large donation in one year instead of spreading it over five years to win more attention and approval. And fluctuations in giving by the wealthy may reflect response to powerful fundraising appeals. So, Auten and Rudney say, the wealthy may be approachable for more frequent solicitation of large gifts.

Emmett Carson addresses black philanthropy, a subject shrouded in myth and fear. Myth has it that black Americans are strangers to the philanthropic tradition, that the growing number of middle- and upper-middle class blacks give less than whites, and that blacks spend less time as volunteers than the majority population. The fears, in the black community itself, are a sort of Catch-22: Whether research discloses that these beliefs are supported or if not (that blacks devote a good proportion of their resources for self-help), government support of the black poor will decline even further.

Carson is the principal investigator of the most extensive research ever done on historical and current patterns of black philanthropy. Under the aegis of the leading black think tank in the United States, the Joint Center for Political Studies, he reports on the historical record and on the results of a Joint Center–Gallup Poll of black and white philanthropic behavior. The review of the past confirms what any reasonably informed student of black history knows—that black philanthropy stretches back more than two centuries in the United States and that black voluntary organizations persisted even in secrecy when they were declared unlawful in southern states. The time blacks devote to voluntary activity is undervalued, as it has been in the general population until recently. Indeed, Carson says, black voluntarism in the civil rights movement "was perhaps the greatest mobilization of charitable activity of any group" in American history. Because previous studies of charitable giving and voluntarism have scanted the black community, the Joint Center–Gallup survey oversampled blacks and questioned them in direct interviews, mainly by black interviewers. Blacks were shown generally to differ little from whites in contributions, in their confidence that charitable organizations will deliver what they promise, and in voluntary participation both with organized philanthropy and in direct assistance to the homeless and to needy neighbors and friends. Higher percentages of blacks than whites in the low-income and $20,000 to $25,000 income bracket make large contributions to charitable organizations. More so than whites, the object of black charitable contributions is churches, although as Carson observes, churches are often the conduits of funds to other charitable activities, such as black colleges. Carson concludes that one-dimensional perceptions of blacks as only recipients of charity are unwarranted.

Not since the Tax Reform Act of 1969 has a piece of federal legislation set off so many tremors across the landscape of organized philanthropy as its 1986 counterpart. Several chapters deal with the known and presumed effects of the 1986 act on individual donors, organized philanthropy, and their nonprofit clients. Jerald Schiff throws up cautionary notes on economic forecasts of dire effects. He points out, for one thing, that most economic analysis has focused on individual giving, largely ignoring foundation and corporate giving and volunteering of time. Also, despite the little research done on the "free rider" concept in donor decision making, he gingerly estimates that the impact of tax policy changes on money donations is about 10 percent less than suggested by past estimates, which largely ignored interactions among donors. He also comments on the resilience of nonprofit organizations in confronting adversity, through improved solicitation, for example, or through sale of services. Although deductibility for charitable donations is regarded as a major positive force in philanthropy, Schiff points to the potentially negative side of other issues associated with deductibility; for example, the disproportionate influence it accords the wealthy. Taking up factors other than deductibility in influencing donations, he discusses why, nonetheless, economists continue to fasten on the "price of giving" phenomenon. Schiff believes tentatively that the 1986 tax reform will have a smaller effect on giving, particularly in the short run, than estimates suggest, although volunteering will fall as well. The upshot, he concludes, is that estimation in this area is very imperfect, and policy makers will find only partial answers. Schiff suggests any number of potentially fruitful areas of research to fill the gaps.

Charles Clotfelter offers a wide-ranging discussion of federal tax policies and charitable giving, including but going well beyond the 1986 act. He reminds us that private giving, if not the only source of the vast charitable expenditures ($129 billion in 1975, $228 billion by 1986), is a major factor (one third). Individual giving, about which there is a notable amount of research and remarkable consistency of result, is the largest component, intimately connected with tax policy, and the preeminently important charitable deduction is the focus of major debate on this policy. Clotfelter reviews various aspects of this debate: the effectiveness of the deduction as an incentive for giving, alleged favored treatment of the wealthy, distribution effects, and the effect on tax progressivity. Besides the deduction, he examines such ingredients as nominal tax schedules and inflation. Clotfelter discusses the implication of the hypothesis that the charitable deduction be eliminated altogether and the possible effects of optimal tax systems on such matters as the comparative effectiveness of government and private nonprofit activity, the social welfare function (the utility of recipients), pluralism, and revenue costs. Clotfelter cites major unanswered research questions: the magnitude of price elasticity for low-income households, how the response of taxes varies according to the type of organization being supported, the precise price elasticity of corporate income and its appropriate measure, and the interactions among donors and between contributors and recipient organizations. He concludes by underscoring the importance of analyzing efficiency and distributional equity as well as the response to tax incentives in judging the desirability of incentives for charitable giving.

The Reagan administration's cutbacks in social spending produced a flurry of speculation and partisan argument over the "replacement" issue. Would state and local government and private philanthropy make up the difference? This is the obverse of "crowdout," the presumption that for a government dollar spent, private donors withhold a dollar. Economists began empirical studies of crowdout before the new federalism took the saddle, and such research understandably increased in the 1980s. Richard Steinberg's chapter reviews this work and offers a theoretical model of the interplay among federal, local, and private philanthropic spending. Simple dollar-for-dollar crowdout is extremely unlikely, he observes. The phenomenon is complicated by a "web of complex and interdependent causalities." Such factors come into play as the particular kind of government contribution, private donor perception of the degree to which government and other donors are filling a need, and tax and income effects. Local government reaction depends on the nature of grants from one level of government to another, targeted versus general intergovernment transfers, the presence or absence of matching requirements, government user charges, and the provision of related services by for-profit firms.

One also considers the greater leverage effect of public spending compared to that of one individual. A dollar-per-person increase in taxes to support social services raises total spending enormously; a $1 charitable donation is just that. Ideology also plays a role—the pure dream of replacing government action on social problems almost entirely. Steinberg says that cumulative research shows that some crowdout occurs, usually between one-half cent and 30 cents on the dollar. Theory commonly suggests that combined local donations, local government spending, and user charges will not increase enough to compensate for federal and state cutbacks on social services. But he concludes that we still lack data that easily confirm this conclusion.

Perhaps the most unique way Americans have chosen to give is the establishment and endowment of private foundations. The boldness of the Carnegies, Rockefellers, Sages, Harknesses, Mellons, et al. in giving birth to foundations is what gives the early twentieth century its luster as a golden age of philanthropy. Significant foundation formation continued, if less dramatically, in the post-World War I period (the Ford Foundation came into existence only in 1936) and a few of the largest foundations have been established in the last quarter of the century. But more than 10,000 foundations have been terminated since the Tax Reform Act of 1960.

Teresa Odendahl, an anthropologist with a decent respect for economics, public policy, and psychology, has taken an audacious look at the reasons foundation forming as a preferred philanthropic avenue by the rich has been waning since mid-century. In addition to scouring historical and Internal Revenue data, she and a research team drew information and insights from the wealthy themselves and from lawyers and others who influence their philanthropic habits, some 240 in all. In extensive interviews, they probed the motivation for their charitable giving and the incentives and disincentives to choosing the foundation approach.

Whereas most of the wealthy interviewed ranked tax incentives lower than several other motivations for establishing a foundation, they ranked them quite high in choosing another charitable option when the Revenue Acts of 1954 and 1964 eroded the tax advantages for foundations relative to other charitable channels. The 1969 act was a heavy blow to foundation formation, not only by limiting deductibility but through the "excess business holdings" provision that prohibits individuals from using a foundation to maintain a controlling interest in any company.

In the late 1940s and early 1950s, large endowments were added to foundations that had been set up earlier. Odendahl doubts that the "acorn" pattern of foundation growth will continue. Yet, foundations continue to offer certain attractions for wealthy people: the accumulation of assets; a vehicle for systematic giving: a counterweight to government monopoly support of educational, scientific, welfare, and cultural activities; control over wealth even if not for personal benefit; and a means of carrying on family tradition. Foundations also are attractive to the moderately wealthy trying to raise their status.

Odendahl offers interesting speculations on generational differences in the source of wealth (manufacturing versus the service and trade sectors of business), the less liquid fortunes of the new entrepreneurs, and the tendency of the young wealthy to avoid the foundation as a symbol of great wealth. Many more foundation donors are "self-made" than heirs to wealth. The children of the self-made rich, however, are less disposed to use the foundation route for their charitable impulses.

Because of the decline in the birth and growth of large independent foundations, the Council on Foundations, which co-sponsored the Odendahl study with the Yale Program on Non-Profit Organizations, has now launched a three-year Private Foundation Project, an aggressive campaign to promote the concept of foundation philanthropy to the wealthy. The argument is that establishing a foundation makes it possible for an individual to do more for the public good than the same money would as a one-time gift made directly to a charity. The council believes that ignorance of the law and confusion among wealthy individuals and their advisors about how philanthropies are created and about the laws governing their operation are to blame for the decline as well as legislative disincentives.[16]

As the growth rate of private foundations has slackened, community foundation formation has quickened. Peter Hall attributes this not only to the tax advantages of giving to the community form of organized philanthropy but also to a residual nostalgia for a localized focus and a reaction to the nationalization of American life. Hall traces the origin of the community foundation from its birth in Cleveland as an outgrowth of the Community Chest movement. Both were vehicles whereby enlightened businessmen hoped capitalism could cure the ills arising from a rapidly urbanizing, rapidly industrializing society. Between 1920 and 1960, a period that saw the birth of almost 60 percent of today's 24,000 foundations, community foundations grew modestly, mainly in the cities of the Midwest. Hall believes the turning point occurred in the 1960s, with public disclosure of private foundation abuses and the rise of new local and regional elites in the West and South. Today's community foundation expansion harks back in part, he says, to a Progressive era notion that organized philanthropy should be limited to the very rich and is in part a rediscovery of hands-on local solutions to social problems. Hall suggests that as private foundations come under greater government regulation, community foundations may be favored even more by philanthropists.

Although most philanthropic giving comes from individuals, a new class of professionals has arisen whose work is to give away others' money.

They include lawyers and advisors referred to in Odendahl's chapter, but the greatest number of those who work full time at it are the employees of the nation's private and community foundations. Elizabeth Boris and her colleagues have done the most exhaustive research on this group in recent years. Boris is well postioned for such work, since under her aegis as vice-president of research, the Council on Foundations conducts biennial surveys of foundation staffing and other management aspects. It is a small industry, estimated at 8000 employees, most of whom work for the 500 largest foundations. Their influence, in the dispensation of over $3 billion a year, is far out of proportion to their numbers. Workers in this vineyard are motivated by an ideology of doing good for others and making a difference, Boris observes. But power and status attach to the work. The boards of foundations are less homogeneous than they were fifteen years ago, but white men still dominate. So, too, in the ranks of chief executive officers of foundations, though the number of women CEOs has risen to 35 percent and members of minorities to 5 percent. The diversity of foundations is measured not only in size and programmatic emphasis but also in the delicate balance between professional staff and board perspectives. Boris outlines four models of this crucial relation. Equally critical are the attitudes and work styles of foundation staff members, ranging from the reputed arrogance of program officers to their unsung work beyond check writing—technical assistance, brokering, and networking, for example. Boris's research also is revealing in the recruitment patterns, career paths, compensation of foundation staffs, and the constraints faced by women in foundations. And, although employment in foundations is highly sought after and job satisfaction is high, she observes that all is not milk and honey. Mobility is limited, and some foundation executives may play paradoxical roles: powerful, if not arrogant, toward grant seekers, but humble toward their boards of trustees; liberal to outsiders, conservative in their posture before their boards.

Is business philanthropy different? The conventional recollection of the nineteenth-century public's attitude toward business is that of a rampaging capitalism that deserved every assault by the muckrakers. But Peter Hall, in a historical review that begins with colonial entrepreneurs' community activities even before corporations existed, reminds us that in the light of widespread civic corruption and the privations arising from industrialization and urban growth, many businesspersons regarded themselves as reformers. "Founders of corporate liberalism" supported many local social services and some engaged in national reform politics, the founding of social research centers, and the reform of higher education. The transition from the philanthropy of an owner-manager to corporate giving ran into shareholder challenges, but the new professional managers picked up the pursuit of social goals in the way they ran their enterprises, through the creation of credit machinery that softened class divisions, through support of colleges and universities and service on their boards, and through creation of the first corporate foundations in the 1920s. Today, Hall points out, business philanthropy—more broadly, the social role of business—is in flux as perhaps never before. Firms are under pressure to tie their giving more directly to company operations and to measure the payoff of their charitable investments. The effect of mergers and acquisitions appear to reduce corporate giving. But the continuing thread, says Hall, is the desire of business to take part in shaping public life; witness the surprising degree of corporate engagement in public education reform efforts.

Hall concludes that welfare capitalism came to an end with the advent of Social Security, increased income taxes, and the developing strength of organized labor. The corporate charitable deduction was passed during this period. Direct corporate giving became especially important after World War II, as businesspersons sought to preserve the private sector in the face of growing government involvement in every aspect of American life. A major breakthrough was achieved when the New Jersey Supreme Court removed legal obstacles to corporate philanthropy.

Corporate officials who decide whether, how much, and how funds are donated to worthy causes are, philanthropically speaking, less professional than their counterparts in private and community foundations, although outstanding exceptions exist—men and women as thoughtful and skilled at giving as anyone. The corporate philanthropist may have the job as a de facto demotion, or at best lateral move, from the "hard" side of corporate life. Or he (or she) may have been promoted to it with extensive responsibility or even recruited from outside the corporate world. E. B. Knauft's research provides an insight into the culture and dynamics of corporate giving programs. Based on a sample of forty-eight major companies, his analysis benefits from his years on the inside.

Corporate giving programs may shrink or expand as the company's fortunes rise and fall, and the CEO plays a major role in determining the level. Knauft traces the relation of size not only to earnings but to the size of stockholder dividends and long-term growth. Also covered is the proportion of company contributions devoted to employees' matching gifts programs; the role of gender (larger programs were managed by men, more generous programs relative to company earnings by women), public disclosure policy, proportion of United Way giving, and many other variables. Particularly interesting is a matrix Knauft constructed from subjective ratings by contributions managers of other companies' programs. Knauft also cites the effect of corporate contributions on several models of corporate culture. He found the character of most corporate giving programs reflected in case-by-case decisions rather than well-developed priorities, reactive rather than "proactive." Such priorities as do exist tend to continue year after year, without being reassessed, but Knauft cites several encouraging exceptions.

Complementing Knauft's analysis of the characteristics and governance of corporate giving, Joseph Galaskiewicz focuses on motivations, from an extensive view of the literature and from his own exhaustive research in one of the meccas of corporate philanthropy, the Twin Cities. Two of the five driving forces he identifies are directly tied to business operations: cause-related marketing (e.g., the giving of a unit of charitable contribution for a given unit of sales) and public relations. The former clearly is reflected in dollar-and-cents yield to the company, but there is little research to show that companies enhance their public relations if they give more money to charity. A third motivation, "enlightened self-interest," seeks a longer-term, more amorphous goal, an environment congenial to business operations; for example, a form of insurance against government regulation or a means of enhancing the well-being of employees. More immediate and straightforward is the use of charitable contributions as a tax strategy, though surprisingly few executives who give cite as a priority the goal of alleviating the tax burden on the company. And it is fairly well known that fewer than one in five companies gave up to the deductible 5 percent of pretax limit. Finally, he examines "contributions as social currency," corporate charity that reflects staus competition among executives. Although little research has been done on this motivation, Galaskiewicz found it strong in his examination of corporate giving in Minneapolis and St. Paul. Companies gave more depending upon the CEO's position in the networks of locally prominent businessmen. He documents how the latter both pressured them into giving money to nonprofit organizations that served members of this elite, and how the elites accorded corporate givers status in return. Galaskiewicz calls on researchers to test propositions derived from class control theory, to do content analysis of the business press, to look at executives overly concerned with bottom-line indicators, and to examine the role of United Fund brokers, fund raisers, and contributions professionals in affecting the distribution of corporate grants.

The volume concludes with examinations of certain consequences of charitable giving. Francie Ostrower analyzes the thorny issue of donor control of a favored channel through which wealth can flow toward public purposes: the charitable trust. The attraction in such instruments is their ironclad protection against succeeding generations tampering with the wishes of the donor. But Ostrower demonstrates that "nothing lasts forever." It is no surprise, given long-held suspicion of "dead hand control," that all manner of legislative, judicial, and informal steps have been devised to alter or adapt charitable trusts to current conditions or unforeseen circumstances. Donor intentions early on were nullified to curb the power of the church and monasteries; but closer to our own time, restrictions also arose from fears of liberal beneficiaries, such as antislavery causes. And, of course, foundations have been scooped up in laws prohibiting donors from exercising perpetual control of corporations. Reviewing the welter of judicial opinion centering on the cy pres doctrine, Ostrower notes that conflicts may be more between the living than between deceased donors and posterity; for example, the fierce struggle over the Buck Trust. She reports that even without resort to courts some universities, foundations, and other nonprofit beneficiaries have altered, if not evaded, donors' intentions. Some institutions have declined bequests with strings that are too restrictive, and

she calls for more research on the boundaries of unreasonableness and the
devices institutions use to cope with them. But, she reminds us that deny-
ing donors the ability to control in perpetuity may divert funds from charity.

Using as cases social change movements of the mid-twentieth century,
from civil rights to the nuclear freeze to consumerism, J. Craig Jenkins
examines a troubling issue in the application of private wealth to social
change: "Is . . . upper class patronage compatible with our conceptions
of democratic government and politics?" In so doing, he quantifies the
distance between the widespread impression of private foundations as major funders of such movements and the rather slim actuality. Thus, although such funding began to rise rapidly in the mid-1960s and peaked at
$25.2 million in 1977, that was less than 1 percent of total foundation
giving! Still, it was enough to stir up storms of controversy—sharp criticism in Congress, chilling moves by the Internal Revenue Service, academic
debate on the motives of wealth embodied in private foundations, and
intense feeling within social change organizations on whether they were
more distorted than helped by donors' assistance. Jenkins delineates the
characteristics of traditional grass-roots organizations in comparison with
the increasingly prevalent "professional movements" (preferred by funders), along with the new genre of technical support centers. As other authors in this volume discuss the motives of individual givers, Jenkins reviews institutional motives, particularly conscience, social control, and
political opportunities. He says his data support the conscience interpretation, a classic case of which was foundation support of the civil rights
movement. On the other hand, he says, the control argument, the desire
to dampen radicalism and militancy, lacks direct evidence and is ultimately
impossible to demonstrate. The politicial motivation, an attempt to gain
an advantage by bringing new players into the political game, enters into
the policies of foundations as well as political leaders, he also concludes.

The bulk of Jenkins's chapter deals with the impact of foundation patronage on social movement organizations, whether it withers their grass
roots, how it affects their internal democracy, and what it means for the
representativeness of the American political system. He says his evidence
refutes the most outlandish conservative criticism—that foundations actually generated the movements; that they stir up baseless grievances and
radical demands and "divert tax dollars toward quixotic political organizing campaigns that misrepresent the interest of their purported beneficiaries." Jenkins's cases, while necessarily brief, afford a vivid scan of the
complexities and inner workings of the events that stirred American society across several fronts for more than two decades. In most instances, the
role of the foundation fit a "channeling intepretation"; that is, foundation
funding did support movement professionalization and squares with several arguments of the cooptation theorists. Yet, he says, it was frequently
constructive, dovetailing with forces within the movements. In channeling
groups into less militant directions, the foundations helped institutionalize

moderate social reform and helped direct middle-class reformers into new issues. Jenkins concedes that foundation funding tended to lead to weaker direct participation and a representational monopoly of groups recognized and licensed by the state. Yet, he says, the real choice may be between professionalization and no advocacy whatsoever. He concludes that professionalization has often strenghtened grass-roots challenges and broadened the range of views in politics. In an important sense, then, foundation patronage has helped "represent the underrepresented," creating a more accessible political system.

No one is better qualified than Robert Bremner to review the history of the attitudes of American society toward children, from the colonial period to the present. Although Bremner is best known within philanthropic circles for his classic, slim volume *American Philanthropy* (University of Chicago Press, 1960), his principal contributions as a historican are works on poverty (*From the Depths*, New York University Press, 1956), and child welfare policy (editorship of the three-volume *Children and Youth in America*, Harvard University Press, vols. 1 and 2, 1971, and vol. 3, 1974). His chapter in this volume blends the two streams of knowledge. Thus, the great diversity of philanthropic giving is woven into every historical period and shift in attitudes and practices toward children, especially children in need. The evangelist George Whitfield stirs a congregation to give nearly $200 for an orphanage on a single Sunday in 1740. An antebellum asylum for blind children raises money from fees, fairs and exhibitions at which its pupils display their accomplishments. The Children's Aid Society, which shipped poor children from New York City to midwestern homes in the 1860s and 1870s, raises money through contributions of a few cents a week from Sunday School pupils. YMCAs around the turn of the century perfect fund-raising drives, whose basic elements remain today. A large bequest leads to the transformation of the Association for the Aid of Crippled Children from a service agency to a foundation that makes grants for research, policy studies, and advocacy projects for children. He also touches on one of the notorious scandals in youth-serving philanthropy—Boys Town's accumulation of a $200 million endowment and excessive fund-raising costs. Bremner points out that many childless rich men were responsible for giving for the education of poor children. Among them was Stephen Girard, a subject of discussion in McCarthy's chapter for other reasons.

Bremner demonstrates that from the earliest organized concern for children in trouble, the private sector enlisted government financing. Thus, in the first half of the nineteenth century, privately established orphanages and reform schools received government funds, and even in the Reagan years the largest single source of revenue for nonprofit agencies serving children and youth was the government (38.4 percent).

And he notes and strongly approves the policy advocacy role of youth-serving nonprofit organizations. For example, the successful effort of the National Society for Crippled Children to persuade Congress to include provisions in the Social Security Act of federal grants-in-aid to states for services to crippled children. We are also reminded that the current concern about the condition of the American family has antecedents stretching back to the earliest days of the Republic.

John Stanfield brings a distinctly qualified view to one historical aspect of private foundation giving that is generally viewed through rose-colored glasses. Foundation assistance to the education of blacks after the Civil War is usually recorded as a noble venture. It was indeed the focus of some of the earliest foundations, including the Slater and Jeanes Funds and Rockefeller's General Education Board. But Stanfield suggests that their work was an accommodation to the racial status quo and implies that at various points they and their successors may actually have impeded efforts toward racial desegregation. He also observes that the majority of foundations ignored race relations and black education altogether. And he finds equivocation in foundation connections with government in this field. From the first, some foundations sought to stimulate government attention to black education, even to the extent of paying the salaries of state employees. In the early days, government officials served on foundation boards. Whether the foundations pushed gently or forcibly for government action, he says, they observed racial etiquette. When they overstepped the status quo, in the post-World War II period, they suffered the wrath of Senator Joseph McCarthy and the Congressional authors of the 1969 Tax Reform Act. Stanfield may be stretching matters in attributing 1950s witch hunting or the investigations of the 1960s to racial backlash, but it would be stretching in the other direction to deny that racial factors weighed in to some extent. Stanfield is not averse to giving credit; for example, to the Rosenwald Fund for helping to build 5000 schools for black children at a time when educating blacks was distinctly unpopular, and to the Ford Foundation and Carnegie Corporation for responding to the beat of the civil rights movement, especially in the development of black scholarly and professional talent. But, essentially, he portrays a set of institutions whose claims to risk taking and concern with social issues are confined to a very small population and always were.

The growth of research on the nonprofit sector generally, giving included, is suggested by the latest (1985–86) edition of Independent Sector's annual Research-in-Progress report, whose 440 entries included 245 entirely new projects. Whatever gaps remain in research on philanthropic giving, it is at least free of a bias criticized in other forms of research on the nonprofit sector; that is, that such research concentrates on agencies that provide services and neglects religious and advocacy institutions. In dealing with aggregates and policies, research on giving makes no such distinctions, and it is worth noting that discussions of nonprofit activities in this volume have dealt as much with advocacy as with services, though hardly at all with religious institutions.

Despite the need for further historical work, a flurry of short-term research on giving will probably be stimulated by the impending change of administration in Washington, as it was eight years ago. This period, marked by sharp reductions in overall federal activity in fields of concern to nonprofit organizations, heightened interest in the ability of philanthropic giving to offset losses. As it turned out, predictably, private giving could not offset the cutbacks, not even by half, despite overall increases in giving. Remarkably, however, philanthropy did catch up with the direct revenue losses nonprofit organizations experienced. By 1986, Lester Salamon and Alan J. Abramson concluded, "the growth in private giving had enabled nonprofits to return to the service levels they had attained as of 1980, but still left them with only limited ability to expand their activities to fill in for government cutbacks." [17]

Future research is also likely to deal with the growing "democratization of giving." Notwithstanding laments about a current mood of self-centeredness, if not narcissism, the long-term trend is toward the democratization of philanthropy, according to Brian O'Connell, president of Independent Sector. He observes:

A far larger proportion and many more parts of our population . . . have successfully organized to deal with a vast array of human needs and aspirations . . . rights of women, conservation and preservation, learning disabilities, conflict resolution, Hispanic culture and rights, education on the free enterprise system, the aged, voter registration, the environment, Native Americans, the dying, experimental theater, international understanding, drunk driving, population control, neighborhood empowerment, control of nuclear power, consumerism. . . . Our interests and activities extend from neighborhoods to the ozone layer and beyond."

The democratization of giving, however, may be threatened by the loss of tax deductions for persons who do not itemize their contributions. As the proportion of taxpayers who itemize their deductions decreases sharply, deductibility will increasingly be regarded as a benefit for the rich, and challenges to the deduction mechanism itself may find more political sympathy.

Furthermore, especially in an atmosphere of deep concern for large government deficits, more credence may be given to the view of some economists (and legislators) that philanthropic giving that is deductible is a "tax expenditure," since the government loses tax revenue equal to the deduction. The corollary is that government should have more control over where deductible funds are used, undermining the basic concept of pluralism and free choice in giving.

Very little research has been devoted to nonprofit activity outside the Untied States, much less individual giving. In part this arises from the premise that giving and volunteering are so pervasive in American society as to render philanthropy elsewhere insignificant, or at least not worth examining in any detail. That this view may have to be modified is illustrated by a recent reminder from, of all places, the Soviet Union. "For 49 years, since the closing of a fund for homeless children that was created in Lenin's memory, the mythology of the all-sufficient state left no room for philanthropy. [But] 1987 has been the year in which charity came in from the cold," notes a front-page article in the *New York Times*. The account describes a wide array of private giving and volunteering in the Soviet Union, from a fund for victims of the Chernobyl nuclear disaster to a fund for Soviet orphans (to which $200 million, including proceeds from a rock music festival, has been donated).[18]

So a wider ambit may be traversed than intended in the recent prediction by the editor of *Daedalus* that "much more attention" will be given to philanthropy by the research community because it is just too important to leave alone: "The values of a society are admirably expressed in what it chooses to support; so, also, are its ambitions and fears."[19]

NOTES

1. F. Emerson Andrews, *Philanthropic Giving* (New York: Russell Sage Foundation, 1950).

2. "Gilt Guilt: Left-Wing Venture Philanthropists," *The New Republic* (December 8, 1986).

3. Richard Kluger, *Simple Justice* vol. 1 (New York: Alfred A. Knopf, 1975), p. 164.

4. Notwithstanding an emphasis on giving by the wealthy, it is worth remembering that most of their estates stay in the family. That is truer of older wealth than new wealth, according to a *Fortune* survey (September 29, 1986). Thus, Eugene Lang is leaving nothing to his three children, planning to bequeath most of his fortune to his foundation.

5. Brian O'Connell, "State of the Sector: With Particular Attention to Its Independence," report to the annual meeting of Independent Sector (October 25–28, 1987), p. 8.

6. Robert N. Bellah, Richard Madsen, Ann Swidler, and Steven M. Tipton, *Habits of the Heart: Individualism and Commitment in American Life* (Berkeley: University of California Press, 1985).

7. Barbara Ehrenreich, "The Moral Bypass," *The Nation*, December 28, 1985/January 4, 1986, p. 717.

8. *American Volunteers 1985: A Summary Report* (Washington, D.C.: Independent Sector, 1986).

9. Ibid.

10. Ibid., p. 5.

11. Gary Becker, "A Higher Cost of Giving Is No Cause for Low Spirits," *Business Week* (August 11, 1986).

12. *From Belief to Commitment: The Activities and Expenditures of Religious Congregations in the United States* (Washington, D.C.: Independent Sector, 1988).

13. One of the most successful professional fund raisers in American history, Arnaud C. Marts, complained about the dearth of research on philanthropic giving: "No one has been able to give me documented answers to the question 'How and why did this vast generosity develop in America and nowhere else?' Libraries have been consulted in the belief that our private gifts . . . for the public good must have attracted the attention of responsible historians and authors. No titles are to be found which trace this cultural force back to its origin. . . . No one, evidently, has been fascinated by this analytical history; no one has spotlighted within the larger framework of historical events.

"Many analytical histories have traced . . . single strands in the fabric of the nineteen centuries of Western civilization in books or magazines or journals. For example: *Rats, Lice and History, The Influence of Naval Power upon History, The Wheel and Civilization*. However, no comparable—nor satisfactory—analytical history of generosity or philanthropy or charity is available about their historical origins . . . for the future, goodwill, and civilized well-being of Western man." Arnaud C. Marts, *The Generosity of Americans* (Englewood Cliffs: Prentice-Hall, 1966).

14. John Gardner, Address to the International Fundraising Conference, National Society of Fundraising Executives, Toronto (March 17, 1982).

15. David Johnson, "Yale Study Discovers Pools of Untapped Funds," *Non-Profit Times* (November 1987), p. 1.

16. Paul Desruisseaux, "Philanthropic Groups Stepping up Efforts to Boost Creation of New Foundation," *Chronicle of Higher Education* (November 5, 1986).

17. Lester M. Salamon and Alan J. Abramson, *Nonprofit Organizations and the FY 1988 Federal Budget,* a report to Independent Sector and the 501(c)(3) Group (August 1987).

18. Felicity Barringer, "Soviet Communism Lets Private Charity Revive a Tradition," *New York Times* (December 25, 1987), p. 1.

19. Stephen R. Graubard in his preface to "Philanthropy, Patronage, Politics," *Daedalus* 116, (Winter 1987), p. v.

I
THE GIVING TRADITION

1

Philanthropic Values

ROBERT L. PAYTON

THE PRESENT SITUATION

A millionaire computer entrepreneur subsidizes the largest rock concert in fifteen years. A New York foundation gives New York's police department a million dollars to improve data on robbery patterns and suspects. An interdenominational group organizes relief programs for victims of the conflict in Lebanon. A twenty-eight-year-old Saudi sheik gives $30,000 to Virginia Beach, Virginia, for city youth programs and a museum. The Willimantic, Connecticut, Elks Club contributes leftover food to a local soup kitchen, one element of the New Haven Food Salvage project that collects surplus food for the poor from supermarkets. A Boston industrialist creates a new foundation with $40 million to help improve the way society looks upon and treats the elderly. The University of San Francisco withdraws from intercollegiate basketball largely because the university's most generous contributor also made under-the-table gifts to basketball players. The state attorney's office in New York issues warnings about charity fraud during the holiday season. A neighborhood center "wrote to 85 large food manufacturers asking them to make donations for food baskets to the elderly. To our dismay, there was no response." An essay in the *New York Times* asserts that "charity needs coercion." CBS gives $250,000 to advance the condition of Hispanics in this country and will contribute $1 million more if matching funds can be generated from other sources. A *Wall Street Journal* article reports that corporate employees resent the pressure on them to contribute to charities. Is it possible to make sense of such diversity? Is it possible from such diversity to determine who ought to be doing what, where, and for whom?

This chapter will offer some definitions of philanthropy and of the values that philanthropic activities express. The list of news items just presented helps us consider our current behavior in the context of the tradition out of which it has grown.

An earlier version of this paper was prepared as the discussion paper for a Wilson Center symposium in 1982.

DEFINITIONS

Values reflect what we think and feel about things and guide the way we act. Our view of the way we would like the world to be is made of both personal and social perspectives. Our standards of performance, our "instrumental" values, are expressions of our sense of morality and of our abilities.[1]

The essential value expressed by the word *philanthropy*—shared by other words such as *benevolent, charitable, humanitarian,* and *altruistic*—is defined as "having or showing interest in or being concerned with the welfare of others." If a value is a preference, then the philanthropic will *always* tend toward the altruistic end of a scale of preference and away from the egoistic.

By a narrower definition, philanthropy has to do with giving money, what Boulding might call the "voluntary, one-way transfer of exchangeables."[2] *Philanthropy* has succeeded *charity* as the embracing term. For our purposes, it is useful to distinguish between two terminal aspects of the value we call philanthropic:

Charitable stresses either active generosity to the poor or leniency and mercifulness in one's judgements of others, but in each case it usually retains in some degree the implications of fraternal love or of compassion as the animating spirit behind the gift or judgement. [Charity is thus personal, an act of mercy and of concern for others.]

Philanthropic and *eleemosynary* also (with *humanitarian*) suggest interest in humanity rather than the individual, but they commonly imply (as *humanitarian* does not) the giving of money on a large scale to organized charities, to institutions for human advancement or social service, or to humanitarian causes.[3]

Philanthropy in its more specialized sense is less personal, more concerned with the betterment of humankind rather than simply with the short-term alleviation of suffering. Although charity and philanthropy are important in other cultural traditions—Islam, for example—the dominant influences on our ways of thinking about and doing philanthropy derive from the formative ideas of European civilization.

It is my contention that the value of philanthropy originated in ancient practice and belief, and that modern philanthropy has lost its conscious grip on that tradition. Although charity and philanthropy reflect the enduring aspirations of Western civilization, their further evolution may be in question.

Policy makers are further removed from the recipients of charity and philanthropy than was the case in simpler, smaller societies. As giving and receiving become increasingly systematized and specialized, organizational values replace personal ones. Detached from religion, charity and philanthropy lose their place to value-neutral terms like *grant making* and *contributions*. Policy decisions are expressed in utilitarian and pragmatic terms. We should begin to deepen our understanding of issues in philanthropy by encouraging scholars, practitioners, and policy makers to study and discuss philanthropic values and their roots.

RELIGIOUS ORIGINS OF CHARITY

"Religion is the mother of philanthropy,"[4] a modern writer has declared. Philanthropy emerged in a primitive culture that is becoming increasingly difficult for the modern imagination to grasp. The ancients lived in a hard world; suffering was a fact of life. Yet, it was a world created by a God who declared that he wanted men to make it a better place. He even began to tell them how: "When you reap the harvest of your land, you shall not reap right into the edges of the field, neither shall you glean the loose ears of your crop; you shall not completely strip your vineyard nor glean the fallen grapes. You shall leave them for the poor and the alien. I am the LORD your God."[5]

On the basis of later passages there is reason to suspect that not everyone adhered to the letter of the divine command. There may have been skeptics or others whose self-interest dampened the powerful sense of religious awe that moved the soul of primitive man. Whatever the case, the passage establishes the principle of the voluntary gift of gleanings from one's surplus, and introduces the radical principle of helping those beyond the persons for whom one is immediately responsible. The charitable act of giving is mandated.

Reciprocity is not mentioned. Perhaps it was simply overlooked in that passage, for it is clear elsewhere that self-interest was present. The reward of charity would be God's mercy, a reflection of the developing relationship between the believer and his God. God would help man in his present difficulties, perhaps in return for honoring God's commandments.

A second radical idea was introduced by the prophets, by linking economic poverty to oppression: someone is to blame for the condition of the poor and the downtrodden.[6] "The LORD opens the indictment against the elders of his people and their officers: You have ravaged the vineyard, and the spoils of the poor are in your houses. Is it nothing to you that you crush my people and grind the faces of the poor?"[7]

Political and economic brutality are the instruments of oppression. The cry is not simply for mercy but for justice: "he shall judge the poor with justice and defend the humble in the land with equity; his mouth shall be a rod to strike down the ruthless, and with a word he shall slay the wicked."[8]

A profound change occurs when one's lot is thought about in terms of suffering that is the product of human action and suffering that is brought on by natural causes. The seed of the idea of righteousness is planted.

The temper of New Testament times is similarly conditioned by oppression. Spiritual well-being is declared to be more important than material comfort. Love itself, the highest form of altruism and benevolence, is the fundamental virtue. Love of neighbor is fused with love of God as the terminal value. More practically, the New Testament provides the charitable agenda for the subsequent two thousand years: "For when I was hungry, you gave me food; when thirsty, you gave me drink; when I was a stranger you took me into your house, when naked you clothed me; when I was ill you came to my help, when in prison you visited me."[9]

There has been debate for a century or more about the extent to which the Christian gospel is a gospel of social action and reform, to what extent charity is to alleviate suffering in this life in anticipation of a richer life beyond, and to what extent the kingdom is to be sought with the end of suffering on earth. In such a debate charity is a factor or, as some would have it, a deception that diverts attention from the causes of the poverty to which charity is only one response. It is clear, however, that Christianity adds a further responsibility to the charitable act: it is necessary to *care* about the welfare of the other, as well as to *give* material help.

SECULAR ORIGINS OF PHILANTHROPY

The Judeo-Christian emphasis on charity and other acts of mercy to relieve suffering is not similarly evident in the classical civilization of Greece and Rome. The classical contribution is the relationship of philanthropy to the community. There is much less concern for the poor in pre-Christian Greek and Roman literature than in the literature of the Bible. The differences of substance are as profound as the differences of style between the Old and New Testaments, on the one hand, and Aristotle and Cicero, on the other. The culture of classical Greece and Rome is a civic culture. Philanthropy and reason are more powerful than religion and faith. Religion is civil religion. The highest aspirations are expressed in terms of honor and dignity rather than humility and salvation. Wealth in the classical world is the means by which honor, power, and privilege are obtained. The calculus of self-interest is measured in mundane rather than celestial benefits. When Aristotle wrote about the virtues he wrote about the vices as well, about excesses of prodigality and stinginess.[10]

This reasoned awareness of the good and bad of philanthropic behavior appears when Cicero advises his son Marcus about being generous: "We should see that acts of kindness are not prejudicial to those we would wish to benefit or to others; second, we should not allow our generosity to exceed our means; and third, it should be proportionate to the merits of the recipient."[11] He continues, stating that some people will go so far as to steal from one to give to another, that people will use their gifts in order to buy advantage for themselves. Cicero explains that philanthropic contributions are used to win public favor, and too much ambition can impoverish a family. The point begins to be made that it is difficult to give intelligently, that it is possible in the process of giving to harm oneself as well as the other.

The uses of philanthropy in Greece and Rome covered a broader range of objects than one associates with the Judeo-Christian tradition. The classical philanthropist made gifts for the dole of grain and oil, and also for games and festivals, for theaters and baths and stadiums. The classical benefactor was also a patron of the arts. On occasion he was also by origin nouveau riche. The prototype of the great American philanthropist is to be found in ancient Rome.

The religious, charitable, tradition is founded on altruism; the secular, philanthropic, tradition is founded on what Aristotle called prudence and what we would call enlightened self-interest. The culture of Israel gave us charity; classical civilization gave us philanthropy. Both are the expression of social as well as personal ideals and values, and the sense of community is terribly important.

By the end of the ancient world the framework of charity and philanthropy was established. The ancient world defined itself in terms of tribe and polis, yet the inclusion of the alien among those whom one seeks to help is as old as the charitable tradition itself. As the economy developed and urbanized, the methods of giving expanded beyond gifts in kind to gifts of money. The ancient world introduced the link between the need for charity and the fact of economic oppression. Finally, it provided the first examples of spiritual assistance and concern for others as the highest expression of charitable action.

INSTITUTIONALIZATION

The evolution of philanthropy is in large part a history of its institutionalization and adaptation. "The beginnings of organized Jewish relief are closely associated with the rise and spread of synagogues." The problem of how charity is collected, and who does it, are aspects of the institutionalization process. For a thousand years, charity and philanthropy were essentially coextensive with the church. The church preserved and elaborated the doctrine of charity and its practical manifestation in almsgiving. St. Thomas Aquinas fitted this particular value carefully into the *Summa Theologiae,* balancing seven corporal works of mercy with seven kinds of spiritual almsgiving. The corporal works are to feed the hungry; give drink to the thirsty; clothe the naked; receive the stranger; visit the sick; ransom prisoners, and bury the dead. The spiritual works are instructing, counseling, consoling, reproving, forgiving injuries, bearing another's burdens, and praying for all.[12]

The religious tradition of Christian Europe proclaims the necessary link between intention and act. Almsgiving must be animated by love. The Anglican Jeremy Taylor, writing four centuries after Aquinas, reaffirmed the sentiment: "Mercy without alms is acceptable. . . . But alms without mercy are like prayers without devotion."[13]

Private voluntary giving by individuals, beyond what was exacted by the church, continued throughout the Middle Ages. The motivation has been criticized as being excessively "self-regarding." We will all be judged in heaven for our acts on earth, and this is "the element of self-interest in the Christian doctrine of charity" that allegedly detracts from its altruistic purity.

The medieval church also perpetuated and extended the idea of stewardship, "based on the recognition that all gifts come from God and must be used to his glory, and applying equally to all types of gifts, whether of money, time or talents."[14] The church claimed the right to be mediator between God and humankind in matters of human claims to use God's property. Stewardship in its narrower usage as the fund raising of the church is an American innovation. Its Protestant origins are found in John Calvin, the ultimate source of modern capitalism (at least in the disputed thesis of Max Weber).

TRANSITION TO THE MODERN ERA

Poor laws mark the watershed between medieval and modern philanthropy. Economic conditions for increasing numbers of people continued to deteriorate. The economic transformation of Europe expanded the cities and improved the standard of living, but also caused dislocation, and temporarily, the population of unemployed poor increased. Relief provided by traditional almsgiving was too limited in scale to ameliorate conditions; the traditional institutions of the nobility and the church lacked both leadership and new vision. Poor Law legislation culminating in the reign of Elizabeth I would suggest that the state had moved in to fill the vacuum; W. K. Jordan argues a different case.[15] He assigns more symbolic than actual importance to the enactment of the Poor Laws. The private sector response to the need for increased philanthropy is what makes the period most distinctive. He notes "the momentous shift from man's primarily religious preoccupations to the secular concerns that have moulded the thought and institutions of the past three centuries" (p. 16). "Calvinism was in England sublimated into a sensitive social conscience that was secular in its aspirations and fruits even when the animating impulse may have been religious" (p. 18). "Two classes of men, the gentry and the newer urban aristocracy of merchants, assumed an enormous measure of responsibility for the public welfare" (p. 18).

Elizabeth I also authorized the creation of the legally sanctioned charitable trust. A persistent failing of Roman philanthropy had been the absence of a mechanism to permit the establishment of permanent endowments. Individuals had only friends and family to rely on to preserve the intention of their bequests. The medieval church provided some assistance, but the modern charitable trust, with its carefully rationalized claims against the future, was definitively created in England in 1601.

The new spirit of private philanthropy described by Jordan was ambitious in its aims as well as practical in its methods. The objects of philanthropy included an attack on poverty, not only by expanding and improving the system of almshouses and assistance for the aged and infirm, but by what Jordan calls *municipal betterments* ranging from fire protection to public parks and the maintenance of bridges. Education at all levels won new attention and emphasis, most significantly in efforts to improve schools for the poor; education was recognized as the best means to escape from poverty. Workhouses were built, loan funds were established. It was a period of great optimism and confidence, and it was followed by periods of failure and disillusionment—even with philanthropy.

For a very long time there had been complaints that philanthropy, usually described as indiscriminate almsgiving, reinforces the very conditions it seeks to eliminate. Giving alms to those who could work to support themselves was thought to encourage them to take the "easy way" and not work. It was believed that help should not be given to the able-bodied; that they should be forced to work and, when necessary, punished for not working. In difficult times, when everyone must contribute his or her share, idleness was thought to be criminal as well as sinful ("the man who will not work shall not eat"[16]).

These same sentiments were voiced in the eighteenth and nineteenth century attacks on Poor Law legislation. There are at least rhetorical parallels between these attacks and the twentieth century attack on public welfare programs.[17] The key questions are whether one can identify and separate out the deserving from the undeserving poor, and whether providing basic needs from philanthropic sources perpetuates poverty.

The sixteenth through nineteenth centuries added to the philanthropic agenda the questions about the growing role of the state, the diminishing role of the church, and the expanding influence of a new commercial class. Direct attention was focused on the link between philanthropy and the maintenance of social order and the growing concern about whether charity really works.

MODERN PHILANTHROPY

The nineteenth century was stimulated powerfully by confidence in the effectiveness of social organization founded on scientific principles. Critics measuring public relief under the Poor Laws by scientific standards revealed a scandal of bureaucratic inefficiency; private charity was declared by other critics to be a scandal of thoughtless and duplicative effort. Responding to criticism, private charity progressed to social service. Charitable activities were to be coordinated in the new Charitable Organization Society, founded in 1869. A key figure in this effort for more than forty years, Charles S. Loch, wrote *How to Help Cases of Distress* (1895), in which he deplored "charity which, for love or pity's sake, seduces the individual from the wise and natural toilsomeness of life . . . is . . . the poor man's greatest foe." Loch summarized five principles of charity:

(1) As a rule, no work of charity is complete which does not place the person benefited in self-dependence. . . . Charity should . . . become a partner, as it were, in the work of thrift.

(2) All means of pressure, such as the fear of destitution, a sense of shame, the influence of relatives, must be brought to bear, or left to act on the individual. He must, as far as possible, be thrown on his own resources.

(3) In deciding whether relief should or should not be given, or what assistance

should be provided, the family must be taken as a whole; otherwise the strongest social bond will be weakened. Family obligations—care for the aged, responsibility for the young, help in sickness and in trouble—should be cast, as far as possible, on the family.

(4) Further, as material charity is only a part, and a small part, of efficacious charity, a thorough knowledge is necessary both of the circumstances of the persons to be benefited and the means of aiding them; and the element of personal influence and control must very largely predominate over the monetary and eleemosynary element. . . .

(5) The relief, to effect a cure, must be suitable in kind and adequate in quantity. Charity must [consider its beneficiaries] not as the recipients of gifts, but men and women whose standard of life has to be raised. The truest charity often lies in the righteous fulfillment of duty, whether personal or public; and next to it must be placed that charity which is vigilant to see duty done.[18]

Loch measures charity by its results, and its results must show the individual standing on his or her own feet. He believed that relief would not work unless it carried with it new behavior and that new behavior could not be accomplished without direct intervention. There is strong emphasis on the inadequacy of the individual, and no question about what constitutes the virtuous life that the poor must be required to follow.

It is perhaps helpful to balance this passage with one from the nineteenth century's preeminent spokesman for social evolution, Herbert Spencer:

People who think that the relations between expenditure and production are so simple, naturally assume simplicity in other relations among social phenomena. Is there distress somewhere? They suppose that nothing more is required than to subscribe money for relieving it. On the one hand, they never trace the reactive effects which charitable donations work on bank accounts, on the surplus-capital bankers have to lend, on the productive activity which the capital abstracted would have set up, on the number of labourers who would have received wages and who now go without wages—they do not perceive that certain necessaries of life have been withheld from one man who would have exchanged useful work for them, and given to another who perhaps persistently evades working. Nor, on the other hand, do they look beyond the immediate mitigation of misery. They deliberately shut their eyes to the fact that as fast as they increase the provision for those who live without labour, so fast do they increase the number of those who live without labour; and that with an ever-increasing distribution of alms, there comes an ever-increasing outcry for more alms.[19]

Spencer sees the consequences of charity as harmful to the larger economy. He shares Loch's goal of a community of self-reliant citizens, but he thinks charity fails to reach it. Charity, in fact, he believes, *pauperizes*, reinforces the poor in their poverty. In spite of such criticisms, the progress made by Loch and others raised charity to a new level—organized, disciplined, purposeful.

What is the next phase of this evolution? Some argue that as charity was outmoded by social service, so social service will be superceded by justice. What has in the past depended on voluntary action should become obligatory of all citizens. Charity, as economist Barbara Bergmann argued in a *New York Times* article, must be coerced. The shift in emphasis is fundamental: from thinking about what someone may choose to give we pass to thinking about what someone is entitled to receive. What one has a right to have someone else may have a duty to provide. The criterion is no longer contribution but need.

Basic needs first, but what is sought is not only material relief but a change in social relations. "Civility towards others," wrote Collingwood, "is . . . inseparably bound up with self-respect. This enables us to distinguish two different kinds of demeanour which are often confused: *civility*, or the demeanour of a self-respecting man towards one whom he respects, and *servility*, or the demeanour of a man lacking self-respect towards one whom he fears."[20] Collingwood was not concerned with the issue of whether basic needs are a right (welfare) or a privilege (charity). Another way of looking at the problem, of course, would argue that there are market solutions, if social organization (especially the family) is maintained.

THE MODERN AGENDA

The modern agenda differs from previous ones in the extent to which government has taken on philanthropic responsibilities. "Public altruism,"[21] as one writer explains, covers far more than transfer payments to reduce poverty. Governments have come increasingly to bear the burden of spiritual alms as well as the corporal works of mercy as Aquinas described them.

The modern era provides a study in private initiatives to reform social institutions. In the realm of the social and cultural, private philanthropic initiatives shaped the duties and reordered the priorities of government. The more recent experiment reverses the process, enabling administrative agencies of government, with broad legislative mandate, to plan and organize the private sector.

Until the emergence of social democracy, philanthropy provided the financial means to transform the biblical priorities into programs of action. In motivation and means, social reform and philanthropy are inseparably linked in recent centuries. The abolition of slavery, the amelioration of conditions in asylums and hospitals, the reform of prison administration and judicial practice, the reduction and eventual elimination of child labor are cases in point. These were charitable and philanthropic endeavors in the broad sense begun in private initiative, largely inspired by religious values, often intended to improve government as well as private practices. Worked out in the process of social reform is a new way of thinking about the source of responsibility for suffering in this world that asks: to what extent is the individual responsible for his or her own condition or to what extent is the person a victim of social structures not of his or her own making?

This question includes such concepts as the Christian idea of sin. That concept is gradually replaced by the concept of the perfectibility of man. The shift from religious to secular is more than a matter of turning from religious to secular institutions. There is a new view of human nature, as dramatic as when the kingdom is declared to be found here on this earth rather than in heaven.

The age of revolution marked a transition in thinking about government, about who it serves and the services it provides. Call this long and complex process the emergence of *social democracy,* which seems a better term than *welfare state* because the latter suggests a narrower range of public concerns. *Social democracy* has come to mean government entering, dominating, and perhaps controlling all types of charitable and philanthropic activity.

The shift is from voluntary, private, largely religiously motivated and directed charity to tax-supported secular humanism (in its pre-Falwell sense), a shift of institutional framework and a shift in values as well. As private philanthropic practices become public, administrative actions, they are deliberately purged of their religious rationale. Philanthropy as a secular practice is then based on utilitarian rather than religious values.

This process of change can also be seen in the extension of modern philanthropy beyond national boundaries. Nineteenth century missionaries brought the philanthropic program with them; along with the Christian religious message came schools and hospitals, literacy and leprosariums. The religious missionary was largely superseded after World War II by secular missionaries, funded by governments, and providing cultural as well as corporal alms. Governments provided art exhibits, symphony orchestras, athletic coaching, libraries of approved books, as well as food and malarial suppressants to impoverished countries.

For what purpose? Public altruism is usually justified in terms of national interest, to the extent that some would have the United States provide food relief to friendly countries. Here, on a global scale, is the ancient practice of helping the stranger. Can it survive in a political environment if it is at root a religious doctrine? Or, is public foreign aid an example of religious influences shaping modern secular society?

American philanthropy in the nineteenth century extended from primitive medical assistance for Africans to advanced medical research for Americans. As private philanthropy expanded, usually well ahead of public initiatives, its list of priorities grew long. Andrew Carnegie's list is indicative; he suggested seven priorities: (1) founding a university; (2) establishing libraries; (3) establishing hospitals, medical colleges, and laboratories; (4) providing land for development of parks, (5) constructing meeting places and concert halls; (6) constructing swimming facilities (presumably we would broaden it to include recreational facilities); and (7) constructing of church buildings.[22] With the exception of the last item, all of these philanthropic activities are now common to the programs of governments at all levels.

Collective philanthropy as it appears in government has grown out of a new concept of the role of the state, and from the inability of voluntary philanthropy to meet the scale of charitable needs. It also emerged in the private sector in the late nineteenth century in the form of grants from business corporations. That practice grew from the individual philanthropy of owners and managers as the corporation developed. The self-interest of the business corporation has not prevented its participation in philanthropy.

The rise of corporate philanthropy is largely an American phenomenon, although recently corporations in other countries have also begun the practice. The acceptance of government-financed and controlled philanthropy is ideologically far advanced in the Western social democracies. On the one hand, government monopoly of support for higher education, for example, has only recently aroused concern about government control. The famous dictum that charity has no seat at the directors' table simply did not take hold. Circumstances developed in which it was clear that corporate interest and community interest would be served by philanthropic activity: building YMCAs along the railroad lines was the beginning. The corporation is viewed as a citizen, and citizens are expected to be generous.

In recent years corporate philanthropy has been criticized by some economists as being inconsistent with the corporate mission to make a profit. Philosophers argue about whether the corporation can be thought to be a moral agent, as the individual is.[23] A related inquiry asks whether a corporation can be said to have a conscience,[24] and, by implication, whether its conscience directs it to pursue philanthropic, altruistic purposes.

Modern values are said to be secular, because the secular state dominates society. The state has sharply reduced the role of the church, and that anticlerical victory is assumed to have reduced the importance of religion as well. The vast improvement in material conditions that has resulted from the economic achievements of the capitalist era has also, it is thought, replaced spiritual values with material ones. Technological advances in information and organization have made possible new ambitions as well as new magnitudes of philanthropic activity. Philanthropy has become collectivized, in the public and private sectors. Activity on such a large scale presumably entails depersonalization. Yet, what are we to make of the fact that some 80 million Americans offer their time as volunteer workers? To what extent does this minimize the depersonalization characteristic of philanthropy on a mass scale?

Given the religious origins and traditions that have been emphasized in this paper, the erosion of religious values in the face of the secularization of modern institutions is a matter of considerable consequence. One set of facts leads to the conclusion that philanthropy has been severed from its religious past. Another set of facts indicates that religion, even so, is still a powerful force in the realm of private philanthropy. Religion received 46.6 percent, or $43.6 billion, of the total of $93.68 billion contributed in 1987.

The objects of religious philanthropy are largely charitable, addressing many of the same concerns as the programs of public welfare. Religion is also the largest private influence in international philanthropy. And some of its work internationally seeks philanthropic ends—social reform, "development" rather than "relief." The effectiveness of the so-called liberation theologies, especially in Latin America, depends directly on religious philanthropy in support of political action. Evangelical theologies, especially in Africa, expand rapidly outside of, and often in opposition to, Socialist and Marxist governments. Evangelical emphasis, however, is focused primarily on individual salvation rather than social reform.

As noted earlier, perhaps religion has triumphed in that government has been infused with religious (charitable) values. The future of private philanthropy does seem inextricably linked to organized religion.

Secular values inhere in secular objects and institutions, those of government and the marketplace, and philanthropy is often compromised by its use of marketplace techniques, as it is bureaucratized by the organizational practices of government. Governments are faulted for insensitivity and rule blindness; markets are faulted for turning even altrustic relationships into self-serving ones. Large complex organizations tend to yield to bureaucratic weakness, whether public or private. Each is capable of turning philanthropy to its advantage, political or economic as the case may be. Philanthropy owes its credibility to its altruistic imperative. It is not that self-interest does not often yield altruistic benefits; what matters is that acts guided primarily by self-interest are called something else.

The interaction of voluntary, not-for-profit, private-sector, public-interest institutions with government and with business can lead to a confusion of roles. Any of the three sectors can be compromised by borrowing too many of the core values of the other. Some overlap is necessary as well as desirable; too much leads to an essential compromise of purpose and method.

Such scale and complexity require specialization. This occurs when philanthropists (such as Andrew Carnegie) list priorities; this occurs when institutions (such as colleges) define their services; this occurs when givers and seekers of funds come together in organizations to improve their effectiveness. We have much less history and tradition to rely upon in these matters of specialization. Those who serve as agents of benefactors and as agents of recipients, for example, claim a very small part in the literature of philanthropy. Those charged with collecting and distributing alms in the Mishnaic period of early Judaism were known as *Gabbai Zedekah* or *Parnasim;* they were highly respected members of the congregation and served without compensation. Deacons and presbyters presumably provided similar services in the early church.[25] The first "professional" collectors and dispensers of alms, however, appeared in the Middle Ages. The *almoner* distributed alms in behalf of the bishop (or in France, in behalf of the king).

The history of appeals for charity is largely that of spokesmen in behalf of the poor and the oppressed. Often the spokesmen included the nobility and, in more recent years, members of the middle classes as well. The prototype of organized appeals for charity, however, is the medieval mendicant, the monk given dispensation to appeal for funds for himself and for the poor, door to door. Abuses of the practice are severely denounced:

medieval mendicants were apparently guilty of some of the same bad practices as those engaged in by some television evangelists more recently.

Abuses of philanthropy are not simply fraud or the misuse of funds; the relationships are psychologically complex. One writer says there are four kinds of gifts: the anonymous, the nonreciprocal, the reciprocal, and the self-help ("the donor shares in the benefits of his own generosity").[26] Giving alms in public can encourage pride in the giver and shame in the recipient. Reciprocity can erase every mark of emotions such as humility and appreciation. Ancient society was aware of the awkwardness of giving and developed the technique of giving in secret. "No; when you do some act of charity, do not let your left hand know what your right is doing; your good deed must be secret, and your father who sees what is done in secret will reward you."[27] The *Lishkat Chashaim* of ancient Judaism was the "chamber of silence" where "the pious, unobserved, left donations to the poor."[28] The poor were thus permitted to be grateful to the community and were not bound by gratitude toward a specific individual.

The perspective of asking is examined less often than the perspective of giving: how does one persuade others to give and to give generously? Manipulation of emotion is practiced on a national and even worldwide scale in our time, but it is at least as old a practice as the art of persuasion itself. Philanthropy, after all, is a product of persuasion, not of logical demonstration. The abuses of rhetorical technique in a good cause are so familiar as to be commonplace: "charity needs coercion"—or at times, seduction, or manipulation, or threat.

The fund raiser for others is not a *schnorrer,* as the Yiddish term describes the professional beggar who engages in a contest with his benefactors for his own benefit.[29] The contest is the same, at times, but the beneficiaries are others (and the fund raiser only incidentally). Most of the actual asking for money is not done by professional fund raisers themselves, of course. Their task is to stimulate and organize others to do so.

There is often a stigma attached to asking for money, even when everyone recognizes the unavoidable necessity to do so. How many campaigns for colleges or hospitals or other important purposes fail because of the failure to ask rather than the failure to give? How much of the failure to ask results from timidity, embarrassment, or unwillingness to "lower oneself" to the subordinate role of petitioner?

In the vast and impressive organization of contemporary American philanthropy, what is at risk are the personal values and direct involvement that characterized philanthropy in earlier centuries. The very progress of society reduces the frequency with which the affluent confront the poor. In rapidly changing social circumstances it is possible that our attitudes become outmoded, our ability to respond, attenuated.

CONCLUDING QUESTIONS AND COMMENTS

Despite the modern movement from private to public philanthropy, private philanthropy continues to be the source of much, and probably most, social and cultural innovation. Ideas are introduced and tested by social and cultural entrepreneurs supported by philanthropic venture capital, and the successful innovations come to constitute a claim on public funds. Entrepreneurs sit on both sides of the philanthropic table, as do bureaucrats. The emergence of new "professionals" in the roles of almoner and mendicant raises new questions about how philanthropic policy is formulated, and whose values it reflects.

"Philanthropic activity can never be understood (or defined) except against the background of the social ethos of the age to which it belongs."[30] Our social ethos is reflected in the news items that opened this paper. Do they reveal a pattern of philanthropic activity and values that we can justify or condemn in terms of the tradition we have inherited? Who should give, how much, to whom, in what way, and, above all, *why* are the persistent questions.

- Should the purpose be taken into account? Is it appropriate to subsidize the world's largest rock concert? To what extent should the marketplace govern the survival of any cultural value?
- Should the distinction between public and private sectors prohibit private philanthropy from making grants to public agencies or institutions?
- Should we limit philanthropy to our own people? What obligations do we have to help Lebanon or Afghanistan rebuild? Is that a job for public foreign aid rather than private philanthropy? Should charity provide funds for relief and leave development to the government?
- Is there such a thing as "tainted money" for a good cause? A Bolivian drug dealer who buys food for the poor, passes out medicines to women, and restores churches, (and Al Capone and Robin Hood) violates the ancient principle not to take dishonestly or forcibly from one to give to another. Some clergy urged that gifts from Andrew Carnegie and John D. Rockefeller should be refused because they were derived from wealth gained by exploitation of the workers.[31]
- Should we accept money from foreigners, especially those whose wealth may have been generated from profits gained at the expense of the American worker or consumer?
- Are there limits to the measures people take to help the poor? Is "scrounging" in trash barrels to salvage surplus food an acceptable charitable practice? What right does an individual philanthropist have to decide how the elderly in our society should be treated? Is it not true that philanthropy "transfers to the hands of individuals what is essentially the special function of men in moral bodies, that is to say, what ought to be institutionalized as the expression of the moral will and sentiment of men in their corporate political capacity"?[32] The abuses of philanthropy are reported regularly in the press. Under the cloak of philanthropy it appears that some institutions have lost their integrity. Is there no way to control the philanthropic bribe or manipulation for personal aggrandizement?
- Is private charity a participant in determining public social welfare policy? Must charity be coerced?
- Is corporate philanthropy a contradiction in terms? Should it be? Is the corporation a new element in the philanthropic tradition that requires a new rationale? Is the corporation a moral agent? "Can a corporation have a conscience?"
- Does the publicity about philanthropy suggest that our society believes that more

good comes from giving in public than from giving in secret? Is the anonymous gift morally superior to the one in which the donor is known to the recipient? Is "enlightened self-interest" moral?

- Can we sustain our vast philanthropic enterprise and retain the values on which it is based? Will the pressure to give become an intolerable irritant?

Modern society seems to assume that it can fashion its values out of its social scientific understanding of people and events. Prevailing philosophy is utilitarian and pragmatic, and there is widespread lack of interest in history and deep skepticism about religion. How well equipped are we, morally and intellectually, to address the problems of philanthropy? It seems reasonable to inquire seriously into the values we offer our children. Are charity and philanthropy among them? Can they be taught? Where will they be taught and by whom? Were they attacked—say, as general education was attacked by some students and faculty fifteen years ago—would our defense of them collapse in confusion and uncertainty, as happened to general education?

Someone once wrote that he had forgotten the religious faith of his father's generation, that it was like a shadow for him. His son's generation, he concluded, would be even further removed from that living faith, that it would be "the shadow of a shadow" for them.[33] Is that the fate of the philanthropic tradition that we pass on? Will it become "the shadow of a shadow"?

Some answers might be proposed, some conclusions offered about the place of philanthropic values in our time.

- The integrity of philanthropic activity rests on the primacy of concern for others. To practice philanthropy is to engage in a constant struggle with the claims of self-interest. Our skill at rationalizing those claims threatens our philanthropic integrity.
- To achieve and maintain a high level of philanthropy requires effort of intellect and effort of will. As individuals and as institutions we have still not accepted how difficult it is to do the right thing. We tend to be guided by an easy sentiment or by narrow rationalism. We will improve our performance only if we seek to understand how philanthropy works—and why, so often, it doesn't.
- Philanthropic activity at its best is an informed discipline, a habit of mind and behavior infused with value. Such habits are acquired by socialization and education, especially in the early years. We should pay more attention to the way that philanthropy is taught.
- Philanthropy thus becomes a sort of First Amendment right, and we should protect the philanthropic entrepreneurs who presume to make our lives better. We should not let our legitimate concern about some silliness or even fraud to obscure the enormous value of charitable activity dispersed throughout the population.
- It is argued here that private initiatives supported or even originated by philanthropy have shaped the social role of government. As "public altruism" has emerged, those initiatives have been seized by bureaucrats drawing on public funds. Governments have often proved to be inept as philanthropic agents. The need to increase the magnitude of funds available for social and cultural pur-

poses has led us to rely on governments to take over philanthropic responsibilities. The answer to our dilemma appears to lie in tax incentives and in public funds administered by private agencies.

* Finally, the habit of philanthropy seems to be best acquired in a social *ethos* that encourages attention to matters of ultimate concern—to religious values. Both the material and the spiritual values of our society are tested in the matrix of philanthropy, and we cannot treat intelligently one without the other. Our secular preference for focusing on the material is not sufficient, we must come to grips with our religious values as well; and philanthropy forces us to do that.

Whether we like it or not.

NOTES

1. These definitions are based on Milton Rokeach, *The Nature of Human Values* (Free Press, 1973), p. 7.

2. Kenneth E. Boulding, *The Economy of Love and Fear* (Belmont, Calif.: Wadsworth, 1973).

3. These definitions are quoted from *Webster's New Dictionary of Synonyms*.

4. Henry Allen Moe, "Notes on the Origin of Philanthropy in Christendom," *Proceedings of the American Philosophical Society* 105, no. 2 (April 1961), p. 141.

5. Leviticus 19:9–10, *The New English Bible* (NEB).

6. Ephraim Frisch, *An Historical Survey of Jewish Philanthropy* [1924] (Totowa, N.J.: Cooper Square, 1969) p. 6.

7. Isaiah 3:14–15, NEB.

8. Isaiah 11:4, NEB.

9. Matthew 25:35–36, NEB.

10. *The Ethics of Aristotle: The Nichomachean Ethics,* trans. J. A. K. Thompson and Hugh Tredennick (Baltimore: Penguin, 1976), pp. 104, 142.

11. Cicero *On Moral Obligation (De Officiis),* trans. John Higginbotham (London: Faber and Faber, 1967), p. 54.

12. St. Thomas Aquinas, *Summa Theologiae, Charity* (2a2ae. 23–33) trans. R. J. Batten O. P. (Cambridge, Engl.: Blackfriars, 1974) vol. 34, pp. 241–43.

13. Jeremy Taylor, *The Rules and Exercises of Holy Living and Holy Dying* [1650] (London: Parker, 1859), p. 232.

14. John Maquarrie, *Dictionary of Christian Ethics* (Philadelphia: Westminster Press, 1967), p. 333.

15. W. K. Jordan, *Philanthropy in England 1480–1660* (New York: Russell Sage Foundation, 1959), pp. 16–18.

16. 2 Thessalonians 3:10, NEB.

17. Martin Anderson, *Welfare* (Stanford, Calif.: Hoover Institution Press, 1978).

18. C. S. Loch, *How to Help Cases of Distress* (London: Charity Organization Society, 1895), pp. v–ix.

19. Herbert Spencer, "The Need for Sociology" [1873], in J. D. Y. Peel, ed., *On Social Evolution* (Chicago: University of Chicago Press, 1972).

20. R. G. Collingwood, *The New Leviathan* (Oxford: Oxford University Press, 1942), p. 308.

21. Morris Silver, *Affluence, Altruism, and Atrophy* (New York: New York University Press, 1980).

22. Andrew Carnegie, "Wealth", in E. C. Kirkland, ed., *The Gospel of Wealth* (Cambridge, Mass.: Harvard University Press, 1962), pp. 33–46.

23. Peter A. French, "Corporate Moral Agency," in Tom L. Beauchamp and Norman E.

Bowie, eds., *Ethical Theory and Business* (Englewood Cliffs, N.J.: Prentice-Hall, 1979), pp. 175–86.

24. Kenneth E. Goodpaster and John B. Matthews, Jr., "Can a Corporation Have a Conscience?" *Harvard Business Review* (January–February 1982).

25. Frisch, *An Historical Survey,* pp. 39–40.

26. Jeffrey Obler, "Private Giving in the Welfare State," mimeographed copy (1979).

27. Matthew 6:3–4, NEB.

28. Frisch, *An Historical Survey,* p. 35.

29. Israel Zangwill, *The King of Schnorrers* [1894] (Mineola, N.Y.: Dover, 1965).

30. A. R. Hands, *Charities and Social Aid in Greece and Rome* (Ithaca, N.Y.: Cornell University Press, 1968), p. 7.

31. Robert H. Bremner, *American Philanthropy* (Chicago: University of Chicago Press, 1988), p. 107–108.

32. W. K. Jordan, *Philanthropy,* p. 215.

33. Mentioned by Will Durant in an interview.

2

The Gospel of Wealth: American Giving in Theory and Practice

KATHLEEN D. MCCARTHY

In 1975, Mrs. Beryl Buck died, leaving a tidy sum for the use of "non-profit charitable, religious or educational purposes" in her native Marin County, California. She left management of the funds, in the form of a trust, to the San Francisco Foundation, a community foundation whose other assets totalled $50 million at the time. At the time of her death, the Buck estate totaled $10 million: not one of the great fortunes of the twentieth century, certainly, but a respectable sum for local charitable and educational needs in one of the country's richest counties. What no one could have foreseen at the time was that her charitable nest egg would grow to almost $400 million in the years immediately following her death. Begun as a simple statement of civic stewardship, the Buck trust became a cause célèbre in modern philanthropic annals.

As nonprofit agencies in adjacent, less wealthy areas began to cast a covetous eye on the Buck Trust's resources, the San Francisco Foundation came under increasing pressure to spread the fruits of Mrs. Buck's beneficence more widely. The legal mechanism for accomplishing this already existed. Under the doctrine of cy pres trustees can modify the terms of a bequest if the original stipulations prove illegal or impossible to enforce. But Mrs. Buck's specifications were technically neither. To break the original tenets, therefore, the San Francisco Foundation would have to prove that trusts such as these should be altered not only when illegal or unenforceable but also if unwise. This it sought to do a decade later and failed.

The Buck Trust case served to highlight one of the leitmotivs of twentieth century philanthropy: the continuing quest to strike a suitable balance between the prerogatives of a donor and local and national needs. Civic stewardship, the notion that citizens owe a dual obligation of time and money to the communities in which they have lived and prospered, is one of our oldest national traditions. With the growth of professionalization (in social work, medicine, curatorial work, fund raising, and so on) and government services, these localistic values often clashed head on with national aims, counterpointing the quest for efficiency, system, centralized planning, and rational reform.

Prior to the twentieth century, it is likely that the arguments of those who wished to alter Mrs. Buck's bequest would have fallen on deaf ears. The philanthropic spirit that Americans inherited from their colonial forbears was often intuitive and highly personal. The act of giving was imbued with equal or greater significance than the gift itself—a means of reaffirming public faith that the links between wealth and virtue, individual prosperity, and community well-being were still intact.

These were imported ideas, formed in the tempestuous climate of the English Reformation. As Tudor reforms divested the church of its capacity to care for the country's educational needs and for the ailing and the poor, philanthropy was secularized, with much of the responsibility transferred to wealthy urban mercantile elites. As Wilbur Jordan notes in his superb study of Tudor-Stuart giving, the doctrine of stewardship that echoed from English pulpits during these years held that men were made rich so that they might benefit the poor. In the words of John Donne, the religious obligations of charity were "a doctrine obvious to all."[1]

These notions were imported to American shores with the first settlers. As John Winthrop sternly reminded his fellow passengers on the wind-swept decks of the Arbella in 1630, every man would be obligated to "afford his help to another in every want and distress" in the New World. "Wee must be willing to abridge ourselves of our superfluities," he urged, "for the supply of other's necessities." Winthrop's "Modell of Christian Charity" represented far more than the idealistic faith of a Jacobean gentlemen. For Winthrop and his fellow emigres, the twin notions of civic and religious stewardship lay at the heart of their wilderness experiment. Virtuous men would labor in their callings here, as in the more settled English communities they left behind. If they found God's favor and worked diligently, prosperity would be their inevitable reward. But prosperity could be a Pandora's Box, breeding jealousy, covetousness and civil unrest. The doctrine of civic stewardship ensured that excess riches would be siphoned back into the community, preserving the spirits of the wealthy while ministering to the needs of the less fortunate. In effect, it helped to hold the settlement together and make it a community in the wilderness, a "city upon a hill" to serve as a beacon to the rest of the world.[2]

Although personal service provided a more common gauge of stewardly impulses during these years, some notable gifts were given. One of the best memorialized was John Harvard's 1668 donation to the college that would henceforward bear his name. Other colonial gentlemen of property and standing later followed John Harvard's example, providing a comparable namesake for Eli Yale and, in Benjamin Franklin's case, a trust to provide loans for young married artisans of good character. Most of these gifts were given to local institutions, in response to local needs.

These practices took on added importance in the decades after the Revolution, both as a means of grafting people's loyalties to new communities and as a check on aristocratic leanings and material excess. If the rich donated their time and money to communal projects, it was reasoned, then class lines would not harden into caste lines and republican virtue would endure.

Within this milieu, men were constantly enjoined to be Christian gentlemen, devoting their cash and leisure to civic purposes. The cultured antebellum gentleman was to serve as a ready proponent for any sort of charitable, educational or cultural scheme to benefit his city. Rather than lavishing the fruits of his good fortune on himself, he was to minister to local needs, helping to provide essential services in an era of limited government. Outstanding stewards and their gifts were often publicly lauded in sermons and the press. Business publisher Freeman Hunt had a particularly good eye for model merchant stewards, regularly highlighting their activities in his periodical, *Hunt's Merchants' Magazine,* and books like *Worth and Wealth.*[3]

Although modest in comparison to the outpourings of the Gilded Age, some of these gifts assumed handsome proportions, fed by a broadening range of financial and mercantile pursuits. For example, in the 1830s and 1840s, John Lowell bequeathed the hefty sum of $250,000 to found the Lowell Institute, Abbott Lawrence donated $50,000 to Harvard, and John Jacob Astor singlehandedly created the library that bore his name. As these examples suggest, these were often highly personal donations, used to found community institutions that faithfully reflected their originators' aspiratons, biases, and norms. Rather than draw upon the advice of "experts," these men fashioned their donations according to their own experiences and firsthand perceptions of communal opportunities and needs.

Women played a far less prominent role as donors during the antebellum years. In part, this reflected the nature of their activities. Conscripted into philanthropic service primarily through churches, they formed their agencies through collective voluntary efforts, fashioning asylums, charities, and agencies for moral and spiritual reform through modest donations, avid fund raising, and gifts in kind. When contributions failed to cover programmatic needs, they often donated their own energy as well, nursing and training asylum inmates, sewing the linen, and even cooking the meals. These practices also reflected their inferior legal status. In an era when most women married, matrons were prohibited from owning property in their own right under the doctrine of *femme couverte.* As a result, until the passage of married women's property acts beginning in the late 1840s, most ladies were proscribed from controlling, much less donating, their dowries, earnings, or worldly goods.

While sermons celebrated women's activities, more quantitative, secular yardsticks were used to measure the stewardship of men, particularly at life's end. John Jacob Astor and Stephen Girard exemplified two extremes in these assessments. Astor was a cunning businessman who built his fortune in the western fur trade. One of the country's richest men, he left the sum of $500,000 to found the Astor Library in 1848. The press was openly disdainful, citing Astor's lack of civic spirit while alive and casting a dubious eye on the elitist institution bequeathed as his memorial. Although cultural institutions were needed, more broadly based activities were needed as well, and Astor was criticized for failing to serve the needs of the larger community more generously, in more perceptive terms.[4]

Girard fared a good deal better. His life was the stuff of legends. Born in Bordeaux, Stephen Girard emigrated to the New World as a cabin boy at the age of fourteen. After a brief stay in the colony of Port au Prince, he settled first in New York then Philadelphia. Through luck, persistence, and shrewd investments, the Frenchman ultimately came to head one of the country's largest banking concerns. But he was known as much for his civic stewardship as for his financial acumen. During the yellow fever epidemic that enveloped Philadelphia in a tide of suffering and death in 1793, Girard abandoned his countinghouse to oversee the city's pesthouse, caring for the inmates himself. Upon his death in 1831, the banker bequeathed the bulk of his fortune to civic ends, ranging from municipal improvements to local charities.

His major donation, however, was for a school for poor, white orphan boys, which received the then unheard of sum of $6 million. It was a quintessential antebellum donation, in that it bore the donor's imprint to an extraordinary degree. Girard's will specified how the institute was to be constructed—and conducted—in painstaking detail. In addition to specifying the building's measurements, he indicated its height, placement on the lot, where the windows would go, and the design of the interior walls. Inmates were to be admitted between the ages of six and ten, with preference given to youngsters of local birth. Distinctive apparel was not to be permitted. Teachers' qualifications and the content of the curriculum were similarly specified. Instructors were to be chosen according to their moral and personal merits, and coursework limited to basic reading, writing, grammatical and scientific skills. "I would have them taught facts and things, rather than words or signs," the practical businessman asserted.[5]

Girard's beneficence was lauded in his obituaries, but his heirs were less enthused. Unwilling to settle for the paltry $140,000 they had inherited, they took the case to court in an attempt to break the will. The basis of their contention stemmed from Girard's obvious anticlerical biases and the fact that his plans for the school were steeped in unabashedly secular aims. Impatient with sectarian rivalries that swirled around him, Girard had indeed stipulated that ministers be barred, not only from the school's board and faculty but from its premises as well.

The contestants' claims were awkwardly couched, charging that the will was too indefinite and vague to be executed. Their attorney, the celebrated orator Daniel Webster, was more to the point. Noting the document's secular tone and anticlerical aims, the statesman questioned whether "this devise be a charity at all." Because it threatened to weaken men's commitment to Christianity, Webster contended that it would foster "mischievous" rather than useful ends. Yet, neither Webster's eloquence nor his concern for the inmates souls' could sway the court, and the injunctions were implemented without revision. Girard's highly individualistic vision of a good and just society prevailed.[6]

The Civil War marked an important turning point in the course of philanthropic individualism, calling spontaneous beneficence into question for the first time. Prior to the war, stewards had been urged to give and encouraged to shape their gifts according to their own often highly personal interpretations of social needs. The war necessitated a more systematic response, stressing planning, organization, and discipline in the collection and distribution of donated funds and goods. The Sanitary Commission that oversaw these operations prohibited donors from earmarking their donations for specific individuals or regiments. In the process, the prerogatives of the steward were tempered and tamed for the good of the whole.[7]

During the Antebellum Era, the giver's tastes and whims were law. The Civil War discredited many of these aims, giving rise to a new interpretation of civic stewardship. In the ensuing decades the rich would be called upon to create new institutions for the "fit," leaving the "unworthy" poor to fend for themselves. The high priest of this Gilded Age Gospel of Wealth was one of the era's quintessential robber barons, as well as one of its greatest philanthropists: Andrew Carnegie. Carnegie was the icon of the age, the archetypal self-made man. Like Girard, the Scotsman first arrived on American shores as an impoverished immigrant. Through a combination of what contemporary authors fondly described as "pluck and luck," he managed to parlay his modest savings into an empire built on railroads and steel. His classic treatise on "Wealth" (1889) reverberated with the prejudices, aspirations, and biases of the fabulously wealthy postwar nouveaux riches he represented.

Unlike the antebellum Christian gentleman, Carnegie had a leering contempt for traditional charities. "In bestowing charity, the main consideration should be to help those who will help themselves," the millionaire intoned, since "neither the individual nor the race is improved by almsgiving." The best means of benefiting the community, according to Carnegie, was

to place within its reach the ladders upon which the aspiring can rise—free libraries, parks, and means of recreation, by which men are helped in body and mind; works of art, certain to give pleasure and improve the public taste; and public institutions of various kinds, which will improve the general condition of the people; in this manner returning their surplus wealth to the mass of their fellows in the forms best calculated to do them lasting good.

Universities, hospitals, medical colleges, laboratories, churches, and even swimming pools exemplified useful benefactions to benefit the industrious poor and sidestep the claims of "irreclaimably destitute, shiftless and worthless" idlers who refused to work. In the process, new opportunities for self-improvement and social mobility would be opened to the aspiring, and millionaires cast as "but a trustee for the poor, entrusted for a season with great part of the increased wealth of the community . . . administering it for the community far better than it . . . would have done for itself." "Thus is the problem of rich and poor to be solved," the industrialist sanctimoniously concluded.[8]

The man of means was to bestow not befriend. Unlike Stephen Girard, few Gilded Age stewards would volunteer to run an urban pesthouse or personally nurse the sick, much less tarry in their cities during times of pestilence or civic unrest. Instead, they expressed their civic spirit in monumental terms. For example, the decades between the Civil War and the turn of the century were punctuated by a series of great gifts for university development, including the creation of Johns Hopkins and Stanford universities and John D. Rockefeller's contributions to the newly reconstituted University of Chicago. Although Gilded Age donors were increasingly insulated from the harsh realities of the changing urban scene by their wealth and social standing, they felt their virtue vindicated by the monuments that bore their names. For men like Chicago's flamboyant meat-packing king, P. D. Armour (the founder of Armour Institute, which later became the Illinois Institute of Technology), these benefactions often provided a personal haven, a home away from home. Armour visited his institution daily, as did New York's Peter Cooper, whose last visit to Cooper Union came a scant three days before his death. For donors such as these, the institutions they so lovingly created provided a narcissistic reflection of themselves, hymning their achievements in lasting monuments of mortar and stone.

Ironically, Gilded Age beneficence and the highly individualistic ethos it embodied were roundly rejected by the generation of stewards who came to prominence around 1900, setting in motion a chain of initiatives to curb donors' prerogatives and loosen their control over their gifts. Washington Gladden's lively "Tainted Money" attack set the tone for subsequent debates. Written a scant six years after Carnegie's dictum, it castigated nonprofit organizations for accepting the fruits of ill-gotten gains. Pointing an accusing finger at industrialists who forced their workers into penury through unscrupulous employment practices, Gladden warned that "money secured by extortion or crime must carry a curse with it to all who, seeing the blood-stains upon it, covet it for themselves." The philanthropists who amassed these fortunes had done so "by the most daring violation of the laws of the land; by tampering with courts of justice; by the bribery of city councils or legislatures, and even of Congress itself; by practices which have introduced into the body politic a virulent and deadly poison that threatens the very life of the Nation."[9]

"Is this clean money?" asked Gladden. "Can any man, can any institution, knowing its origins, touch it without being defiled?" His answer, of course, was resoundingly negative. Educators and clerics who entered into partnerships with "corruptionists and extortionists" jeopardized their institutions as well as their souls. They bartered their freedom of independence, though, word, and deed. In the process, he predicted, the critical voice of dissent was forever stilled.[10]

Gladden's muckraking article marked a major turning point in American philanthropy, for it added a discordant note of skepticism for the first time. Donors had traditionally been categorized among the Elect: favored citizens who had prospered to the benefit of the community rather than themselves. From Winthrop to Carnegie, the donor's intentions and the wisdom of his or her designs were seldom questioned. And when they were, as in the case of the Girard will, public opinion and the weight of the law sided with the donor rather than the rapacious souls who sought to turn the fruits of their beneficence to more selfish ends. The act of giving seemed at the heart of the public good.

The Progressive Era marked a sharp turn in these attitudes, as donors' benefactions, business dealings, motives, and morals were increasingly subjected to public debate. In the process, the unquestioned association between wealth and virtue, philanthropy and the public weal was irrevocably sundered.

Gladden's criticisms resonated through a spate of periodicals, newspapers, and books in the decades prior to World War I, as muckrakers castigated the nation's "robber barons" for their inhumanity, their questionable business dealings, and their mountainous wealth.

The links between corporate power, civil unrest, and growing skepticism about philanthropy were most vividly illustrated in John D. Rockefeller's quixotic quest to secure a national charter for the foundation that was to bear his name. The idea was hardly without precedent, since agencies such as the American Historical Association had been so chartered. The oilman's intentions in seeking a national charter seem to have been quite innocent, a means of increasing the agency's public accountability by making it a national resource. The press and Congress, however, saw more insidious aims. Rockefeller's motives were vigorously questioned, and the specter repeatedly raised that the new foundation would come to rival the resources of the government itself. The bill was popularly caricatured as a ploy to usurp "The Political Sovereignty Now and Heretofore Lodged in the People of the United States." Critics ominously prophesied that the foundation's trustees would use "their delegated power to reward or punish, to build up or destroy, those who were subservient to them, or those who dared oppose them." In the process, political autonomy would be subverted and the citizenry enslaved. Stung by congressional intransigence and the vehemence of the public's response, Rockefeller withdrew his request and incorporated the foundation in New York in 1913.[11]

The project received a second round of criticism two years later, when the Senate Committee on Industrial Relations launched an inquiry into Rockefeller's business and philanthropic dealings, including a foundation-financed study of a violent strike at the Colorado Fuel and Iron Company, which was one of Rockefeller's commercial interests. The findings listed a litany of ills, charging the foundation with "creeping capitalism" and a

conspiratorial attempt to subvert the nation's educational and social services and suggested that the government should confiscate its funds. The
matter was ultimately dropped with the advent of World War I, and the
foundation survived, albeit with a lingering aversion to public scrutiny.

Individual philanthropists, as well as foundations, received a healthy
measure of criticism during these years. In one of the more balanced commentaries, the revered Chicago settlement worker Jane Addams typified
railroad magnate George Pullman as "A Modern Lear." Like Armour,
Pullman had sponsored a monumental solution to Chicago's social ills in
the form of the model industrial suburb that bore his name. Nineteenth
century reformers were fond of linking poverty and unrest to the evils of
dilapidated housing. Pullman, who had made a fortune by providing lux-
urious accommodations for the nation's railroad passengers, sought to
provide equally pleasant surroundings for his workers. Some of the intended recipients chafed at his magnanimity from the outset, complaining
that "We are born in a Pullman house, fed from the Pullman shop, taught
in the Pullman school, catechized in the Pullman church, and when we die
we shall be buried in the Pullman cemetery and go to the Pullman hell."
The company's ill-considered decision to lower salaries but not rents, during the Depression of 1893, sparked one of the nation's most vicious industrial uprisings.[12]

Addams traced the root of the strike to Pullman's paternalism and his
insensitivity to his employees' wishes, aspirations, and needs. Cut off from
the daily lives of his employees, Pullman "cultivated the great and noble
impulses of the benefactor" but lost "the power of attaining a simple human relationship" with them. He ceased to measure the usefulness of his
benefactions "by the standards of men's needs." "In so far as philanthropists are cut off from the influence of the *Zeit-Geist,* from the code of ethics
which rule the body of men . . . so long as they are 'good to people'
rather than 'with them' they are bound to accomplish a large amount of
harm," Addams sadly concluded, and very little good.[13]

Rather than standard-bearers of public virtue and well-being, philanthropists were now limned as corporate conspirators, plotting to coopt the
nation's schools, the media, the clergy, the government, and indeed its very
soul through their cleverly constructed, self-perpetuating funds. The individual steward no longer seemed best qualified to determine the nation's
needs or the uses to which philanthropy should be applied. As a result, a
new definition of philanthropy began to emerge. Formulated by middle-
class experts and specialists in social reform, the Progressive reassessment
called for more decentralized services accessible to the slums; a greater
reliance on research, rather than impulse, to define emerging needs; flexibility, rather than seemingly permanent solutions; a working partnership
between donors and professionals; and a movement away from older, institutional forms. Rather than creating monuments to last the ages, philanthropy was now to test new models for dealing with emerging needs, turn
the best over to government, and then move on to fresh fields.

These dicta reflected a broader "search for order" that lay at the heart of the evolving national experience. During the Antebellum Era, federal intervention had been minimal. After the Civil war, this system began to erode, undermined by transportation and communications innovations that increasingly called for national, centralized decision making. The public greeted these trends with a good deal of apprehension. Indeed, much of the outcry against corporations stemmed from a growing fear of monopoly, of forces outside the community that increasingly touched upon the quality of local life.

Philanthropy served as both ballast and spur to these trends. Despite a rather rocky beginning, government increasingly came to rely on foundations for research and aid in implementing national reforms. Donors whose fortunes were made nationally also began to look beyond their native communities in dispensing their largesse: hence the University of Chicago, founded by a Cleveland businessman, and foundations such as those founded by Carnegie, Rockefeller, and Russell Sage, devised to serve national and international aims. At the same time, local giving and voluntarism counterpointed these trends. Although the prerogatives of the donors were beginning to be hedged in by the growing ranks of professional policy makers, those with the necessary money, leisure, or vision could still make an impact on their communities beyond the political sphere.[14]

Progressive reformers sought to harness these energies and place them at the service of more systematic ends. In effect, they adopted the role of mediators and social engineers, defining the problems to be addressed and marshaling the behavior of both donors and recipients. Traditional and new communications media provided their primary tools. As the upper and middle classes increasingly insulated themselves from city slums, photographs became a persuasive medium for defining the scope of urban need. In the hands of a masterful reformer such as Jacob Riis, photographs documented the harrowing details of slum life while obviating the need for firsthand observation. Publications further amplified audiences, both through the burgeoning periodical press and books such as Jane Addams' *The Spirit of Youth and the City Streets* and *Twenty Years at Hull-House*. Films played a similar role, conveying new ideas among reformers and the slum dwellers who flocked to urban nickelodeons in the years before World War I. Social surveys added insights, quantitatively charting the dimensions of poverty, while settlement workers stumped the lecture circuit with their findings, carrying the gospel of reform to churches and clubs across the nation.[15]

While settlements and social workers redefined the needs of the poor, professionalization gave rise to new levels of expertise in science, medicine, museum management, and fund raising. During the 1920s, the cult of science reigned supreme, coloring a variety of measures in the philanthropic, social, and professional arenas of American life. Thus, John Price Jones helped to spearhead the professionalization of fund raising. Like contemporary social workers, Jones' firm predicated its campaigns on careful research, basing its queries on the Behaviorist faith that giving could be systemized into predictable patterns of stimulus and response. A whole range of techniques were generally prescribed to tap the interest—and pocketbooks—of potential donors. For example, workers in a University of Chicago development campaign were coached in the implications of recently minted tax laws that had allowed charitable deductions for the first time in 1917. Dinners and lectures were planned, as well as a media campaign to " 'sell' the University to the public in Chicago and throughout the country." In each forum, the university's practical contributions to science and the "solution of world problems" were stressed. Media, science, and publicity helped to define the university's needs, encouraging donors to render the necessary funds for research.[16]

Like the Progressive reform movements, Jones' appeal lay in the ability to systematize the philanthropic impulse and harness it to often intangible ends. The scientist in a laboratory, the visiting nurse pursuing rounds in city slums, the social worker carefully tallying the scope and nature of urban want—all were steeped in professional techniques designed to foster the public weal but offering little in the way of opportunities for lay participation. The donor was increasingly marginalized in the process, assigned to junior status in the philanthropic partnership. As a result, efforts to channel support into a more reliably fluent stream became appealing, whether through the high-powered media blitz of a John Price Jones campaign or the emotional appeals of the United Way.

In the words of one scholar, "an anonymous public" was now asked to support "an anonymous machinery" to serve often anonymous ends. The decades bracketing World War I also witnessed a growing emphasis on curbing the spectral reach of the "dead hand." Initiated in Cleveland in 1913, the Community Trust movement sought to widen the gap between donor and gift by reducing the threat of "mortmain"—control from beyond the grave. Donors were encouraged to set up small foundations under community trust auspices, to be administered by experienced trustees. Besides the obvious advantage of having someone else do the administrative work, promoters stressed the plan's flexibility. Although bound to honor the donor's general aims, whether aiding children, medical charities, or the arts, the trust's administrative framework permitted the grants themselves to be calibrated to contemporary needs. Community trusts ensured that future generations of managers, rather than earlier generations of laypersons would decide the destiny of the gift.[17]

Chicago philanthropist Julius Rosenwald went a step further by suggesting that even foundations should not be allowed to live beyond their prime. Rosenwald had an aversion to perpetual endowments and gleefully collected examples of trusts gone awry. Among his favorites were Ben Franklin's apprentice fund, Bryn Mawr's baked potato endowment, and an 1851 endowment to aid weary pioneers passing through St. Louis on their westward trek. By the 1920s, the last fund had almost $1 million in its coffers, where it languished for lack of suitable grantees. In Rosenwald's estimation, perpetually endowed trusts and foundtations were a "blight." As he explained, "Human conditions are changing so rapidly that a project which today is entirely commendable may be not only useless but viscious tomorrow." To emphasize his point, he stipulated that his own foundation, the Rosenwald Fund, be self-liquidating. That is, it would be required to spend both its interest and endowment within the span of a single generation and dissolve.[18]

During the forty-year interim that separated Carnegie's confident "Gospel of Wealth" and the bleak days of the Great Depression, the notion of civic stewardship was revised considerably. Neither the donor's motives nor the donor's decisions were beyond question after the turn of the century. Instead, growing ranks of professionals worked to systematize the benevolent impulse and channel it toward responsible ends. Science, rather than simple good will, infused the philanthropic ethos of the Jazz Age. In the process, donors were increasingly asked to pay for projects they often could neither touch nor comprehend, deferring to the wisdom of reformers and managers of every stripe and hue.

Nevertheless, American donations steadily climbed during the first three decades of the twentieth century. In 1905, estimates placed the level of annual contributions at $135 million, with colleges, hospitals, and dispensaries heading the list of favored recipients. Two decades later, the figure neared the $2 billion mark, a threshold first crossed in 1925. By 1928, American donations totalled over $2 billion, topped by $60 million in bequests to cultural organizations and medical institutions from Payne Whitney.[19]

The spiral ended abruptly on a bleak October afternoon in 1929. Like the Civil War and Washington Gladden's ill-tempered article on "Tainted Money," the stock market crash cast the donor's role in a new light.

Herbert Hoover's political downfall stemmed, certianly in part, from his determination to preserve the prerogatives of the local donor. Local citizens initially sought to heed his call, forming emergency funds for what many viewed as a short-lived economic downturn. As one observer optimistically noted, "charity is straining every sinew to keep up with the aching need [and] indications are that its coffers will be able to meet every appeal."[20]

As the Depression wore on and breadlines lengthened, the initial optimism soon faded to grim concern. By 1933, giving in the nation's six largest cities had dropped by 19 percent, with education and religious giving taking the brunt of the cuts. As donors' capacities to support local services began to wane, critics subjected their activities to sterner scrutiny, complaining that private charities had failed to keep pace with public needs, that the rich had grown insensitive and ignored their responsibilities toward the needy, necessitating more draconian reforms.

The underlying reality, of course, was that the country's charitable needs had far outpaced the capacities of private largesse. In New York alone, estimates placed the number of families on relief as high as 500,000 at mid-decade, and the numbers continued to climb. As New Deal measures were drafted to address these needs in the euphoric first months of Roosevelt's administration, thoughtful observers began to ponder the implications of the president's designs. Some viewed the prospect of federal aid with an almost audible sigh of relief. Government, it seemed, would provide a sensible alternative to misguided donors who squandered their funds on studies of "the causes of pessimism in the Middle Ages" while leaving more basic needs untouched. It also promised to avoid the pitfalls of emotional, sentimental charitable schemes that encouraged the dependence of the "shiftless," only to leave them "high and dry when [the money] is spent." The *New York Daily News* went so far as to suggest that the government should tax foundations, since private trusts diverted money from the public till and reduced needed revenues.[21]

Others, such as Nobel Laureate Nicholas Murray Butler of Columbia University, regarded such notions as an appalling affront to "the public interest and the public service. Neither Communism nor Fascism could do more." If government encroachment were not kept in check, Butler prophesied that private philanthropy would be "crippled or destroyed," supplanted by "the halting, imperfect, and often incompetent hand of government." Abraham Flexner urged a more moderate approach as well, since "in a vast and diverse nation, the highest interests—social or intellectual—cannot safely be left either to government alone or to private agencies alone." "Complete governmental control involves the dangers of repression, red tape, partisanship," he explained, "and these dangers become greater as the area of the country enlarges—witness Russia, Germany, Italy. On the other hand, uncontrolled private initiative has developed abuses that are absolutely incompatible with a high social and moral sense." Thus, he concluded, "unless private means continue to exist, the role hitherto played in our history by private enterprise in education, in art, and in philanthropy is doomed to shrink in importance, and perhaps ultimately to disappear."[22]

The rise of federal funding cast private giving in a new light. "Until recently the superiority of private philanthropies to publicly controlled, tax-supported institutions has been widely assumed," observed a writer in *The Atlantic Monthly*. "But now the excellent work of many state-supported institutions raises the question whether there is any inherent objection to the government carrying a greater part of the burden of education and welfare than it has done hitherto."[23]

The black and white distinctions these writers saw—institutions based exclusively on public or private giving—proved illusory. As the welfare state expanded, so did the tendency to blend support from a variety of sources, be they individual donors, foundations, corporations, or government agencies. What did change was the way in which private giving was regarded. Increasingly, those who sought to diminish the philanthropist's prerogatives would cloak their arguments in the notion that private giving diverted money from the public till. Thus, although tax incentives undeniably helped to inspire charitable giving, they also laid the groundwork for these debates. By the 1950s, the parameters of public and private giving were increasingly blurred, and the philanthropist's distinctive role increasingly questioned.

Many of the discussions centered on private foundations. The immediate post-World War II years witnessed a remarkable upsurge in the number of these institutions, fueled by general prosperity and favorable tax laws. Over 5,000 were in operation by the mid-1950s, and their numbers continued to climb, reaching well over 20,000 in the 1980s. Congressional interest rose correspondingly, as did the number of regulations governing their management.

In 1950, nonprofit organizations were prohibited from running businesses as tax-exempt concerns. The law traced its origins to a spaghetti company owned by New York University. In 1949, the school was audited by the IRS, which argued that NYU owed $1.5 million in back taxes on the company's profits. The IRS won the resulting lawsuit with the argument that noodle making was neither a charitable, literary, scientific, nor educaitonal endeavor but a taxable business. Several of the post-mortem commentaries scornfully assayed the perquisites of private giving. "In the name of Sweet Charity and her equally gentle sister, Higher Education, a tremendous lot of money which otherwise would to go the United States Treasury, is sticking to the fingers of lawyers, promoters . . . foundations and universities," one observer sneered.[24]

Fed by evidence of genuine abuses, attacks such as this continued throughout the Eisenhower, Kennedy, and Johnson years. A 1954 commentary estimated that fully $120 million of more than $4 billion donated annually was given to "out and out frauds." Similarly, an article on the Ford Foundation's reorganization along national and international lines in 1950 suggested that the foundation had initially been created as a means of maintaining control of the family business while sidestepping high inheritance taxes.[25]

Although the Cox and Reece congressional investigations of the early 1950s, redolent of the McCarthy witch hunt era, failed to produce any substantive legislation, the very wildness of their charges—that the Rockefeller Foundation had precipitated the fall of China into Communist hands through its support of the Peking Union Medical College or that grants for social science research constituted a Socialist plot—demonstrated that the bulk of the American public knew so little about this brand of giving that almost anything seemed plausible. Although the Cox Committee's final report was surprisingly temperate, exonerating most of the nation's grant makers from the taint of Communist associations, the Reece Committee had a virtual field day, cataloguing a litany of abuses and ills. The majority report complained that private foundations exercised enormous control over American life and foreign policy, using their ill-merited gains for vaguely conspiratorial ends.

While Cox and Reece attacked foundations from the right, the investigations conducted by Texas Congressman Wright Patman in the 1960s took a different tack. Rather than deriding grant makers for their supposed ideological leanings, Patman assumed a more populistic stance, citing evidence of fiscal abuses that had diverted an estimated $7 billion from the public till. In Patman's scenario, foundations were born primarily of a desire to sidestep personal, inheritance, and business taxes. Rather than disinterested philanthropy, their efforts were rooted in the lust for gain.

Patman's lone crusade opened the door to more substantial inquiries. A Treasury Department investigation initiated at his behest did, indeed, unveil some of the abuses Patman had suggested. Spurred to action by these findings, the House Ways and Means Committee launched yet another inquiry, this time under the chairmanship of Arkansas Congressman Wilbur Mills. Drafted in 1969, the resulting tax legislation included a number of salutary measures, such as prohibitions against self-dealing, as well a payout clause that set a minimum on foundation disbursements and an excise tax to cover the costs of IRS regulation, set at a rate many considered abusive.

Individual tax laws have also been periodically redesigned to curb or inspire new gifts, reflecting the political temper of the times. Even some formerly autonomous trusts have fallen under the government's vigilant scrutiny. Ironically, Girard's bequest was ultimately overturned by the courts in the 1960s, with the argument that its exclusionary mandate violated civil rights laws.

National attitudes about donors and their gifts have shifted considerably since Girard first wrote his will. The first half of the nineteenth century was the heyday of philanthropic individualism. Men and women were urged to contribute time and money to their communities, and their efforts often provided essential services when few other alternatives existed. Civic stewardship was viewed as an essential check on materialism and unrest, an index of individual virtue and a means of retaining an aura of social cohesion while mitigating social needs.

The value of spontaneous beneficence was first called into question with the Civil War. Faced with the monumental task of systematizing the contributions of an anguished nation and distributing them impartially within the field, Sanitary Commission workers placed themselves squarely between the donor and the object of his or her largesse. After the war, self-made philosophers such as Andrew Carnegie would condone a newer, less sentimental and more managerial role. Carnegie's Gospel of Wealth urged aspiring donors to forge institutional bonds with those they sought to aid and to earmark their gifts for the "fit." Rather than personally ministering to the needs of the poor, donors of Carnegie's generation would use their managerial skills and surplus profits for the benefit of their communities.

Carnegie and his self-made peers increasingly came under fire from Gilded Age and Progressive Era muckrakers. Washington Gladden's tainted money critique set the stage for two decades of intensive debate about the virtues and vices of America's stewards. By the time the first American soldiers embarked to take their place in the trenches of World War I, a new definition of philanthropy had been forged. Research, flexibility, decentralization, a partnership between donors and experts, and noninstitutional solutions to social needs were now celebrated as the earmarks of enlightened giving. In the process, donors were increasingly cast in a secondary role, bowing to the perceptions of growing ranks of professionals and often asked to fund new programs carried out beyond their range of vision. Community Chests and Community Trusts, professional fund-raising campaigns and Julius Rosenwald's crusade against perpetual endowments—all these were designed to streamline the giving process and adapt it to changing needs.

With the advent of the Depression, government entered the funding arena as well, sponsoring projects that had traditionally been the province of private largesse. As government services expanded, so did the dimensions of federal funding needs. Henceforward, philanthropists would have to justify their efforts as an alternative to public giving.

The brouhaha over Mrs. Buck's donation represents yet another chapter in the twentieth century quest to harness the philanthropic spirit and make it more responsive to managerial definitions of local, regional, and national ends. At stake was the fundamental issue of control, whether the donor's wishes should prevail no matter what the circumstances, or whether professional analyses should be paramount.

Beyond its legal implications, the Buck Trust case tested the resilience of one of America's most enduring practices: civic stewardship. Mrs. Buck had hoped to leave her fortune for the good of her community, to don the steward's role. The fact that the case was decided in her favor reveals the continuing power of this ideal.

With the growth of federal government responsibilities after the turn of the century, the practice of civic stewardship assumed new meaning, providing an important safety valve for the spirit of American individualism in an increasingly bureaucratized milieu. As such, it has continued to serve as an important counterweight to government expansion, offsetting the need for national planning, management, and control with grass-roots autonomy and individual initiative. In the process, individual giving served both Jeffersonian and Hamiltonian ends, paving the way for the expansion of government services while simultaneously humanizing and balancing these trends. As such, American philanthropy constitutes one of the quintessential earmarks of American pluralism, enabling individual citizens to pursue their vision of a just and equitable society beyond the range of government control.

NOTES

1. Wilbur K. Jordan, *Philanthropy in England, 1480–1660: A Study of the Changing Pattern of English Social Aspirations* (London: George Allen and Unwin, 1959), p. 182. For

a fuller discussion of the notion of civic stewardship, see Kathleen D. McCarthy, *Noblesse Oblige: Charity and Cultural Philanthropy in Chicago, 1849–1929* (Chicago: University of Chicago Press, 1982).

2. John Winthrop, "A Modell of Christian Charity" [1630], in Jack P. Greene, *Settlements to Society, 1584–1763* Vol 1 (New York: McGraw-Hill, 1966), p. 68.

3. Freeman Hunt, *Worth and Wealth: A Collection of Maxims, Morals, and Miscellanies for Merchants and Men of Business* (New York: Stringer and Townsend, 1856).

4. For a discussion of commentaries surrounding Astor's gift, see Sigmund Diamond, *The Reputation of the American Businessman* (Cambridge, Mass.: Harvard University Press, 1955), Chapter 2.

5. Quoted in Freeman Hunt, *Lives of American Merchants* (Vol 1 (New York: Hunt's Merchants' Magazine, 1856), p. 268.

6. Daniel Webster, *A Defense of the Christian Religion and of the Religious Instruction of the Young* (New York: Mark H. Newman, 1844), pp. 11, 12, 20.

7. For an excellent discussion of the impact of the Sanitary Commission, see George M. Fredrickson, *The Inner Civil War: Northern Intellectuals and the Crisis of the Union* (New York: Harper Torchbooks, 1965).

8. Andrew Carnegie, "The Gosepel of Wealth," in Edward C. Kirkland, ed., *The Gospel of Wealth and Other Timely Essays by Andrew Carnegie* (Cambridge, Mass.: Belknap Press of Harvard University Press, 1962), pp. 27, 28, 31.

9. Washington Gladden, "Tainted Money," *The Outlook* 52 (November 30, 1895), p. 886.

10. Ibid.

11. Washington Gladden, "A Questionable Benefaction," *The Independent* 68 (June 23, 1910), pp. 1406–1407; A. J. Portenar, "Looking a Gift Horse in the Mouth," *The Independent* 68 (June 23, 1910), p. 1388.

12. Quoted in Ray Ginger, *Altgeld's America: The Lincoln Ideal versus Changing Realities* (Chicago: Quadrangle Books, 1958), p. 149.

13. Jane Addams, "A Modern Lear," *Survey* 29 (November 2, 1912), pp. 131–37, reprinted in Christopher Lasch, ed., *The Social Thought of Jane Addams* (Indianapolis: Bobbs-Merrill Co., 1965), pp. 111, 112, 118.

14. For an excellent study of the symbiotic relationship between foundations and government, see Barry D. Karl and Stanley N. Katz, "The American Private Philanthropic Foundation and the Public Sphere: 1890–1920," *Minerva* 19 (Summer 1981), pp. 236–70.

15. Jane Addams, *The Spirit of Youth and the City Streets* (New York: Macmillan Co., 1909); *Twenty Years at Hull-House* (New York: Macmillan Co., 1910).

16. John Price Jones, "Program of Publicity" (August 28, 1925), p. 18, in Harold Swift mss., Regenstein Library, University of Chicago, Box 73. For an excellent account of the history of the professionalization of fund raising see Scott M. Cutlip, *Fund Raising in the United States: Its Role in American Philanthropy* (New Brunswick, N.J.: Rutgers University Press, 1965).

17. Roy Lubove, *The Professional Altruist: The Emergence of Social Work as a Career, 1880–1930* (Cambridge, Mass.: Harvard University Press, 1965), p. 172.

18. Julius Rosenwald, "Principles of Public Giving," *The Atlantic Monthly* 143 (May 1929), reprinted in Brian O'Connell, ed., *America's Voluntary Spirit* (New York: The Foundation Center, 1983), p. 126; "The Burden of Wealth," *Saturday Evening Post* 201 (January 5, 1929), 12.

19. "The Benefactions of the Year," *The World's Work* 12 (August 1906), p. 7816; "Billions for Practical Piety," *Literary Digest* 100 (January 26, 1929), p. 28.

20. "Charity Keeping up with the Demand," *Literary Digest* 107 (December 6, 1930), p. 22.

21. "Where Philanthropy Spends Its Millions," *Literary Digest* 115 (February 4, 1933), p. 21; "Better than Bread Lines," *Literary Digest* 108 (February 14, 1931), p. 19; *New York Daily News* (June 9, 1937).

22. Nicholas Murray Butler, quoted in "Revolution by Taxation," *School and Society* 44

(December 26, 1936), pp. 854, 855; Abraham Flexner, "Private Fortunes and the Public Future," *The Atlantic Monthly* 156 (August 1935), p. 219, 222.

23. John Crosby Brown, "Private Giving and Public Spending," *The Atlantic Monthly* 161 (June 1938), p. 815.

24. A. G. Mezerik, "The Foundation Rocket," *The New Republic* 122 (January 30, 1950), p. 11.

25. Jerome Ellison, "Who Gets Your Charitable Dollars?" *Saturday Evening Post* 226 (June 26, 1954), p. 27; "The Ford Foundation," *The Nation* 171 (October 7, 1950), p. 300.

II

INDIVIDUALS: THE GREATEST SOURCE

3

Patterns of Giving

ARTHUR H. WHITE

The Old Testament is very clear on the obligation to be charitable: "And if thy brother be waxen poor and his means fail; then thou shalt uphold him: as a stranger and a settler shall he live with thee. Take thou no interest of him or increase; but fear thy God; that thy brother may live with thee. Thou shalt not give him thy money upon interest, nor give him thy victuals for increase" (Leviticus 25:35–37). Moreover, the relationship between religious observance and giving is very powerful. Why then do more than half of all Americans give less than 1 percent of their incomes to charities? After all, if we add the fundamentalist Protestants, Orthodox Jews, and traditional Catholics together we have millions of Americans who follow the Bible's tenets strictly on many matters. Why don't more of them follow the commandments on giving more faithfully? The answer may lie in a discovery researchers made a few years ago.

In 1983, Yankelovich, Skelly, and White inquired into the relative clarity of Americans' thinking on a wide variety of subjects. They developed a "mushiness" index that showed very clear opinions and reporting on some subjects but very "mushy" and unreliable results on others. It may be surprising to many to learn that decisions on giving and responses to questions on giving turn out to be high on the "mushiness" scale. Why? There are a flock of causes of ambiguity:

Guilt: Individuals have given less than they feel they should have given and their reporting of gifts is importantly affected.

Habitual inaccuracy in reporting and discussing their charity: Large numbers of givers never tell the truth about their giving for one or more of the following reasons: they are not able to respond to all requests for giving and resort to untruths to get rid of some requests; they consistently understate their giving because they fear that the better known their willingness and ability to give, the more likely it is that they will become the targets of requests for funds; and some exaggerate the extent and nature of their giving to impress or make people think better of them. So, for the many who rarely (if at all) think or talk accurately about giving, mushiness of response to survey questions on the subject is common.

Confusion about giving: the rapid growth in charity appeals at business locations, at home, via television, mail, and so forth causes many to have difficulty in remembering how much and where they give.

Need to know more before they give more: What may be most important is the widely held belief that individuals lack information, which leads to uncertainty and "mushiness." Although much information is given to potential donors about needs and uses of gifts, there is considerable evidence to support the conclusion that more is needed if giving is to increase and if mushiness is to decrease.

Acknowledging mushiness is not a reason, however, to disclaim or doubt the value of studying or attempting to influence charitable behavior. Rather, it is intended to suggest that thorough, extended effort is necessary to obtain reliable results.

KEY FACTS ABOUT GIVING

A number of impressive facts stand out when one considers philanthropic giving in the United States in the 1980s. A major source of these facts is the Yankelovich, Skelly, and White Study on Charitable Behavior of Americans carried out for the Rockefeller Brothers Fund in 1985.[1] (See the Appendix to this chapter for a description.)

- Virtually everyone claims to give (nine out of ten people reporting in surveys carried out in the 1980s). But ambiguity and mushiness set in right here because the variation in extent of giving is very great. Half of all Americans gave 1 percent or less of their incomes to charities, another one-fourth gave 3 percent or more. So, giving is hardly uniform, fixed, and following generally accepted rules such as those laid out in Leviticus.
- By far the most favored charity is a church or religious organization (almost half of every dollar in 1985). But Catholic, Jewish, and Protestant groups have widely varying rates of success in solicitation, which are detailed later in this chapter.
- Giving is not correlated directly with income. Surprising to many is the fact that those with household incomes of less than $10,000 are found to give the highest percentage of income (3 percent to charity. Giving generally grows with age. Persons under 30 give the smallest percent (1.6) of their income, and those from fifty to sixty-four years of age give the highest percent (3.0). Persons over sixty-five, although they have the lowest average household income—just under $15,000—still are among the most generous givers, donating an average 2.7 percent of their incomes. And lest charities lean toward giving up on the under thirty-five cohort as stingy and self-centered, it can be argued, as will be shown, that giving is a cultivated, growing process and much can be done in the early years to assure a better crop in later years.
- Giving tends to grow with education. Those who have done postgraduate work report an average of almost one-and-one-half times the level of giving (3.0 percent of income) as those with less than a high school diploma.
- But before we conclude that the simple game plan for charities should be to go after the religious, older, educated Americans, take note of the key importance

of several other variables: discretionary income, worry about money, and volunteering. Individuals of the same age, income, and education are found to give more if they perceive that they have a moderate or large amount of discretionary income, do not worry about money, and do volunteer work. The remainder of this chapter deals with all these factors in greater detail.

VIRTUALLY EVERYONE CLAIMS TO GIVE

The most heartening finding with respect to Americans and giving is that an astonishing nine out of ten people questioned report giving. Individuals differ a lot in how, when, and where they give, but virtually everyone claims to give and, therefore, the challenge is to increase the amount of giving. This is an easier job than the common marketing problem, which is to convert those who are not believers or customers. In approaching givers with the purpose of increasing their level of giving, it is helpful to consider each element in a widely used model of consumer behavior that shows how awareness and identification are needed to produce favorable attitudes, which then lead to the desired behavior (in this case, increased giving).

Awareness of the needs of charities is high in the United States. Only 5 percent assert that they "don't give more because they don't think there are any charities that deserve their support." The public information programs of charities, the government, and religious organizations, using the heavy program of "space" donated to charitable funding campaigns (valued at $1.2 billion in 1986 by the Advertising Council) has clearly made Americans aware of the need for charitable giving.

An important barrier to increased giving is apparent, however, in the area of *identification*. Identifying an organization or charitable group amounts to having a perception of it—believing you know it and have a basis for judging it. There is considerable evidence to support the conclusion that charities that make their activities better known and understood, therefore perceived more clearly by more people, will receive more gifts. The proof that improving identification or perception pays off is shown in these facts. Analyses of giving regularly reveal that the United Way, religious, and educational organizations tend to receive the most support. These institutions are the institutions that have the most extensive information programs. Volunteers give one-and-one-half times as much to charity as those who do not volunteer. One of the main reasons is that they become more familiar with an organization's work. Americans under thirty-five years of age are found to give less to charity than those thirty-five and above. Intensive study of these younger Americans has revealed that one of the key distinctions of the typical "new values," post-World War II generation, is a desire for choice.[2] Choice requires information. Organizations that give these individuals information are more likely to encourage the individuals to give to them. America in the 1980s is characterized by a competitive environment. President Reagan has placed great emphasis on a policy of "deregulation" and allowing products and services to compete in the marketplace for the consumer's favor. This emphasis on "consumer choice" is applied by many to charitable gifts. It is the stimulus for many United Ways to offer and emphasize "donor choice" programs. Here again, if the choosing process is to meet people's needs and desires they must be given more information.

The next step in the process of encouraging more giving is to develop a positive *attitude* toward giving among potential givers. Here, there is considerable reason for optimism. Eight out of ten Americans beleive it is the responsibility of people "to give what they can to charity."[3] Almost four in ten people queried admit that they are not giving as much as they should be giving.[4] In addition, charitable organizations consistently receive higher marks in assessments by the American people than do businesses, government, or labor organizations.

Given this analysis, how can we get more Americans to alter their *behavior*—to give more? If, as shown, they have high awareness of the needs, and large majorities are positive in their attitudes toward charitable institutions, what stands in the way of increased giving? One barrier is in the area of identification-perception: a perceived lack of sufficient knowledge about individual organizations, about how the money is used, and about what results are achieved. But an even more important barrier is the decision that one does not have enough money left over (to give) after paying for the basic necessities. The second leading indicator of a person's proclivity to give more generously (number one is weekly church attendance) was found, in the YSW/RBF Study on charitable giving, to be whether the person perceives that he or she has a moderate or large amount of discretionary income. But the measurement of discretionary income is a very inexact science. It has apparently received relatively little attention in our primary or secondary education programs. Most Americans find it very difficult to explain how they arrived at their determination of *discretionary income*. There is little consensus on how much money is needed for current expenses, future investments, saving for "rainy days," and so forth. And, even if Americans were used to computing their discretionary income more exactly in the past, new factors in the present era call for new computations. These include

1. The basic necessities have changed importantly. For example, Social Security and pension programs have given large numbers of Americans the confidence that they will have enough resources to live in dignity after they retire. Moreover, Medicare, Medicaid, and private medical-hospitalization insurance plans have led many to feel less need to put aside money to be used if they become ill. Thus, the amount to be saved can be reduced.
2. Attitudes toward what is necessary for day-to-day living have also changed importantly. One of the biggest changes in attitudes of young people in the 1960s and 1970s was in the concern about money. Throughout that period rapidly increasing numbers reported less need and interest in material possessions. For example, in the 1969–1973 period, the trend went as follows:[5]

| | Noncollege | | College | |
	1973	1969	1973	1969
Less emphasis on money	74%	54%	80%	72%
Doing things for others	64%	55%	56%	51%
People should save money regularly	80%	89%	71%	76%

3. In the late 1970s and early 1980s, we have seen some modification of the "new values" that led young people to substitute or give priority to self-fulfillment (creativity, leisure, pleasure, and so on) over the need to work and make money. Many found that they needed to give their jobs more time and effort if they were to earn the incomes necessary to enjoy the "full rich life" they desired. But money still plays a less important role than it did as the underpinning of the traditional American test for success that stressed material possessions.

As noted earlier, the most important indicator of who gives in America is the frequency of attendance at church. But before anyone decides that the best way to increase giving is simply to encourage more frequent church attendance, attention should be given to some complexities. All churches have not done equally well in developing member giving. For example, in the YSW/RBF Study, Protestants, on average, were found to contribute almost twice as much to charity as a proportion of household income (2.9 percent) than Catholics (1.6 percent).[6] Did the Protestants give more than the Catholics because they earned more? No, because the Catholics' average household income was slightly higher ($27,500) than that of the Protestants ($26,400). Among the Protestants there is evidence that Presbyterians and Episcopalians are the most generous givers. Also, many studies have revealed that Jews are also better than average givers, particularly in the higher age brackets.[7] It is important to note that in addition to giving more than Catholics,[8] Protestants also tend to give a significantly larger proportion of their gifts to religious charities than Catholics. Specifically, Protestants give three-fourths of their gifts or more to religious institutions, the Catholics only about two-thirds.

Why has giving to religion persisted even as total church membership and other forms of religious participation have declined markedly in the last twenty years? One important answer is that the support is being given to many new religious groups that have grown rapidly in the last decade. These new groups include fundamentalist Protestant groups, the Havura of Judaism, and nontraditional Catholic fellowships. Another reason for the persistent success of religious organizations in obtaining gifts is that most churches and synagogues are well organized in their solicitations. In many cases, the hat is passed up and down the aisles. And even when the appeal is made "off the premises," it is made by leaders (lay or clergy) who are likely to be known to the donors.

Before proceeding to the final key factor, age, it should be noted that some demographic factors are not found to be major determinants of giving, among them sex and occupation. This may come as a surprise to some who point to the impressive contributions of women to social service and charity in the United States or who might expect professional and business people to give more than those in other occupations. Neither is so.

Table 3.1   The Facts on Giving by Age

Age	Average ($)			As Percent of Income			Average Family Income ($)
	Total	Religious	Other	Total	Religious	Other	
All	650	470	180	2.4	1.7	0.7	27,300
Under 30	380	290	80	1.6	1.2	0.3	23,300
30–34	500	350	150	1.7	1.2	0.5	29,000
35–49	910	660	260	2.6	1.9	0.7	35,400
50–64	880	570	300	3.0	2.0	1.0	29,100
65 +	400	340	70	2.7	2.3	0.5	14,900

SOURCE: YSW/RBF Study, p. 5.

AGE AS A DETERMINANT OF GIVING

Age is a critical factor in the challenge to increase giving but not the whole story. There are several reasons why age as an explanation for giving behavior may be exaggerated:

1. There have always been important differences among young and old in giving but the differences are probably greater now than ever before.
2. The younger age groups will determine giving patterns for the rest of this century and into the next.
3. Information on behavior by age groups is more readily available than information on other demographic or attitudinal segmentations. The Census Bureau and opinion surveyors have no problem in obtaining reliable responses to questions about age, and they provide much of their analysis in terms of age groups.

As Americans grow older, the higher the percentage of their income is given to charity, and also the greater the percentage of giving goes to religious charities. The greatest contrast is between those under thirty-five ("baby boomers") and those over thirty-five. For example, one out of three persons younger than thirty-five gave more than 2 percent of his or her income to charities, but 51 percent of those between fifty and sixty-four gave over 2 percent (see Table 3.1). So, age clearly is an important discriminator for giving. Those seeking charitable gifts stand a better chance of getting them if they target those over thirty-five years of age. But it is even more important for charitable institutions to recognize that age is not the whole story. Young people who attend church weekly and believe they have more than enough money to pay for basic necessities (that is, have a moderate amount or a lot of discretionary income) are well worth their attention.

Although this examination of the patterns of giving does not suggest that we are meeting the biblical injunctions, there is reason for satisfaction and hope that giving can be increased in the future, for several reasons:

1. Of Americans, 70–80 percent consistently say that help should be given to those in need.
2. Even as they voted by a six to four ratio for the reelection of Ronald Reagan, most Americans rejected his view that a safety net was in place and that government programs for the poor can be reduced. This rejection was made in their responses to opinion polls throughout the 1984 election campaign, and thereafter as well, and in their votes on tax, health services, and benefit questions submitted to voters in the 1981–1986 period.
3. Almost nine out of ten people questioned claim to be giving, and there is growing evidence that they are giving and volunteering more than they have in previous eras.
4. After a slump in the 1960s and early 1970s, religious belief and support, the primary factor in giving, is showing signs of growth, again particularly for groups that offer choices in worship forms, frequency, and so forth.
5. Many of the obstacles to increased giving lie in a lack of adequate information on which to base choices. The United Way and many other organizations are impressively showing that this lack of information can be effectively overcome by aggressive communications programs.

APPENDIX

The YSW/RBF Charitable Behavior of Americans Study was conducted in 1985 among 1151 Americans eighteen years of age or older. The sample was designed to be large enough to allow statistically reliable analyses of individuals with a broad variety of demographic, geographic, and attitudinal characteristics. Almost 4000 copies of the study report have been sold, and it was a major contributor to the Independent Sector's 1987 "Daring Goals for a Caring Society," which set the five hours a week and 5 percent of income goals for America's volunteering and giving.

NOTES

1. "Charitable Giving of Americans" (a study by Yankelovich, Skelly and White, for the Rockefeller Brothers Fund), hereafter called the YSW/RFB Study, p. 26 (Washington: Independent Sector, 1986).
2. Daniel Yankelovich, "New Rules: Searching for Self-Fulfillment in a World Turned Upside Down" (New York: Random House, 1981).
3. YSW/RBF Study, p. 1.
4. Ibid., p. 26.
5. Daniel Yankelovich, *The New Morality—A Profile of American Youth in the '70's* (New York: McGraw-Hill Book Co., 1974), p. 93.
6. YSW/RBF Study p. 5.
7. Studies for the American Jewish Committee, N.Y., 1970–1980.
8. YSW/RBF Study, p. 40.

The Variability of the Charitable Giving by the Wealthy

GERALD AUTEN and GABRIEL RUDNEY

The charitable contributions of high-income individuals represent a significant share of total charitable giving in our society. The top 1 percent of taxpayers by income provide 20 percent of total charitable contributions and 45 percent of gifts of property.[1] This disproportionate amount of giving reflects the fact that these individuals on the average give a higher percentage of their income to charities. This study of the charitable giving of high-income individuals from 1971 to 1975 provides new insights about the variability of giving behavior of such individuals.

The use of panel data over a five-year period[2] rather than the cross-sectional data for a single year used by previous studies to analyze high-income giving behavior, allowed us to examine the variability of an individual's giving over the period as well as the variability in generosity among individuals.[3] Thus, it provided substantially more information than cross-section data, which incorporate the necessary assumption that the observed behavior for giving and income in a particular year reflects equilibrium or "normal" relationships. This may not be the case if either income or giving vary significantly over time.

Two basic samples were used in this study: a permanent income sample that included returns with a permanent (average) income of at least $100,000 over the five years, and an annual income sample that included returns with income of at least $100,000 in one or more years during the five-year period.[4] These samples limited the study to approximately the top 1 percent of taxpayers. Expanded income was used as the basic definition of income because it is a better measure of economic income than Adjusted

The authors wish to acknowledge the valuable comments of Charles Clotfelter and Jerry Schiff, and the editorial assistance of Janet Auten. The research was funded in part by the Ford Foundation, the J. M. Foundation, and Yale University's Institution for Social and Policy Studies (Program on Non-Profit Organizations) and received computer support from Bowling Green State University. The views expressed are those of the authors alone and do not represent the views or policy positions of these institutions or individuals.

Table 4.1 Numbers of Returns by Income Class under Alternative Definitions of Income, 1971–1975

Income Class (in $1,000)	Permanent Income	Annual Income
$100–200	93,270	546,340
$200–500	18,670	128,522
$500–1,000	3,290	18,066
$1,000 and over	948	7,655
Total	116,178	700,583

NOTES: *Permanent income* is defined as average expanded income over the 1971–75 period in 1973 dollars. The permanent income column includes individuals with permanent income of $100,000 or more and for whom data are available for all five years. The annual income column includes all returns with incomes of $100,000 or more in any one year. In the annual income column, an individual taxpayer may appear up to five times, once for each time income is in the income classes shown. Individuals may appear in the annual income sample without being included in the sample based on permanent income. Expanded income is defined as Adjusted Gross Income plus the excluded portions of capital gains and dividend income plus preference income reported under the minimum tax.

Gross Income (AGI), which is the basic measure of income used in the tax system.[5]

As shown in Table 4.1, there were 116,178 taxpayers in the permanent income sample. Fewer than 1,000 taxpayers reported a permanent expanded income of $1 million or more. The annual income sample included over 700,000 tax returns. There were more than five times as many returns in the annual income sample as in the permanent income sample because the annual income sample included individuals whose income rose to over $100,000 once or twice but whose permanent income was under $100,000. It is interesting to note that although 7655 returns reported income of over $1 million, only 948 taxpayers had permanent income of this amount. Therefore, it appears that quite a few taxpayers had income that rose to the $1 million level only temporarily.

HIGH-INCOME GIVING: A STATIC VIEW

Annual and Permanent Giving

Table 4.2 provides useful perspectives on the giving patterns of the wealthy during 1971–1975. The average contribution increases from $5,305 in the $100,000–200,000 income class to $132,805 in the $1 million-and-over income class. As people move up the income scale, average giving increases faster than income. The average propensity to give rises from 3.93 percent of income in the $100,000 income class to 6.44 percent of income in the highest income class. This compares to giving of about 2 percent of income for all taxpayers. Therefore, the average giving of high-income families is more generous than that of middle-income families.

Table 4.2 Annual High-Income Giving, 1971–1975

Expanded Income Class ($1,000)	Number of Returns	Average	Median	Distribution of Giving	
				Lower Quartile	Upper Quartile
		Amount of Annual Giving			
$100–200	546,340	$5,305	$2,101	$785	$4,456
$200–500	128,522	13,610	2,678	872	10,981
$500–1,000	18,066	39,653	5,400	743	37,224
$1,000 and over	7,655	132,805	16,264	3,006	119,021
Total	700,583	9,108	2,254	813	5,405
		Propensity to Give			
$100–200	546,340	3.93%	1.61%	0.60%	3.54%
$200–500	128,522	4.69	0.99	0.33	3.98
$500–1,000	18,066	5.92	0.86	0.10	5.86
$1,000 and over	7,655	6.44	0.96	0.12	7.47
Total	700,583	4.15	1.48	0.57	3.65

NOTES: This table includes returns in the annual income sample. The propensity to give is the amount of charitable contributions as a percent of expanded income in each year of the sample. Individual taxpayers may appear up to five times in this table for the different years of the panel.

Since the use of averages can provide a misleading impression of generosity when there is great variability in giving, Table 4.2 also provides information on median giving. The median amount of giving increases from $2,101 in the $100,000 income class to $16,264 for those with incomes of $1 million and over. Note that the average is 2.5 times the median in the $100,000 income class and 8.2 times the median in the highest income class. The high ratios of average to median giving indicate large variance in the distribution of giving within the income classes. There is quite a bit of variability in the willingness to support charitable institutions.[6] The distribution is most unequal in the highest income group. This evidence of considerable variation in giving is bolstered by looking at the interquartile range of giving, which provides a perspective on the distribution of giving. The upper quartile shows the amount of contributions at the twenty-fifth percentile and the lower quartile shows giving at the seventy-fifth percentile. Half of all contributors make contributions within this range, while the other half make larger or smaller contributions. At the $100,000 income level, half of contributions were between $785 and $4,567, a ratio of approximately six to one. For the highest income class, the interquartile range is from $3,006 to $119,021 so that giving at the top quartile is nearly twenty times as high as the bottom quartile. Clearly the variability in levels of giving is greater at higher-income levels.

It is also useful to examine long-run or permanent giving over a period of time. Table 4.3 provides the same distributional information on giving as Table 4.2, but shows it over the five-year period of the panel. Whereas the general pattern is the same as with individual-year giving, the variation in permanent giving is somewhat less. For example, whereas the annual

Table 4.3   High-Income Giving over a Five-Year Period, 1971–1975

Expanded Income Class ($1,000)	Number of Returns	Average	Median	Distribution of Giving	
				Lower Quartile	Upper Quartile
Amount of Giving per Year					
$100–200	93,270	$5,336	$2,475	$1,163	$3,432
$200–500	18,670	14,201	4,364	1,097	14,012
$500–1,000	3,290	35,172	6,337	2,211	44,129
$1,000 and over	948	151,272	62,891	11,681	209,147
Total	116,178	8,952	2,777	1,196	6,825
Propensity to Give					
$100–200	93,270	3.97%	1.97%	0.94%	4.51%
$200–500	18,670	4.85	1.70	0.44	5.10
$500–1,000	3,290	5.29	0.83	0.44	5.70
$1,000 and over	948	9.96	4.05	0.53	14.98
Total	116,178	4.21	1.90	0.81	4.70

NOTES: This table includes taxpayers in the permanent income sample. The data in this table show average giving over a five-year period or "permanent" giving. The propensity to give is the amount of charitable contributions as a percent of expanded income over a five-year period.

propensity to give of the highest quartile is more than sixty times the propensity of the lowest quartile, the ratio is less than thirty for permanent giving.[7] Thus, it appears that some of the variation in giving observed in individual years is evened out in giving over time. Another explanation why there is less variation in permanent giving than in annual giving is that the annual-giving table includes individuals with temporarily high-incomes whose giving reflects a lower permanent income. To illustrate, the median propensity to give for individuals with annual incomes of $1 million and over was 0.96 percent as compared to 4.05 percent for those with permanent incomes of $1 million and over.

The U-Shaped Curve of Giving

Observers of charitable giving have long noted that the average propensity to give is greatest for the low- and high-income groups and lowest in the middle-income groups. In previous studies, the U-shaped curve has been shown in terms of Adjusted Gross Income (AGI) or family income.[8] Table 4.4 and Figure 4.1A show the U-shaped curve of annual giving in terms of both AGI and expanded income. The U-shaped curve relationship is clearly observable in terms of the average propensity to give, although it is less pronounced when expanded income is used. Giving is 5.48 percent of AGI in the lowest-income class, and 10.36 percent of AGI in the highest-income class, but only 2.22 percent of AGI in the $20,000 to $50,000 income class. The same pattern holds when income is measured in terms of expanded income.

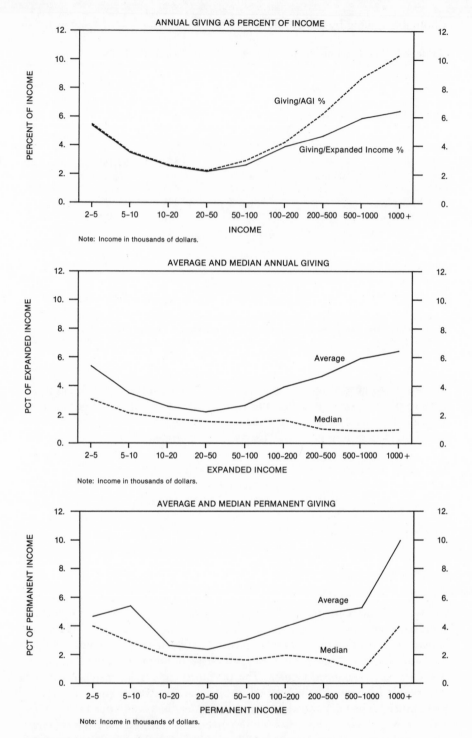

Figure 4.1 The U-shaped curve of giving. *A:* Giving as percent of annual income. *B:* Average and median giving based on annual income. *C:* Average and median giving based on permanent income.

Table 4.4 The U-Shaped Curve of Giving, 1971–1975

Income Class ($1,000)	Annual Giving as Percent of				Permanent Giving as Percent of Permanent Income	
	AGI		Expanded Income			
	Average	Median	Average	Median	Average	Median
Less than 5	5.48%	3.07%	5.37%	3.07%	4.66%	3.99%
5–10	3.51	2.08	3.49	2.09	5.41	2.86
10–20	2.60	1.73	2.58	1.71	2.63	1.88
20–50	2.22	1.53	2.18	1.50	2.35	1.76
50–100	2.96	1.50	2.65	1.42	3.04	1.62
100–200	4.23	1.72	3.93	1.61	3.97	1.96
200–500	6.26	1.43	4.69	0.99	4.85	1.70
500–1,000	8.72	1.59	5.92	0.86	5.29	0.84
1,000 and over	10.36	1.66	6.44	0.96	9.96	4.05

NOTES: The first four columns use the annual income sample and the last two columns use the permanent income sample. Permanent income and giving are defined as a five-year average of expanded income and giving.

Because of the wide variation in the propensity to give at each income level, it is important to examine the U-shaped curve in terms of median as well as the average propensity to give. Although the median propensity to give is high in the lowest income groups and declines in the middle-income groups, there is no significant increase in the propensity to give in the high-income classes (Table 4.4 and Figure 4.1B). Although there is a small increase in the propensity to give in the $1 million-and-over income class, the propensity to give is still well below that of middle-income givers. Therefore, it appears that the reputation of the wealthy for generosity is largely the result of exceptional generosity on the part of a minority of high-income givers rather than widespread generosity among the wealthy. Since giving choices may be based more on permanent income than current income, Table 4.4 and Figure 4.1B also show the U-shaped curve between permanent giving and permanent income over the five-year period. The U-shaped curve relationship for average permanent giving and income is similar to that for annual giving. However, in contrast to annual giving, the majority of individuals whose permanent income is $1 million or more are significantly more generous than individuals in the middle- and upper-middle-income groups. The apparent reason for this difference is that individuals with temporarily high incomes do not increase their giving to the level of those with permanent high incomes.

EXPLAINING THE VARIATION IN GIVING

Economists view charitable giving as a consumption decision in support of a socially worthwhile cause from which the donor derives benefit in the form of personal satisfaction or utility. As with other consumption decisions, the main factors that affect giving are the income and wealth of the donor, the price of giving, and various individual characteristics, such as age, sex, and institutional affiliations, that affect tastes and preferences. Since charitable contributions are deductible in calculating taxable income for taxpayers who itemize deductions, the net after-tax cost of giving a dollar to charity is less than a dollar. The "price" of giving equals one minus the income tax rate of the donor. As with other consumption goods, it is expected that higher income and wealth and lower prices (which result from higher income tax rates) will be associated with increases in giving.

In the following sections we examine the influence on giving of two of these factors: the price of giving and age. The relation between giving and income was discussed in the previous section.

The Price of Giving

By reducing the after-tax cost or "price" of giving, the deductibility of charitable donations on income tax returns provides an incentive to give. Moreover, because of the progressivity of income tax rates, the tax incentive is more favorable to high-income donors. For example, for a taxpayer in the 50 percent tax bracket (the top bracket before the 1986 tax reform), a deductible gift of a dollar would reduce taxes by $0.50. The taxpayer's net after-tax cost or price of a $1 gift is therefore $0.50. An additional impact of the deduction is to increase the after-tax income of contributors, thereby providing additional resources that could be used for giving.

Historically, the donation of stock or other property that has appreciated in value has been even more advantageous than giving cash. This is because with a gift of appreciated property, the taxpayer not only obtains a deduction for the full amount of the gift but also escapes the capital gains tax that would be paid on the asset if it were sold. It is better to donate the appreciated property than to sell it and donate the proceeds.[9]

Since, in the 1970s, the highest federal income tax rate was 70 percent, the basic price of giving in the highest tax bracket was $0.30 per dollar of gift. Not all of the high-income taxpayers were in the highest tax bracket, however. Some taxpayers with large deductions or large proportions of capital gains or preference income had taxable income that was only a small fraction of their expanded incomes. Therefore, they were in lower tax-rate brackets and faced higher prices of giving. Table 4.5 shows the distribution of the price of giving for high-income taxpayers in the 1971 through 1975 period who made cash gifts of 1 percent of their income.[10] Almost all of the high-income taxpayers had a price of giving of less than 0.50, indicating that the government tax subsidy would be more than half of the value of the gift. More than half of the taxpayers with incomes over $1 million dollars had a price of 0.30, indicating that they were in the highest marginal tax bracket. In addition, the table shows that the average and median price of giving declines as income rises.

The effect of price on giving behavior has been studied extensively, and

Table 4.5 The Price of Giving Cash for High-Income Givers, 1971–1975

Price of Giving	Percent of Returns in Income Class (by $1,000)				
	$100–200	$200–500	$500–1000	$1,000 and over	Total
1.0	0.5	1.3	1.5	2.1	0.7
0.60–0.99	6.0	4.2	1.3	2.0	5.5
0.50–0.59	14.8	12.0	4.4	2.7	13.9
0.40–0.49	60.9	34.5	21.3	8.2	54.5
0.31–0.39	17.4	42.4	43.6	29.3	22.8
0.30	0.3	5.6	27.8	55.6	2.6
Total	100.0	100.0	100.0	100.0	100.0
Mean Price	46.2	42.0	37.9	35.9	45.3
Median Price	46.3	40.0	35.4	30.0	45.0

NOTES: This table includes returns in the annual income sample. The price of giving is calculated in terms of the price of cash gifts of 1 percent of income. It is calculated using a tax liability calculation program that takes into account the various features of the federal income tax code.

the consensus of the research is that the reduced price of giving has a strong positive effect on the amount of giving. The incentive effect is so strong that the increase in charitable giving is greater than the loss of revenue to the government due to the deduction. The charitable deduction is therefore viewed as highly efficient in terms of stimulating charitable giving.[11]

A simple way to see the effect of price on giving is to look at the giving of high-income taxpayers with the highest and lowest prices of giving. Table 4.6 shows the propensity to give for the quartiles of taxpayers with the highest and lowest prices of giving. The propensity to give of the quartile with the lowest price of giving is 60 percent greater than the quartile with the highest price of giving. Similar relationships are found in each of the income groups. Thus, it is clear that at least some of the variation in giving is related to differences in the price of giving.

Table 4.6 Giving Propensities of High- and Low-Price Givers, Annual Giving as a Percent of Income, 1971–1975

Income Class ($1,000)	Average Propensity		Median Propensity	
	High Price	Low Price	High Price	Low Price
$100–200	3.86	4.74	1.16	1.48
$200–500	2.53	8.22	0.55	2.25
$500–1,000	2.93	11.91	0.62	3.89
$1,000 and over	3.21	8.83	0.64	1.61
Total	3.55	5.67	0.97	1.60

NOTES: This table includes returns in the annual-income sample. The high price givers were the quartile of returns in each income class with the highest price of giving. The cutoff levels for the four income classes (lowest first) were .50, .48, .42, and .362. The low-price givers were the quartile of returns in each income class with the lowest price of giving. The cutoff levels for the four income classes were .40, .34, .30, and .30. The price of giving is calculated based on a standard amount of giving of 1 percent of income and includes the effect of giving an average amount of appreciated property for each income class.

Giving and Age

Although income and price explain some of the giving behavior, other factors, economic and noneconomic, account for a much larger portion of giving behavior. These include wealth accumulation, savings and investment decisions, discretionary consumption of luxuries, family financial commitments, age, and other economic, social, and moral traditions and values. Of these, the panel provided information only on the relation between giving and age.

The relation between giving over the five-year period and age is shown in Table 4.7. The propensity to give increases dramatically with age. The median level of permanent giving for taxpayers under age forty is less than 1 percent of permanent income. The propensity to give increases with age to 4.6 percent of income for taxpayers over age seventy. The propensity to give increases with income as well as with age. Table 4.7 shows the propensity to give stratified by income as well as by age. For each income level, the propensity to give is more than four times higher for those over age seventy than for those under age forty.

The evidence in this and previous studies (Clotfelter 1985b) is consistent with the hypothesis that the increasing propensity to give is the result of people becoming more generous as they get older. Although greater generosity among the high-income elderly may be attributable to economic factors such as high-incomes and accumulated wealth, social factors such as the important influence of religion and institutional commitment may be equally important. Jencks (1987) offers another hypothesis, however: the observed increases in the propensity to give by age may be the result of generational differences with respect to giving habits. He suggests that the generation born in the late nineteenth and early twentieth century were more generous throughout their lifetimes than those born subsequently. This generation would appear in the panel as those who are age sixty and older. If his thesis is valid, it implies that historical experiences of certain cultural, social, and economic environments and events affect the giving of individuals throughout their lifetimes. Further investigation is necessary to distinguish between Jenck's hypothesis and the conventional view. An appropriate test would be to examine giving data by age for at least two years that are separated substantially separated in time. For example, the

Table 4.7 Propensity to Give and Age (median giving as a percent of income over a five-year period)

Age	Total	Permanent Income Class	
		$100,000–499,999	$500,000 and Over
Less than 40	0.93%	0.90%	1.90%
40–59	1.73	1.73	1.87
60–69	2.24	2.20	3.09
70 and over	4.61	3.72	9.37

NOTE: This table includes taxpayers in the permanent income sample.

age and giving results in this panel could be compared with a panel in the 1980s. Thus, the giving behavior of cohorts could be observed over time.

Steuerle (1987) has investigated the timing of giving by the wealthy during their lifetimes and at death. Although the evidence indicates an increasing propensity to give with age, Steuerle's findings imply that wealthy older persons should actually be giving even more. His study shows that older people tend to postpone their giving until death even though their tax savings would be greater by giving during their lifetime. He offers the hypothesis that top wealth holders prefer to accumulate and control wealth until death rather than consume their wealth or give their wealth away to family and to worthy causes during their lifetimes.

THE VARIABILITY OF GIVING OVER TIME

The analysis in the previous sections showed the considerable variability in the propensity to give among high-income taxpayers. The analysis, however, was static in that it looked at a cross section of taxpayers and their giving in a given time period. The advantage of a panel data set is that it allows examinations of changes in the giving of individual taxpayers over a period of time, in this case from 1971 through 1975.

It is commonly believed that donors have a stable pattern of giving or a commitment to a certain level of giving over time. Our analysis shows, however, that high-income giving is quite volatile, implying that much giving by the wealthy involves periodic discretionary choices for how much to give rather than regular or "committed" giving.[12]

One way of looking at the dynamics of giving is to look at year-to-year changes, as shown in Table 4.8. Only about 12 percent of the high-income

Table 4.8 Changes in Year-to-Year Giving, 1971–1975

Change in Giving	Percent of Returns
Decrease:	
More than 30%	26.1%
15–30%	10.6
5–15%	7.9
Total with Decrease in Giving	44.6
Stable (less than 5% change)	11.7
Increase:	
5–15%	7.9
15–30%	7.7
More than 30%	28.1
Total with Increase in Giving	43.7
Total	100.0%
Number of Returns	526,829

NOTE: This table includes tax returns in the annual-income sample with $100,000 of expanded income in the first year of each pair of years in the 1971–75 period. Taxpayers may appear up to four times in this table.

Table 4.9   Largest and Smallest Amounts of Giving over Five Years, 1971–1975

Ratio of Largest to Smallest Annual Giving	Percent of Returns	Percent of Returns in Income Class ($1,000)			
		100–200	200–500	500–1000	1000+
1.0–1.5	13.2%	14.4%	9.4%	6.5%	3.5%
1.5–2.0	17.8	19.2	14.0	5.2	7.0
2.0–4.0	30.3	31.9	26.0	12.2	21.7
4.0–10.0	19.3	19.5	18.7	17.1	22.9
10.0 or more	19.4	15.0	31.9	59.0	45.0
Total	100.0	100.0	100.0	100.0	100.0
Percent of Returns	100.0	79.6	16.6	3.0	0.9

NOTE: This table includes taxpayers in the permanent income sample.

donors had stable giving; that is less than a 5 percent increase or decrease in giving from year to year. The vast majority increased or decreased giving by more than 5 percent. The most striking finding is that more than half of the annual changes involved increases or decreases of more than 30 percent. Thus, not only are there changes in giving over time, but large changes are also the rule rather than the exception.

Another way of looking at variability over time is to look at the largest and smallest annual giving of each high-income individual over the five-year period. Table 4.9 shows the ratios of the largest to the smallest annual giving for individuals over the sample period. The highest annual giving was 3.3 times the lowest for the median high-income donor. For 19 percent of the high-income donors, the largest giving was more than ten times the smallest giving. Only 14 percent of high-income taxpayers had less than a 50 percent differential between the largest and smallest annual amounts of giving. These results show that regular or habitual giving is not the standard giving behavior among high-income people.[13]

To provide a perspective on the variability of giving by income level, Table 4.9 shows the ratio of the largest to smallest annual giving by income class. The proportion of relatively stable givers, those whose largest giving was less than twice the smallest giving, declines sharply as income increases. One-third of those with $100,000–200,000 permanent income were stable givers, as compared to only 10 percent of those with $1 million of income. By contrast, the proportion of irregular givers—that is, individuals whose largest giving was more than four times their smallest giving—was more than two-thirds of the highest income class and only one-third of the $100,000–200,000 income class.

ECONOMIC VARIABLES AND GIVING OVER TIME

Possible explanations for the substantial fluctuations in high-income giving include a host of economic and noneconomic factors. Economic factors include changes in individual income and wealth, and changes in the tax incentive to give, operating and capital requirements of a business, and changes in savings and investment alternatives and in luxury consumption expenditures. Changes in the preferences of individuals themselves may cause variability. These preferences arise from a multitude of motivations, including planned giving over time, family and institutional obligations, and plans for the disposition of wealth at death. Finally, another influence is the effort of fund raisers to encourage major gifts. This section examines the relation of changes in income and tax incentives to the variability of giving.

Income

As shown in Table 4.10, large year-to-year changes in income are commonplace among the high-income population. For example, 46 percent of the annual changes in income exceeded plus or minus 30 percent. An additional 21 percent of the changes were between 15 and 30 percent. During the 1971–1975 period, more than half of the annual changes were income declines of more than 5 percent. The apparent reason was the onset of a major recession in 1974 and 1975, and a more than 25 percent decline in the stock market from the 1972 peak to the 1974 bottom.

Comparing the variation in giving with the variation in income, Table 4.11 shows that for those whose giving declined by more than 30 percent, 72 percent had declines in income. Similarly, 50 percent of those whose giving increased by more than 30 percent reported an increase in income. The surprising result is that over 38 percent of those who increased giving by 30 percent actually had a decline in income that year. Clearly factors other than annual income are highly influential in determining giving.

Table 4.10 Annual Change of Income during Five-Year Period, 1971–1975

Change in Income Over Previous Year	Percent of Returns
Decline:	
More than 30%	33.2%
15–30%	10.7
5–15%	9.9
Total with Decline	53.8
Stable (less than 5% change)	12.5
Increase:	
5–15%	9.9
15–30%	10.7
More than 30%	13.1
Total with Increase	33.7
Total	100.0%

NOTE: This table includes tax returns in the annual income sample with $100,000 in the first year of each pair of years in the 1971–1975 period. Taxpayers may appear up to four times in this table.

Table 4.11   Year-to-Year Giving and Income Change, 1971–1975

	Percent of Returns	
Income Change	Giving Decreased by More than 30%	Giving Increased by More than 30%
Decrease:		
More than 30%	50.6%	23.2%
15–30%	10.1	7.1
5–15%	11.1	8.2
Total with Decrease	71.8	38.5
Stable (less than 5% change)	8.7	11.3
Increase:		
5–15%	5.9	11.3
15–30%	4.7	20.1
More than 30%	8.8	18.7
Total	19.4	50.1
Total	100.0%	100.0%
Number of Returns	137,334	148,255

NOTE: This table includes tax returns in the annual income sample with $100,000 of expanded income in the first year of each pair of years in the 1971–1975 period. Taxpayers may appear up to four times in this table.

Price

Before looking at the relation between giving and price it is useful to examine year-to-year variation in the price of giving. Since there were no major changes in income tax rates that would have had a major impact on the price of giving during the 1971–1975 period, the changes in price primarily reflect the year-to-year changes in taxpayer characteristics such as income, other deductions, and marital status. As shown in Table 4.12, 43 percent of the changes in price were 5 percentage points or more, large enough change to have an impact on giving. However, in a third of the observations there is little or no change in price from year to year. This may appear to be paradoxical in view of the substantial variation of income of these same observations. But it is explainable by the fact that the rate brackets are wider at the high-income levels and the top bracket is open-ended.

Table 4.13 shows the relation between large giving changes and changes in price. For individuals with large decreases in giving, 41 percent had an increase in price and another 25 percent had little or no change in price. Predictably, a majority of those whose giving declined sharply experienced either an increase or no change in price. But, surprisingly, one-third of those whose giving fell by over 30 percent from the previous year had a decrease in the price of giving. Changes in price do not work as well as changes in income in explaining large year-to-year increases in giving. Of

Table 4.12 Year-to-Year Change of Price of Giving among High-Income Individuals, 1971–1975

Year-to-Year Change of Price	Number of Returns	Percent of Returns
Decrease:		
More than 10 pts.	36,444	6.9%
5–10 pts.	37,218	7.1
2–5 pts.	51,145	9.7
Total with Decrease		23.7
Stable (less than 2 pts. change)	169,716	32.2%
Increase:		
2–5 pts.	78,720	14.9
5–10 pts.	62,122	11.8
More than 10 pts.	91,464	17.4
Total with Increase		44.1
Total	526,829	100.0%

NOTE: This table includes tax returns in the annual income sample with $100,000 of expanded income in the first year of each pair of years in the 1971–1975 period. Taxpayers may appear up to four times in this table.

those who increased giving by at least 30 percent, only 24 percent enjoyed a decrease in the price of giving and 45 percent had an increase in price.

Although these findings seem inconsistent with the idea that giving depends on price, they are consistent with the large portion of unexplained variation in econometric studies. These studies typically find that price and income account for less than half of the variation in giving among taxpayers. However, it is possible that taxpayers may be responding to longer-run price and income factors. There may be lags in taxpayer responses to

Table 4.13 Year-to-Year Giving and Price Change, 1971–1975

	Percent of Returns	
Price Change	Decrease by More than 30%	Increase by More than 30%
Decrease:		
More than 10 pts.	15.9%	6.4%
5–10 pts.	8.8	7.3
2–5 pts.	8.5	10.5
Total	33.2	24.2
Stable (less than 2 pts. change)	24.9	30.5
Increase:		
2–5 pts.	9.5	21.1
5–10 pts.	10.5	11.1
More than 10 pts.	21.9	13.0
Total	41.9	45.2
Total	100.0%	100.0%
Number of Returns	137,334	148,255

NOTE: This table includes tax returns in the annual income sample with $100,000 of expanded income in the first year of each pair of years in the 1971–1975 period whose giving increases or decreases 30 percent or more form the previous year. Taxpayers may appear up to four times in this table.

changes in income and in the price of giving, and the full response may only occur after several years (Clotfelter 1980, 1985b).

GENEROUS AND UNGENEROUS GIVERS

Of special interest in the study of high-income giving are individuals who are unusually generous givers. We define generous givers as those whose giving averages at least 20 percent of their income over the five-year period. This standard of generosity is about four times the median giving for the high-income group. By this standard, about 4 out of every 100 high-income individuals are generous givers. But these generous givers account for more than 35 percent of all giving by the high-income groups. The most generous 10 percent account for 58 percent of total giving. The top 1 percent of givers alone accounts for 15 percent of high-income giving.

The proportion of generous givers rises with income, as shown in Table 4.14. For the $1 million-and-over income class, over 18 percent contributed more than 20 percent of income, accounting for nearly 60 percent of the giving of that group.

Table 4.14 also shows some of the characteristics of the generous givers. For contrast, the table also shows the characteristics of ungenerous givers, those whose contributions averaged less than 1 percent of income. This standard is less than one-fourth of the median level of giving among high-income groups. About 30 out of every 100 individuals are nongenerous givers. Although there were eight times more ungenerous givers than generous givers, the former accounted for less than 3 percent of the giving of all high-income taxpayers.

The generous givers are likely to be the elderly and those who have high tax rates. More than three-fourths of the generous givers were over age 60 as compared to less than one-third of the ungenerous givers. The price of giving is considerably lower for generous givers, 29 cents per dollar gift as compared to 42 cents per dollar, calculated on a standard amount of giving of 1 percent of income.

It is interesting that the difference in tax savings on the actual amounts of giving by generous and ungenerous givers are much smaller than the differences in the tax prices calculated for a standard amount of giving. This is attributable to the fact that generous givers contribute to such an extent that the after-tax cost of their last dollar of giving increases to a level close to the price of the ungenerous givers.

Although the analysis examined those who are generous over a five-year period, there are additional taxpayers who meet the 20 percent standard in only one or a few years. Table 4.15 shows that there are about three times as many taxpayers who meet the generosity standard at least once over the five-year period. About 47 percent of them meet the generosity standard only once and an additional 19 percent meet it twice. Only about 13 percent are generous givers every year.

Table 4.14 Generous and Ungenerous Giving over a Five-Year Period (medians)

Expanded Income ($1,000)	Variable	Giving over 20% of Income	Giving under 1% of Income
100–200	Giving	$40,599	$731
	Propensity	.289	.005
	Price of Giving	.395	.436
	Tax Saving as % of Giving	58.7	55.3
	Number of Returns	2,078	20,233
	Percent of Class Total	2.8	27.1
200–500	Giving	$86,431	$1,058
	Propensity	.282	.004
	Price of Giving	.279	.396
	Tax Saving as % of Giving	66.8	59.0
	Number of Returns	1,053	5,927
	Percent of Class Total	6.8	38.1
500–1,000	Giving	$191,756	$2,010
	Propensity to Give	.284	.003
	Price of Giving	.225	.368
	Tax Saving as % of Giving	69.0	62.3
	Number of Returns	188	1,434
	Percent of Class Total	6.7	51.0
1,000 and over	Giving	$434,761	$4,872
	Propensity to Give	.286	.003
	Price of Giving	.214	.351
	Tax Saving as % of Giving	68.8	62.4
	Number of Returns	149	224
	Percent of Class Total	18.5	27.8
Total	Giving	$83,590	$889
	Propensity to Give	.285	.005
	Price of Giving	.288	.419
	Tax Saving as % of Giving	65.6	56.9
	Number of Returns	3,468	27,825
	Percent of All Income Returns	3.7	29.7
	Percent over age 60	75.9	30.6

NOTE: This table includes taxpayers in the permanent-income sample. Generous givers are defined as those whose contributions were more than 20 percent of their expanded income over the five-year period. Ungenerous givers were those who gave less than 1 percent of their expanded income over the five-year period. The tax saving as a percent of giving is the tax saved by the amount of actual giving. The price of giving is calculated for a standard amount of giving 1 percent of income.

Another way of looking at generosity is in terms of the absolute size of gifts. Approximately 2700 individuals, or about 3 percent of high-income individuals, reported contributions of over $100,000 in one or more years during the period. Of these, almost half made such a large gift only once during the five-year interval and only about 13 percent made such gifts in all of the years. About 250 individuals made at least $500,000 in gifts, which translates into gifts of about $1 million when adjusted to current dollars.

Several implications arise from these findings. The irregular nature of

Table 4.15 Number of Times Giver Was Generous in Five-Year Period

Number of Times	Number of Individuals	Amount of Giving (millions of dollars)
One Time	5,491	$101.2
Two Times	2,256	71.4
Three Times	1,353	78.5
Four Times	1,025	78.4
Five Times	1,577	178.2
Total	11,702	$507.7

NOTE: This table includes taxpayers in the permanent income sample who gave at least 20 percent of expanded income in at least one year in the five-year period.

large gifts may be the result of either carefully planned and periodic large gifts or unplanned responses to fund-raising appeals. The planned bunching of large gifts in particular years may reflect a rational approach to maximizing a donor's own benefit. Giving a large sum of money once every five years may provide a donor more recognition and influence than giving the same amount over five years. In either case, the variability has implications for fund-raising appeals. These discretionary donors who apparently have responded periodically to appeals for large gifts may be approachable for more frequent solicitation of large gifts.

Another implication concerns the effects of the 1986 tax reform provisions relating to the minimum income tax on gifts of appreciated property. The minimum tax is most restrictive on the giving of large gifts of highly appreciated property. But evidence of irregular giving of large gifts suggests that many wealthy taxpayers could avoid or reduce the impact of the minimum tax by spreading large gifts over several years.

CONCLUDING OBSERVATIONS

This study has dealt with three areas of high-income giving: the extent of variability in giving, the incidence of generous and ungenerous givers, and new perspectives on the well-known U-shaped curve of giving. The major finding is the high degree of variability in high-income giving. In any given year, there is tremendous variation in the propensity to give. For example, in the $1-million-or-more income class giving in the top quartile is more than 60 times higher than giving in the bottom quartile. Even more significant is the substantial variation in the giving of individuals over time. One illustration is that the largest annual giving by an individual is more than ten times that person's smallest annual giving over a five-year period for 45 percent of the individuals in the highest-income group.

Another significant finding is that a small proportion of high-income givers who are exceptionally generous account for a large proportion of total high-income giving. On the other hand, a large proportion of high-income individuals give less than 1 percent of income. These findings on variability and generosity provide new insights into the U-shaped curve of giving. A considerable part of the "curve" in the U-shaped curve at the higher incomes is due to the exceptional generosity of a relatively small group of high-income people.

The findings of this paper have implications for research, for prediction of the effects of tax policy changes, and for fund raising. The substantial fluctuation in the giving of individuals over time implies that there may be biases in research using data on giving for one year. Future research should make use of panel data to examine the dynamics of giving and long-run giving patterns. The substantial fluctuations in giving also suggest that much of high-income giving is discretionary and that additional research into the determinants of committed and discretionary giving would be worthwhile.

The finding of variability also has implications for using the results of research to predict the effects of tax policy changes on giving. The substantial year-to-year variability in the incomes and the price of giving for individuals that occur even in the absence of policy changes suggests that the giving effects of tax policy changes could be swamped by the regular annual changes. For example, the year-to-year variability in giving can provide opportunities for avoiding the impact of the new minimum tax on gifts of appreciated property.

Fund raisers should also be interested in the wide fluctuations in high-income giving over time. It suggests that the pattern of giving of such individuals can be influenced by effective fund-raising and development efforts. Fund raisers may be able to assist taxpayers in better tax planning with respect to gifts of appreciated property in order to avoid or reduce the minimum tax.

APPENDIX

The data set used in this study is from a special tabulation of federal income tax returns for the period 1971–1975 called the Sales of Capital Assets Panel. It is a stratified random sample that oversamples high-income taxpayers in order provide adequate representation. The tabulations in this paper use sample weights to obtain population totals. The data set includes selected data from the tax returns of approximately 11,000 individuals over the time period, including information on types of income, deductions, and certain information provided on tax returns such as marital status. For about 85 percent of the sample, the returns were present for all five years. All identifying information has been removed from the data in order to prevent disclosure.

NOTES

1. See Internal Revenue Service, *Statistics of Income. Individual Income Tax Rates, 1984.*
2. The data set is from a special tabulation of a sample of federal income tax returns of

about 11,000 individuals. A more detailed description of the data set is provided in the appendix to this chapter.

3. For exceptions to this see Clotfelter (1980) and Clotfelter and Steuerle (1981).

4. The samples were further limited to taxpayers who itemize deductions. Permanent income is defined as average expanded income over the 1971–1975 period in 1973 dollars. In the annual income sample, an individual taxpayer may appear up to five times, once for each time income is at least $100,000. Individuals may appear in the annual income sample without being included in the sample based on permanent income. Expanded income is defined as Adjusted gross Income plus the excluded portions of capital gains and dividend income plus preference income reported under the minimum tax. For a detailed discussion of the definition of expanded income see the Internal Revenue Service, *Statistics of Income, Individual Income Tax Returns, 1984*, pp. 103–105.

5. For example, AGI excludes part of long-term capital gains, up to $200 of dividend income, and several other types of preference income. Under the tax reform law enacted in 1986, the full amount of long-term capital gains will be included in AGI starting in 1987.

6. The highly skewed distribution of giving can be seen by comparison to the ratio of average to median income in the United States. The ratio of average to median family income in the United States is only 1.3. See U.S. Bureau of the Census, *Current Population Survey*, Series P-20.

7. In Table 4.3, the propensity to give in the most generous quartile of givers was 14.98 percent of income or thirty times the propensity to give of the least generous quartile, which gave 0.53 percent of income.

8. See, for example, Clotfelter and Steuerle (1981).

9. For a more detailed discussion of the price of giving see Auten and Rudney (1986) or Clotfelter (1985a).

10. The price of giving is calculated using a tax calculation program that takes into account the various provisions of the federal income tax code. Depending on the amount of appreciation and what the taxpayer would otherwise have done with the asset, there would be a lower price of giving for gifts of appreciated assets.

11. See for example, Feldstein (1975), Feldstein and Clotfelter (1976), Clotfelter and Steuerle (1981) and Clotfelter (1985b).

12. Dennis, Rudney, and Wyscarver (1983) made the distinction between committed and discretionary giving in the analysis of giving behavior.

13. Unpublished panel data indicate substantially less variation over time in giving by lower- and middle-income individuals. This suggests less discretionary giving and more committed giving by these individuals.

REFERENCES

Auten, Gerald, and Gabriel Rudney. 1986. "Donating Appreciated Property after Tax Reform," *Tax Notes* (October 20).

———. 1985. "Tax Policy and Its Impact on the High Income Giver." In *1985 Spring Research Forum Working Papers*. Washington, D.C.: Independent Sector and United Way Institute, pp. 525–47.

Clotfelter, Charles. 1985a. "The Effect of Tax Simplification on Educational and Charitable Organizations." In *Economic Consequences of Tax Simplification: Proceedings of a Conference Sponsored by the Federal Reserve Bank of Boston*. Boston: Federal Reserve Bank of Boston.

———. 1985b. *Federal Tax Policy and Charitable Giving*. Chicago: University of Chicago Press.

———. 1980. "Tax Incentives and Charitable Giving: Evidence from a Panel of Taxpayers." *Journal of Public Economics* 13.

Clotfelter, Charles, and Eugene Steuerle. 1981. "Charitable Contributions." In H. J. Aaron and J. A. Pechman, eds. *How Taxes Affect Economic Behavior*. Washington, D.C.: Brookings Institution.

Dennis, Barry, Gabriel Rudney, and Roy Wyscarver. 1983. "Charitable Contributions: The Discretionary Income Hypothesis." PONPO Working Paper, No. 63.

Feldstein, Martin. 1975. "The Income Tax and Charitable Contributions: Part I—Aggregate and Distributional Effects." *National Tax Journal* 28 (March).

Feldstein, Martin, and Charles Clotfelter. 1976. "Tax Incentives and Charitable Contributions in the U.S.: A Microeconomic Analysis." *Journal of Public Economics* 5 (January/February): 1–26.

Feldstein, Martin, and Amy Taylor. 1976. "The Income Tax and Charitable Contributions." *Econometrica* 44. (November): 1199–1216.

Jencks, Christopher. 1987. "Who Gives to What?" In Walter W. Powell, ed. *The Nonprofit Sector: A Research Handbook*. New Haven, Conn.: Yale University Press.

Steuerle, Eugene. 1987. "The Charitable Giving Patterns of the Wealthy." In T. Odendahl, ed. *America's Wealthy and the Future of Foundations*. New York: The Foundation Center.

U.S. Department of the Treasury. Internal Revenue Service. 1984. *Statistics of Income: Individual Income Tax Returns*. Washington, D.C.: Government Printing Office.

5

The Evolution of Black Philanthropy: Patterns of Giving and Voluntarism

EMMETT D. CARSON

Debate is quickening about the need for the black community to take a more active role in the development and implementation of strategies aimed at helping poor blacks. A variety of factors have provided an impetus for this discussion, including (1) the belief that increasingly fewer routes provide the poor with upward socioeconomic mobility; (2) the growing income disparity between the black middle class and the black poor; and (3) a belief by many that the current national budget deficit will effectively prevent any substantial expansion of federal domestic programs to address the needs of the poor. Although each of these concerns can be credited for having raised the consciousness of the black community about its responsibility to aid the poor, neither "liberal"[1] nor "conservative"[2] proponents of black self-help have made specific suggestions as to how this self-help should occur.

Rather than attempt to sort out the respective responsibilities of the government and the black community, this chapter will take as given that both entities share a joint responsibility to aid the poor. This paper will focus on black philanthropic activity as one mechanism available to the black community to help it meet its responsibility to the black poor. Although black philanthropic activity is not new, as will be discussed shortly, the establishment of a solid black middle class and the growing number of upper-income blacks (consisting of business owners, entertainers, and athletes) present blacks with new opportunities to harness the professional skills and the financial resources of these groups to support charitable activities aimed at helping poor blacks. This is not to suggest that these groups have been lax in their support of charitable activities. On the contrary,

An earlier version of this paper was presented at the annual meetings of the National Conference of Black Political Scientists in Atlanta, Georgia, April 22–25, 1987. This article represents the views of the author and not the Joint Center for Political Studies or any of its sponsors. The study has received support from the Edward W. Hazen Foundation, the Ford Foundation, the Charles Stewart Mott Foundation, and the Gannett Foundation.

one of the principal research questions is to determine the extent to which the black middle class is engaged in charitable activity relative to blacks in other income groups.

First, it would be useful to define what is meant here by the phrase *black philanthropic activity* and to distinguish it from two related concepts: black self-help and black initiative. Simply stated, *black philanthropic activity* refers to charitable giving and voluntarism by blacks. The voluntarism component is important because in many instances, American society has appeared to undervalue the importance of voluntarism relative to charitable giving. For example, until recently, cash contributions to charity were deductible for nonitemizing tax payers, whereas the giving of time to charity has never been deductible for anyone.

As used here, *self-help* refers to a wide range of activities that are designed to promote the interests (for the purposes of this discussion, socioeconomic interests) of an individual or a particular group. Black philanthropic activity and black initiative are opposite ends of a continuum of self-help. In this context, discussions of personal initiative inevitably rest on the slippery slope that the beliefs, attitudes, and moral values of the individual or group in question are, at least in part, responsible for their socioeconomic conditions. Although this hypothesis may be correct, personal initiative is very difficult to measure objectively.

On the other hand, an analysis of philanthropic activity leads to discussions about the total resources devoted to charitable activity and how these resources are allocated among competing claimants. Unlike research on personal initiative, philanthropic activity can be measured accurately and objectively. Within this framework, proponents of black self-help who recommend that blacks increase their financial support (that is, charitable giving) to traditional black institutions and that blacks participate in Big Brother and Big Sister programs (that is, voluntarism) to provide positive role models for disadvantaged black youth are, in effect, suggesting that blacks become more involved in philanthropic activity.

This chapter is organized into three sections. The first examines the historical roots of black philanthropic activity. The second uses a 1986 Joint Center for Political Studies—Gallup survey to analyze black attitudes toward philanthropic activity as well as black giving and volunteer behavior. The last section discussed the implications of these findings for the future direction of black philanthropic activity.

ROOTS OF BLACK PHILANTHROPIC ACTIVITY

The French chronicler of nineteenth century American life, Alexis de Tocqueville, is often quoted in writings on philanthropic activity as saying:

Americans of all ages, all conditions, and all dispositions, constantly form associations. They have not only commercial and manufacturing companies, in which all take part, but associations of a thousand other kinds, religious, moral, serious, futile, general or restricted, enormous or diminute.[3]

Without doubt, Tocqueville's observation was equally descriptive of black life during this period. As William Lloyd Garrison noted in 1852: "I am encouraged to find not only among yourselves [black Bostonians], but in other cities, a disposition to form societies, both among men and women, for mutual improvement and assistance."[4]

When Tocqueville and Garrison made their observations, organized black philanthropic activity was already six decades old. Organized charitable giving and voluntarism has existed in the black community for over 200 hundred years and is, in large part, responsible for the development of the first black schools, black banks, and black insurance companies. It is not surprising that much of black philanthropic activity began or centered around the black church. Traditionally, the black church has been the center of the black community, addressing all of the community's economic, social, and spiritual needs.

A review of the history of black Americans indicates that blacks historically have relied on the philanthropic resources of their own community to provide for the black poor as well as to supply the people and financial resources to sustain virtually every black protest movement throughout history. One of the earliest recorded mutual aid societies was the fraternal society known as African Lodge No. 459, founded in Boston by Prince Hall.[5] The lodge formally received its charter from the Grand Lodge of England in 1787, although it had been in operation several years earlier. African Lodge No. 459 is widely recognized as the beginning of black Masonry in the United States. Far from simply administering to the needs of its members, the society was active in its activities to aid Boston's black poor in general. During the winter months the society disbursed free firewood and food to the needy.[6] In addition, the lodge was active in the abolitionist movement.

In many respects, the many mutual aid societies and fraternal organizations that came after the African Lodge were equally committed to community service. For example, the African Society founded in Boston in 1796 was so widely recognized for its community service and charitable activities that in 1808, 200 blacks attended a commemoration program in the organization's honor.[7] During the great plague that struck Philadelphia, Pennsylvania in 1793, the Free African society provided the city with an extensive array of voluntary aid including nursing and burial services.[8] Black literary societies such as Baltimore's Young Men's Mental Improvement Society for the Discussion of Moral and Philosophical Questions of All Kinds and the Philadelphia Library Company of Colored Persons were widespread, providing books and establishing libraries.[9] The Brown Fellowship Society, founded in 1790, provided education for free black children in Charleston, South Carolina.[10]

There is strong evidence that the financial needs of mutual aid societies and fraternal organizations led to the development of the first black banks and black insurance companies. As Harris states:

No fewer than twenty-eight banks were organized by Negroes from 1899–1905. Nearly all of these institutions were created to serve as depositories for Negro fraternal insurance orders. Thus, the rapid organization of independent Negro banking followed in the wake of the expansion of fraternal insurance and burial societies that took place in the first twenty years of the post Civil War period.[11]

One measure of the leadership and services that mutual aid societies and fraternal organizations provided blacks is what was done to eliminate them. By 1835, several states including Virginia, Maryland, and North Carolina had laws that banned fraternal organizations and mutual aid societies. Specifically, blacks were prohibited from having "lyceums, lodges, fire companies, or literary, dramatic, social, moral, or charitable societies."[12] Laws such as these contributed to the need for many of such organizations to operate in secrecy.

In addition to providing charitable services to the poor and cultural programs for socioeconomic advancement, black philanthropic activity has been also used as a mechanism by which to implement and sustain protests against the larger society. David Mathews has suggested that philanthropic activity is a political force that provides, among other benefits to the political process, public power. *Public power* is defined by Mathews as the ability of philanthropic activity to raise the public's consciousness to a state where it can influence public decision making.

We are impressed nowadays with formal power—the power of position, money, and law. We tend to lose sight of more basic political power: the power of ideas and ideals, the power of relationships, and the power of commitments. But such power has great force as has been vividly demonstrated in events ranging from neighborhood revitalization to protests against the Vietnam War.[13]

An early example of black philanthropic activity being utilized as public power was the underground railroad. In recounting the accomplishments of the underground railroad, few acknowledge the extent to which it relied on black charitable giving and voluntarism for much of its success. The underground railroad sponsored national and international fund-raising campaigns, used volunteers as railroad "conductors," used the homes of supporters as railroad "stations," and lastly, provided escaping slaves with food, clothing, shelter, new identity papers, and money. Many of the black mutual aid societies and fraternal organizations were actively involved in supporting all aspects of the railroad's operation.

A more recent example of blacks using the public power of philanthropic activity to challenge social injustice is the civil rights movement. It would not be much of an overstatement, if at all, to suggest that the civil rights movement was perhaps the greatest mobilization of charitable activity of any group to be witnessed to date in America. Between 1957–1968, the various civil rights organizations assembled thousands of individuals into a national protest movement that raised money, collected and disbursed food, and recruited volunteers to participate in boycotts, sit-ins, and marches across the country. It would be hard to overemphasize the high level of volunteer participation, which often included children, that was required for the success of countless demonstrations.

These examples illustrate the long and prestigious tradition of philanthropic activity that has existed in the black community. In light of this history, it is interesting to note that current discussions about black self-help have evolved in such a manner as to suggest that philanthropic activity is itself a "conservative" approach for achieving black economic progress. It is not. At various times in history, black philanthropic activity has served as the centerpiece of liberal, conservative, and nationalistic strategies for black socioeconomic progress.

THE JOINT CENTER–GALLUP SURVEY

Despite the historical importance of philanthropic activity in the black community, little contemporary research has been done in this area. Although there have been several major efforts to study charitable giving and voluntarism in general, none of these studies has collected sufficient data to examine the philanthropic activities of blacks. The 1986 Joint Center–Gallup national random survey in which blacks were oversampled overcomes many of these problems. The survey was completed in four days during the week of August 8, 1986. In face-to-face interviews, 916 whites and 868 blacks who were surveyed.[14] In addition, almost 70 percent of all blacks in the survey were interviewed by blacks.

There is at least one significant limitation to these data: the inability to determine whether the individual giver perceived that his or her contribution was to a "black" or "white" charitable organization. Although this is a shortcoming, it is not a critical one. To the extent that blacks have become increasingly integrated into all aspects of American society, the question as to what degree are blacks the recipients of black charity is more difficult to discern. For example, there is no way to determine if blood donated through the Red Cross or goods contributed to the Salvation Army are ultimately received by blacks. Further, this limitation is less severe when taking the first step of estimating and examining the total charitable resources of the black community.

Attitudes toward Charitable Activity

One of the first questions that must be examined before embarking on a strategy of self-help that rests on increasing the philanthropic activity of blacks is, How do blacks feel about charitable giving to poor and charitable organizations in general? The fact that blacks have historically engaged in philanthropic activity is no guarantee that today's blacks have any desire to continue this tradition.

Table 5.1 "What Group Has the Greatest
Responsibility for Helping the Poor?" by Race

	Black	White
Churches	26.3%	24.7%
Other Private Charities	3.9	6.0
Federal Government	31.4	24.6
State or Local Government	13.3	14.7
Relatives of Poor People	3.4	7.6
Poor Themselves	3.8	12.7
Someone Else	.9	.4
Don't Know	17.1	9.3
Number of Observations	868	916

NOTE: The responses in this table have a sampling error of 4 percentage points for each group.

The responses of blacks and whites regarding what group has the greatest responsibility to aid the poor are shown in Table 5.1. The largest number of blacks, 31 percent, believe that the federal government has the greatest
responsibility to aid the poor, whereas 26 percent felt that churches had
the greatest responsibility. Whites were evenly split, with 24 percent believing that the federal government has the greatest responsibility and 24
percent feeling that the church had the greatest responsibility.

Although a greater percentage of blacks than whites in the sample feel
that the federal government has the greatest responsibility for helping the
poor, the difference is less than the 8 percent bound of error. In addition,
roughly the same percentage of both blacks and whites responded that
state and local governments had the primary responsibility to aid the poor,
13.3 percent and 14.7 percent, respectively. These data appear to refute
the claim that blacks are somehow excessively more reliant on the federal
government than other racial groups.

Another crucial question that must be resolved before initiating a concentrated effort to increase the charitable activities of blacks is to determine how much trust the black community has in charitable organizations.
This is important because black charitable organizations traditionally have
been informal groups that operated in relatively small communities, where
there was a high probability for the donor and recipient to know each
other either directly or indirectly. If blacks have strong misgivings, however accurate or inaccurate, about the ability of charitable organizations
to deliver services to the poor, they may be disinclined to participate in
any effort to increase their level of charitable activity.

The responses of blacks and whites to the question of how much trust
they have in charitable organizations is shown in Table 5.2. Whereas a
relatively small percentage of blacks and whites responded that they had a
lot of trust in charitable organizations, 18 percent and 16 percent, respectively, nearly one-half of both groups, 47 percent of blacks and 54 percent
of whites responded that they had some trust. Only one-fifth of either
blacks or whites have very little trust, and less than 7 percent said they
had no trust at all. These findings suggest that the majority of blacks are

Table 5.2 "How Much Confidence or Trust Do
You Have in Charitable Organizations to Help the
Poor?" by Race

	Black	White
A Lot	18.7	16.4
Some	47.2	54.7
Very little	22.0	20.1
None	6.0	3.5
It Depends	3.9	3.2
Don't Know	2.3	2.1
Number of Observations	868	916

NOTE: The responses in this table have a sampling error of 4 percentage points for each group.

not overly concerned that charitable organizations will fail to deliver
promised services to the poor.

Black Charitable Giving

One issue that has been central to discussions of black self-help is the
degree to which the black community (in particular, the black middle class)
is active in contributing money to help the poor. There is a widespread
opinion among many that blacks do not contribute to charitable organizations. This view is unsupported by the data. The total contributions of
blacks and whites to all charitable organizations, controlling for differences in income, can be seen in Table 5.3. With the exception of the $250–
$499 category, a greater percentage of blacks than whites with incomes
under $12,000 make larger contributions to charitable organizations. Sim-
ilarly, among blacks and whites with incomes between $12,000–25,000,
a higher percentage of blacks than whites, with the exception of contributions less than $50, made greater total contributions to charity.

Of blacks and whites with incomes greater than $25,000, a higher per-
centage of blacks than whites made contributions less than $100 while a

Table 5.3 Total Contributions to Charitable Organizations
Percentage of Blacks and Whites by Income

	Less than $12,000		$12,000–25,000		Greater than $25,000	
	Black	White	Black	White	Black	White
Less than $50	14.7	12.5	7.4	11.6	8.4	2.9
$50–99	8.5	6.4	11.2	6.1	5.9	4.0
$100–249	17.2	16.6	23.6	16.1	14.8	14.8
$250–499	8.1	9.0	12.6	12.2	13.3	16.6
$500–999	6.2	3.5	15.8	11.6	13.6	22.5
More than $1000	3.1	4.7	6.5	11.6	20.2	20.2
No Contribution	28.6	17.2	14.0	9.6	16.4	7.5
Unknown	13.4	30.2	9.0	21.2	7.4	11.6
Number of Observations	387	204	260	304	170	352

Table 5.4 Where Black and White Charitable
Contributions Go

	Black	White
Own Church	68.0%	59.4%
Religious Organizations	7.4	8.3
Educational Organizations	3.8	6.0
Social Welfare Organizations	6.2	7.7
Hospitals and Medical Centers	6.6	7.3
Political Organizations	1.1	2.7
Social and Fraternal Organizations	1.3	1.6
International Aid Organizations	2.8	1.5
Other Charities Not Mentioned	2.8	5.5

greater percentage of whites than blacks made contributions between $250 and $999. An equal percentage of blacks and whites with incomes over $25,000 made contributions over $1000, 20.2 percent, respectively.

Table 5.4 shows the percentage distribution of the total amount that blacks and whites in the sample contributed to various charities in 1985. The overwhelming percentage of charitable funds were received by the donors' own church with blacks contributing 68 percent to their own church and whites contributing 59 percent. Other religious organizations received 7 percent of all black charitable dollars, whereas social welfare organizations and hospitals and medical centers received 6 percent.

Before drawing conclusions about the implications of these data, at least two limitations should be noted. As mentioned earlier, the black church has historically sought to provide for many of the needs (spiritual, social, and economic) of the black community. It is not unreasonable to assume that in many cases the church serves as a conduit to channel charitable donations received from members of the congregation to other organizations; for example, black institutions of higher education.

The other limitation to interpreting the data is that there may be some ambiguity with regard to how respondents interpreted making charitable contributions to several of the organizations listed. For example, if one made a contribution to a social or fraternal organization that was engaged in fund raising on behalf of the United Negro College Fund, it is unclear whether the respondent would report having made a charitable contribution to a social-fraternal organization or to an educational organization.

The findings suggest that, overall, blacks and whites contribute similar percentages of their total charitable contributions to the same charitable organizations. Other research using these data have suggested that there are few differences between the average contributions of blacks and whites to these types of charitable organizations when controlling for differences in income.

Black Voluntarism

One of the most neglected aspects of philanthropic activity is voluntarism. However, for a variety of individuals—the elderly, the young, and the poor—

Table 5.5   "For Which of These Organizations, If Any, Did You Personally Do Volunteer Work in 1985?" by Race and Income

	Less than $12,000		$12,000–25,000		Greater than $25,000	
	Black	White	Black	White	Black	White
Church	30.0%	27.0%	40.8%	36.1%	38.4%	37.9%
Religious Organization	3.4	2.9	6.0	10.8	7.7	4.5
International Aid						
Organization	0.3	2.0	2.0	2.6	5.4	1.9
Educational Organization	8.1	4.7	11.2	10.0	22.0	19.4
Hospitals	1.8	3.8	9.0	5.3	6.9	7.2
Other Health						
Organization	4.1	4.7	5.3	8.4	5.1	11.3
Social Welfare	3.5	5.2	6.0	7.3	6.9	9.6
Social and Fraternal						
Organization	2.3	2.3	4.0	5.5	7.7	12.4
Political Organization	0.9	2.0	3.3	8.4	11.3	7.8
Other Charity	6.2	6.1	3.6	15.3	13.3	16.7
None Mentioned	56.6	60.5	45.7	42.2	40.7	39.2
Number of Observations	387	204	260	304	170	352

it may be easier to engage in voluntarism than charitable giving because of limited incomes or greater amounts of leisure time. In addition, voluntarism is important because, as noted earlier, some efforts, particularly protest demonstrations, rely on the ability to motivate large numbers of supporters to volunteer their time to participate in scheduled events designed to influence the political process by demonstrating large scale public support for ideas that are believed to be unpopular.

The percentage of blacks and whites in the sample who responded that they performed some type of volunteer work for different types of charitable organizations, controlling for differences in income, is shown in Table 5.5. Consistent with the data on attitudes and giving, the church is again the dominant institution among both blacks and whites for volunteering time. Over one-fifth of both blacks and whiter reported that they engaged in volunteer work for their own church in 1985. In addition, at every level of income, blacks are more likely than whites to perform volunteer work for both their own church and educational organizations.

It is important to note that a greater percentage of blacks with incomes between $12,000 and $25,000 and blacks with incomes over $25,000 performed volunteer work for every charitable organization mentioned than blacks with incomes less than $12,000. Again, these findings provide firm evidence to refute suggestions that blacks are not actively engaged in voluntarism, or that the black middle class is somehow less active in voluntarism than the white middle class or other blacks.

Nonorganized Individual Philanthropic Activity

Thus far, this chapter has examined organized philanthropic activity; that is, the charitable giving and voluntarism through established charitable or-

Table 5.6 "In 1985, Did You Give Money, Food or Clothing, or Perform Some Other Service for Any of the People Listed Here?" by Income

	Less than $12,000		$12,000–25,000		Greater than $25,000	
	Black	White	Black	White	Black	White
The Homeless	13.5%	9.3%	17.5%	14.3%	22.3%	11.9%
Needy Neighbor	16.0	20.6	28.1	25.0	29.2	23.1
Needy Relative	31.2	18.9	34.3	30.6	35.5	35.7
Needy Friend	33.2	26.7	37.8	35.4	39.6	32.0
Needy Individual	17.3	17.4	23.9	26.1	32.7	25.8
None–Don't Know	35.9	39.2	26.2	25.7	15.9	25.3
Number of Observations	387	204	260	304	170	352

ganizations. However, an exclusive focus on organized philanthropic activity ignores the efforts of people who engage in charitable activity on their own, independent of any organizatinal support.

The percentage of blacks and whites, controlling for differences in income, who reported having given money, food, clothing, or performed some service for the homeless, a needy neighbor, a needy friend, or a needy individual is shown in Table 5.6. In general, although there are a few exceptions, a greater percentage of blacks than whites at each level of income reported having given or provided some type of charitable service to one of the needy groups listed. For example, a greater percentage of blacks than whites at every level of income reported having aided a needy friend and the homeless. The findings suggest that blacks are no less active in acts of personal philanthropy than they are in supporting organized philanthropic activity.

CONCLUSION

The current discussions about black self-help have raised serious questions about the extent to which blacks participate in efforts to help themselves. Although it is not clear that any other ethnic group has been put to a similar litmus test, the stakes are too high for blacks to refrain from the debate. Part of the reluctance to engage in this debate stems from two sources. First, many both inside and outside the black community have unsupported beliefs that blacks have not been engaged in activities to help themselves. Second, others believe that an examination of the capacity and the resources available to blacks will ultimately lead to a justification of a reduction in government services to blacks. Neither of these arguments is very compelling.

The idea that blacks have not been engaged in efforts to help themselves is simply untrue and ignores a continuing tradition that began over 200 years ago. This paper has documented both the history and the immense participation of blacks at all levels of income in all aspects of philanthropy, ranging from charitable giving and voluntarism to individual acts of helping the needy. None of the data presented suggest that blacks are less active than whites in their support of charitable activities nor do they suggest that the black middle class has been particularly negligent.

This paper began by accepting the proposition that the black community and the government have a dual responsibility to aid the poor. Although there are those who will maintain that black charitable resources should be substituted for government services, these arguments are less persuasive in light of the research findings presented here that blacks at all levels of income are doing their fair charitable share. In addition, these arguments are weakened by the disproportionate resources available to the government compared with the black community. Instead, these data argue that despite the difficult socioeconomic circumstances faced by some blacks, one-dimensional perceptions of blacks as only recipients and not givers of charity are unwarranted.

NOTES

1. See Committee on Policy for Racial Justice, *Black Initiative and Governmental Responsibility* (Washington, D.C.: Joint Center for Political Studies, 1987).

2. Glenn C. Loury, "Internally Directed Action for Black Community Development: The Next Frontier for 'The Movement,'; " *Review of Black Political Economy* (Summer–Fall 1984): 33–46.

3. Alexis de Tocqueville, *Democracy in America,* reprint, (New York: Mentor-NAL, 1956), p. 198.

4. Abram L. Harris, *The Negro as Capitalist* (College Park, Md.: McGrath Publishing Co., 1968), p. 21.

5. August Meier, *Negro Thought in America* (Ann Arbor: University of Michigan Press, 1983), p. 14.

6. James Oliver Horton and Lois E. Horton, *Black Bostonians* (New York: Holmes and Meier, 1979), p. 28.

7. Ibid.

8. Benjamin Quarles, *The Negro in the Making of America* (New York: Collier Books, 1979), p. 98.

9. Ibid.

10. E. Franklin Frazier, *The Negro Church in America* (Liverpool: University of Liverpool Press, 1963), p. 44.

11. Harris, *The Negro as Capitalist,* p. 46.

12. Lenone Bennett, Jr., *et al,* Eds., *Pictorial History of Black America.* Nashville, Tenn.; *Ebony,* The Southwestern Co. Volume 1. 1971. p. 183; and Raymond S. Wilmore, *Black Religion and Black Radicalism.* New York: Orbis Books. 1983. page 25.

13. David Mathews, "The Independent Sector and The Political Responsibilities of the Public," an address to the Spring Research Forum of Independent Sector (March 19, 1987), p. 6.

14. Face-to-face interviews are less likely than telephone interviews to underrepresent low-income blacks. However, the sampling methodology that was used to oversample blacks may have disproportionately sampled black middle-class households.

III
EFFECTS OF GOVERNMENT POLICIES

6

Federal Tax Policy and Charitable Giving

CHARLES T. CLOTFELTER

The United States is distinctive in the degree to which it subsidizes the nonprofit sector through its tax system. Its provisions for the deductibility of charitable gifts in addition to the tax exemptions accorded to nonprofit institutions are unparalleled in scope. Although the relations that have evolved between government, nonprofit institutions, and the legal structure are the result of hundreds of years of complex social development, it seems by no means accidental that this special reliance on nonprofit institutions and these favorable tax provisions have developed side by side.

In recent years, however, there has been evidence of increasing concern about the vitality of the nonprofit sector and the adequacy of federal tax provisions affecting charitable giving. One source of concern has been the standard deduction, introduced as a simplification into the income tax system over forty years ago but blamed for reducing the incentive to make contributions. Public commissions in the 1960s and 1970s investigated the role of tax policy in philanthropic giving and made their recommendations to Congress. One of those, the Commission on Private Philanthropy and Public Needs, known as the Filer Commission, began its report by recommending several basic changes in the tax treatment of contributions (Commission on Private Philanthropy and Public Needs 1977, pp. 3–21). For its part, Congress responded in the Economic Recovery Tax Act of 1981 (ERTA) by providing for the gradual extension of the charitable deduction to nonitemizers over a five-year period. However, Congress reversed direction in the Tax Reform Act of 1986. Although Congress rejected tax reform proposals that would have eliminated the charitable deduction altogether, the 1986 law reduced tax incentives to make contributions by ending the new nonitemizer deduction and, for the first time, taxing capital gains on some gifts of appreciated property. Combined with the Reagan admin-

Adapted and updated from *Federal Tax Policy and Charitable Giving* by Charles T. Clotfelter. Copyright © 1985 by The University of Chicago Press. Reprinted by permission.

istration's cuts in federal spending for social programs, these tax changes raised concern about the adequacy of revenue sources for the nonprofit sector at the very time those same cuts increased the demand for many services provided by nonprofit organizations. Rising labor costs and other developments within the nonprofit sector combined with slow growth in private support have caused one commentator to conclude that the sector as a whole "is in serious and growing difficulty" (Nielsen 1979, p. 3). Needless to say, such concerns have heightened interest in the role of the tax system in influencing the level and distribution of private support for charitable and other nonprofit organizations.

This paper examines one important aspect of the relationship between the tax structure and the nonprofit sector: how federal taxes affect charitable giving. Specifically, it examines the effect of tax provisions on contributions by individuals, corporations, and estates and on grants by foundations. The focus is on the connection between policy variables and behavior as observed in econometric analysis and other empirical study.

CHARITABLE GIVING AND THE NONPROFIT SECTOR

Table 6.1 presents data for major categories within the nonprofit sector based on returns for tax-exempt organizations in 1975. The organizations are divided according to whether contributions made to them are generally deductible in calculating federal income taxes. Of the 220,000 nonprofit organizations filing returns in 1975, the largest single group was in fact charitable organizations. Often referred to by the Internal Revenue Code section applying to them, such 501(c)3 organizations include religious, educational, cultural, scientific, and social-welfare organizations. This category represented over a third of all nonprofit organizations, based on number of returns, and over a half of total receipts of the sector, although these figures are probably underestimates since some religious groups do not submit returns. Charitable organizations represented an even larger share of contributions received—some 83 percent based on only 1975 returns and almost 90 percent counting all organizations.[1] The most important other category, based on receipts, was civic clubs such as Lions and Rotary. Other significant categories included voluntary employee beneficiary associations, labor and agricultural groups, business groups, and life insurance associations. As numerous as these other nonprofit organizations were, however, Table 6.1 makes clear that charitable organizations account for a sizable portion of the entire nonprofit sector.

Over the period 1955–1985, the real level of total charitable giving more than doubled, from $28 to $72 billion in 1982 dollars. As a percentage of national income, this giving has varied around 2.4 percent with no apparent trend. Over this period, individual donations have averaged about 80 percent of the total. Contributions by corporations have declined slightly

Table 6.1 Tax-Exempt Organizations, 1975

| Type of Organization | Number of Returns | Receipts (in millions of dollars) | | Applicable Code Section |
		Total	Contributions, Gifts, and Grants	
Tax-Deductible Contributions Generally Allowed				
Corporations Organized under Act of Congress	665	527	11	501(c)1
Charitable, Religious, Educational, and Scientific Organizations	82,048	65,544	17,110	501(c)3
Cemetary Companies	1,518	255	5	501(c)13
War-Veteran Organizations	1,921	130	7	501(c)19
Tax-Deductible Contributions Generally Not Allowed				
Title-Holding Companies for Exempt Organizations	3,263	490	23	501(c)2
Civic Leagues, Social-Welfare Organizations, and Local Associations of Employees	28,064	19,558	681	501(c)4
Labor, Agricultural, and Horticultural Organizations	28,258	5,028	120	501(c)5
Business Leagues, Chambers of Commerce, and Real Estate Boards	17,530	3,890	230	501(c)6
Social and Recreational Clubs	18,228	2,535	32	501(c)7
Fraternal Beneficiary Societies	12,066	2,134	46	501(c)8
Voluntary Employees' Beneficiary Associations	4,285	6,806	1,926	501(c)9
Domestic Fraternal Societies	4,674	507	21	501(c)10
Teachers' Retirement-Fund Associations	49	100	6[a]	501(c)11
Local Benevolent Life Insurance Associations	4,975	3,725	17	501(c)12
State-Chartered Credit Unions	1,610	2,259	1[a]	501(c)14
Mutual Insurance Companies or Associations	864	59	0[a]	501(c)15
Farmers Cooperative Organized to Finance Crop Operations	36	54	18[a]	501(c)16
Supplemental Unemployment-Benefit Trusts	496	959	244	501(c)17
Employee-Funded Pension Trusts	42[a]	13[a]	7[a]	501(c)18
Other Organizations[b]	9,605	309	62	—
Total	220,197	114,890	20,565	

SOURCE: Sullivan and Coleman 1981, pp. 7–8, Figure 1; p. 10, table 1.
[a] Estimates based on small samples.
[b] Organizations not specified included trusts for prepaid group legal services (covered in section 501(c)20), black lung trusts (501(c)21), religious and apostolic associations (501(d)), farmers' cooperative associations (521(a)), cooperative hospital service organizations (501(e)), and cooperative service organizations of operating educational organizations (501(f)). Contributions to the last two types of organizations are generally tax deductible.

Table 6.2 Estimated Charitable Giving by Source, Selected Years

Year	Individuals	Corporations	Bequests	Foundations	Total	Total in 1982 Dollars	Percentage of Total as National Income
Amounts (in billions of dollars)							
1955	6.75	0.42	0.24	0.30	7.70	28.31	2.3
1960	9.16	0.48	0.57	0.71	10.92	35.34	2.6
1965	11.82	0.79	1.02	1.13	14.75	43.64	2.5
1970	16.19	0.80	2.13	1.90	21.02	50.05	2.5
1975	23.53	1.20	2.23	1.65	28.61	48.25	2.2
1980	40.71	2.36	2.86	2.81	48.74	56.87	2.2
1985	65.94	4.40	5.18	4.90	80.42	72.32	2.5
Percentage of Total Giving							
1955	87.7	5.5	3.1	3.9	100.0	—	—
1960	83.9	4.4	5.2	6.5	100.0	—	—
1965	80.1	5.4	6.9	7.7	100.0	—	—
1970	77.0	3.8	10.1	9.0	100.0	—	—
1975	82.2	4.2	7.8	5.8	100.0	—	—
1980	83.5	4.8	5.9	5.8	100.0	—	—
1985	82.0	5.5	6.4	6.1	100.0	—	—

SOURCES: *Giving U.S.A.*, 1988 Edition, © 1988 AAFRC Trust for Philanthropy, p. 11 used by permission; U.S. Council of Economic Advisers, *Economic Report of the President,* 1988, Tables B-3 and B-24.

Table 6.3   Receipts and Expenditures of Philanthropic Organizations, 1980 (billions of dollars)

Receipts			Expenditures	
Sales		$60	Purchase of Goods and Services	$43
To Businesses	$ 4			
To Households	30			
To Government	26		Salaries	75
Subsidies		69	Capital Costs (including	
Private Donations	45		rental property)	11
Government Grants	8			
Investment Income	7			
Rental Value of Property	9			
Total	$129		Total	$129

SOURCES: Estimates of Gabriel Rudney, "Toward a Quantitative Profile of the Nonprofit Sector," in Program on Non-profit Organizations 1981, p. 3.

in importance, and foundation grants have dropped significantly. Bequests have fluctuated over time, being particularly sensitive to large gifts.

The dollar amounts in Table 6.2 do not directly reflect contributions made on fiduciary income tax returns for trusts and estates. Representing, for the most part, gifts not otherwise reflected on personal income or estate tax returns, contributions by fiduciaries were some $600 million in 1974, or about 2 percent of total giving.[2] Because of their small size these contributions by fiduciaries are not covered in the present study.

Although charitable giving is an important source of support for operating nonprofit organizations, it is by no means the only source. Table 6.3 presents a summary of aggregate receipts and expenditures in 1980 for the entire "philanthropic sector," those nonprofit organizations eligible for tax-deductible contributions. Total expenditures for this sector were $129 billion.[3] To cover this amount, organizations raised $60 billion through sales of goods and services and $69 billion from subsidies, of which private donations constituted the most important component—about $45 billion. For this group of nonprofit organizations, therefore, charitable contributions represented about 35 percent of total support. The two most important sources of private support were direct gifts from individuals and transfers from umbrella fund-raising organizations, each with 8 percent of total support in 1978. Bequests and transfers from trusts and foundations were also important sources, together accounting for about 14 percent of support. Not only is charitable giving an important source of support for nonprofit organizations, therefore, but many organizations receive this support by means of intermediaries, such as trusts, foundations, and umbrella fund-raising bodies.

SUPPORT BY RECIPIENT GROUP

It is often important to go beyond aggregate measures to observe the relative importance of giving for the major recipient groups within the non-

Table 6.4 Charitable Giving by Recipient Group, 1960 and 1986

Recipient Group	Giving (in billions of dollars)		Percentage of Total	
	1960	1986	1960	1986
Religion	5.01	40.90	45.8	46.9
Education	1.72	12.73	15.7	15.0
Human Services	1.63	9.13	14.9	10.5
Health	1.35	12.26	12.4	14.1
Arts, Culture, and Humanities	0.41	5.83	3.8	6.7
Public/Society Benefit	0.31	2.38	2.8	2.7
Other	0.50	3.99	4.6	4.6
Total	10.93	87.22	100.0	100.0

SOURCES: *Giving U.S.A.*, © 1988 AAFRC Trust for Philanthropy, 1988 Edition, p. 15, used by permission.

profit sector. Probably the best summary is a simple tabulation of total contributions by major groups, as is given in Table 6.4 for the years 1960 and 1986. The most striking aspect of the distribution for either year is the large share of contributions that go to religious organizations. Between 1960 and 1986, the largest relative increase was recorded by organizations concerned with the arts and humanities. Increases in share also occurred in the health group. The share of groups concerned with human services and education fell during the period.

The relative importance of charitable gifts as a source of support also varies by subsector. Table 6.5 presents a distribution of funding sources by major recipient group in 1974. This tabulation shows that, for the nonprofit sector as a whole, private contributions and government funds each provided about 30 percent of total support, with the remaining 40 percent coming from dues, sales, and endowment income. Among the major recipient groups, religious organizations were most dependent on charitable gifts for support, receiving some 94 percent of all revenues from contributions.

Table 6.5 Distribution of Support for Nonprofit Organizations, by Recipient Group, 1974

Recipient Group	Philanthropy	Service Charges and Endowment Income	Government Funds	Total
Religion	94	6	—	100
Health	11	47	42	100
Education	32	56	12	100
Other	31	35	34	100
Total	31	40	29	100

SOURCE: Report of the Commission on Private Philanthropy and Public Needs, cited in Sumariwalla 1983, p. 195.

Given the lack of fees or government funding for religious activities, this dependence is not surprising. Health organizations, in contrast, showed the least dependence on contributions, with almost 90 percent of their revenues being derived from service charges, government support, or endowment income. Education and other nonprofit groups received about 30 percent of their revenues from contributions. Religious institutions aside, nonprofit organizations receive a sizeable part of their funding from self-generated revenues and government support.

PHILANTHROPY AND TAX POLICY

Two cornerstones underlie U.S. tax policy toward charitable activity: the deductions for contributions allowed in major federal taxes (the personal income tax, the corporate tax, and the estate tax); and the tax-exempt status generally accorded nonprofit institutions. The size of individual giving suggests that the charitable deduction in the personal income tax is of preeminent importance. Adopted in 1917, four years after the enactment of the individual tax itself, the provision allows the deduction of individual contributions of cash or other assets made to eligible organizations up to certain limits. Since 1917 the deduction has been modified in two principal ways. First, the introduction of the standard deduction in the 1940s as a major simplification measure effectively eliminated the charitable deduction for a majority of taxpayers. Second, the nonitemizer deduction contained in the 1981 tax act briefly gave all taxpayers an opportunity to deduct contributions. The effect of provisions such as these cannot be evaluated without reference to the overall structure of the income tax and its tax rates. In addition, state income taxes usually allow a deduction for gifts similar to the federal deduction and thus are another influence on contributions.

The debate over these provisions has pitted those who think the deduction is an effective and appropriate incentive for charitable giving against those who believe that simplification or equity would be better served with less favorable provisions for contributions. These issues were as relevant in the discussions that preceded the passage of the Tax Reform Act of 1986 as they were in the debate over the standard deduction in 1941. Between 1982 and 1986, for example, dozens of proposals were put forward to reform the income tax, many of which were "flat-rate" tax schemes that would have eliminated the charitable deduction altogether. Levels of giving may also be influenced by the structure of taxes and tax rates. The debate over provisions affecting charitable deductions is therefore framed by normative questions of their equity, the importance of tax simplification, and the comparative value of public and nonprofit provision of services as well as by the factual question of how taxes affect contributions.

Besides the charitable deduction in the individual tax, two other deductions and a separate set of related provisions affect charitable giving directly. First, charitable bequests made as part of the disposition of estates are deductible without limit in calculating the federal estate tax. Individuals wealthy enough to be subject to the estate tax may choose between making deductible contributions during life or deductible charitable bequests at death. Second, contributions made by corporations are deductible up to a limit in calculating the corporate income tax. In addition to these provisions, the tax law allows individuals to set up foundations or charitable trusts and deduct the value of gifts made through them. As with the income tax, state taxes are generally similar to their federal counterparts in how charitable gifts are treated.

TAX POLICY AND GIVING BEHAVIOR

The bulk of econometric analysis and attention in economic studies has been directed toward individual giving, which seems appropriate given the large share of total gifts accounted for by individuals. Contributions by individuals vary widely by income level and age as well as among individuals within those classifications. The major tax policy instrument affecting individual giving is the charitable deduction allowed in the calculation of taxable income for taxpayers who itemize their deductions. As a result of this tax treatment, there are two major tax effects on individual giving: the tax liability affects the after-tax income from which taxpayers can make contributions; and the deduction reduces the net price per dollar of contribution made. For example, a taxpayer facing a marginal tax rate of 30 percent will reduce his or her tax liability by 30 cents for every dollar contributed; the "price" or net cost of giving a dollar is thus 70 cents. The econometric analysis of individual giving implies that the income tax has a strong effect on giving. This is not to say, however, that taxes are the only or the major influence on individual contributions, only that they are a significant factor.

Taken as a whole, the empirical work on tax effects and individual giving is notable for the number and variety of studies in the area and the consistency of the findings. Studies of charitable contributions have used aggregated and individual data, data from tax returns and survey data, and foreign as well as U.S. experience. Following the practice in economics, the magnitudes of these effects are usually given in terms of elasticities, or the ratio of percentage changes. The consensus of studies in this area is that the price elasticity for the population of taxpayers is probably greater than 1 in absolute value. This result implies that a 10 percent increase in the price of giving (say, from 50 to 55 cents per dollar) would lead to a decrease of more than 10 percent in the amount given. There are certainly estimates in the literature both larger and smaller than 1, but the range of most likely values appear to be about 0.9 to 1.4 in absolute value. Taxes also influence giving through an income effect, with most estimates of the income elasticity falling between 0.6 and 0.9, classifying contributions as a "normal" expenditure that rises with income.

In order to appreciate the implications of these findings, it is necessary to consider the specific hypotheses, different uses of data, and qualifications that apply to the studies themselves. For example, one maintained hypothesis is that itemization status and marginal tax rate work together through the price effect to affect giving, and that there is no separate "itemization effect." Separate tests of such an effect, in fact, confirm this maintained hypothesis. Another important question is whether the price elasticity varies by income level. The extensive analysis on this question has failed to provide a definitive answer, but it appears that the elasticity rises in absolute value with income. It is reasonable to conclude, however, that the price elasticity is significantly less than zero even for low-income taxpayers. A question of particular importance for evaluating the impact of tax policy is whether taxpayers respond immediately to changes in price and income. Evidence on this question suggests that there are substantial lags in giving behavior, with the result that short-run responses are much less complete than those in the long run. One other question related to the impact of fiscal policy on contributions is whether increased government spending "crowds out" private giving. The econometric evidence on this question shows little if any effect of this sort in spite of the apparent relationship observed among nations in the size of government and the strength of private giving. Throughout this empirical literature certain econometric issues have had to be dealt with, in particular the high correlation between price and income. Based on attempts to correct for possible biases as well as the variety of data and models used in these studies, it appears that these econometric problems are not a major factor in explaining the pattern of estimates.

There is also a sizable econometric literature on the effect of taxes on corporate giving. The evidence suggests that the corporation tax has both a price and a net-income effect on corporate giving. Such behavior by firms would be consistent with a number of models other than pure profit maximization. The estimates of the income-effect elasticity using a cash flow measure of income are close to 1, suggesting that contributions are proportional to after-tax income. An important question remains, however, regarding the proper specification of this income measure. Qualitatively similar results are obtained using after-tax net income. The estimated price elasticities appear to be smaller than those estimated for individual contributions, but the estimates presented here leave some doubt due to the difference in results using marginal and average tax rates, respectively. Taken together, these results suggest that the price elasticity is less than 1 in absolute value. Finally, there is evidence that corporations time their gifts in order to take more deductions during years in which tax rates are higher.

Tax effects are also apparent in bequest giving and foundation activity. The econometric evidence on bequest giving produces estimates subject to substantial variation. Nevertheless, these estimates imply that the deduction in the estate tax has quite a strong effect by and large. Most estimates of the price elasticity are greater than 1 in absolute value. Bequests also rise with estate size, but the elasticity of estate size is substantially smaller than 1. On estimates obtained for the very important group of the wealthiest decedents, those with net estates over $1 million, the estimated price elasticity was greater than 2 in absolute value, and the income elasticity exceeded 1. In any assessment of the aggregate effect of estate tax changes on charitable bequests, the largest estates are of paramount importance because the account for most bequest giving. No comparable econometric evidence on foundation activity has as yet been produced. The limited information available suggests, though, that the provisions in the Tax Reform Act of 1969 related to private foundations had the effect of raising payout rates without threatening the existence of foundations.

IMPLICATIONS FOR TAX POLICY AND
THE NONPROFIT SECTOR

The major conclusion arising from this empirical work is that federal taxes, especially tax provisions affecting charitable giving, have important effects on the size and distribution of giving. The deductions in the individual, corporate, and estate taxes are of course most imortant, in the sense that no other tax changes with comparable revenue effects would influence charitable giving as much as the elimination of these deductions. But other, more general tax provisions and changes also have profound effects on giving. Probably the most important of these effects arise from the combination of the standard deduction, nominal tax schedules, and inflation. The effect of inflation has been to erode the value of the standard deduction, causing an increase in the proportion of taxpayers who itemize their deductions. This in turn affects the price of giving. Another important set of tax changes not directly related to charitable giving has been revisions in the rate schedule itself. In particular, the decline in the marginal tax rates applicable to the top income classes from 91 to 28 percent over the last three decades has had a profound impact on the net cost of giving by the wealthy. A tax change such as the 1986 tax act combines several changes likely to affect charitable giving. Simulations based on estimated models of individual giving suggest that the combined effect will be a reduction in total giving in the long run on the order of 15 percent below what would otherwise have been contributed.

Similarly, econometric evidence implies that federal taxes will affect other forms of giving as well. The 1981 changes in the corporate tax resulting in an increase in the number of firms with no tax liability tended to discourage corporate giving by raising its average net price, but the 1986 act appears to have added firms back into the tax rolls. At the same time, the 1986 act cut top corporate marginal rates from 46 to 34 percent. Based on available econometric evidence, the resulting increase in the net cost of making gifts will result in a modest decline in corporate giving, on the order of 5 percent. The implications of the empirical analysis of bequests are similar to those applying to individual contributions.

One of the most important implications of existing empirical work is that tax policy can affect the distribution as well as the level of contributions. Since donors at various income levels differ markedly in their propensities to make gifts to various kinds of charitable organizations, tax changes that affect the distribution of giving among income classes will tend to affect the distribution of support to various parts of the philanthropic sector. For example, the 1981 and 1986 tax acts had the effect of significantly reducing marginal tax rates for taxpayers in the top income brackets. If the effect of such price changes outweighs the influence of changes in net income, which they in fact appear to do, these tax changes are likely to cut the relative share of giving undertaken by the wealthy. Already there have been significant declines in average contributions in the highest income classes following the cut in top rates in 1981. This would imply a decline in support for institutions such as colleges, universities, cultural institutions, and private foundations and toward religious organizations and certain health and welfare groups. Simulations of the 1986 tax law, for example, suggest that percentage declines in contributions to colleges and universities will be twice as large as percentage declines in religious giving. It is important to emphasize, however, that implications such as these are based on price and income effects and do not account for changes in underlying behavior by donors or charitable organizations.

The econometric estimates also have implications for proposed or hypothetical tax provisions. Simulations in *Federal Tax Policy And Charitable Giving* examine several proposals that involve changes in the charitable deduction or general tax rate revision. *Probably* the largest effect would be observed if the charitable deduction were eliminated altogether. Such a change would have important effects on the distribution as well as the level of contributions, with gifts by wealthy taxpayers falling the most. Substituting a tax credit for the present deduction, depending on the rate used, would have the effect primarily of redistributing the pattern of gifts between low- and high-income groups. Similarly, any reinstatement of the deduction for nonitemizers would shift the distribution of giving toward lower- and middle-income taxpayers and the institutions they tend to support. Smaller changes would come about as a result of less sweeping revisions, such as the constructive realization of appreciated assets given as gifts. Each of the proposals noted here would affect overall tax revenues, and it is important in simulating their effects to adjust for this. Similar effects could be calculated for bequest giving, with the elimination of the deduction in the estate tax having much the same kind of effect.

Unanswered Questions

Even though it encompasses many different studies, the econometric literature linking taxes to charitable giving still leaves a number of important
questions unresolved. Some of these could in principle be answered within
models such as those that have been estimated. In order to answer others,
it would be necessary to employ more general models. Within the context
of the models that have been estimated, questions remain in every major
area of charitable giving. We still do not have a precise idea, for example,
of the magnitude of the price elasticity for low-income households. This is
an important policy question because of the introduction of the new deduction for nonitemizers and because of the distributional implications of
general tax changes. A second unanswered question is how the response
to taxes varies according to the type of donee organization being supported. Although it might be difficult to estimate separate price and income elasticities by detailed donee class, it might well be possible to determine whether religious giving is affected differently from other types of
contributions. And, although there is a fair amount of consistency among
studies of corporate giving, there remains considerable uncertainty as to
the precise price elasticity and the appropriate measure of corporate income. Within the context of the models estimated, however, information
is required on one additional question. If the deduction for charitable contributions were limited or eliminated, corporations would have the incentive to substitute other deductible expenditures for corporate gifts. Because
of this substitutability, the price elasticities based on the current regime of
full deductibility would not be applicable. Concerning charitable bequests,
the instability of elasticity estimates in several studies suggests that our
knowledge about the tax effects is not as good as we would like.

Other questions left unanswered by existing empirical work would require broader models than have been used in previous work. The models
underlying virtually all empirical work on charitable contributions are partial equilibrium in nature. They ignore interactions among various kinds
of giving as well as interactions between donors and donee organizations.
It seems reasonable to suppose that changes in the tax treatment of contributions in one tax could affect contributions made subject to another.
For example, a restriction in the deductibility of charitable bequests might
well increase lifetime giving. Except as between volunteering and lifetime
gifts, there is no evidence on interactions of this kind. More generally,
most of the empirical analysis of charitable giving subject to a given tax
assumes that the tax base itself is given. A more general analysis would
recognize the possibility of endogenous changes in the tax base. The models
employed are also inadequate in their failure to reflect interactions among
donors and interactions between contributors and charitable organizations. Contributions by peers may increase or decrease an individual's contributions, and this relationship has important consequences for tax policy
effects. Charitable organizations, for their part, may respond to changes

in tax policy by varying their solicitation efforts. As long as effects such as these are not reflected in econometric models, projected effects based on those models must be seen only as conditional statements.

NORMATIVE CRITERIA FOR EVALUATING TAX POLICY TOWARD CHARITABLE GIVING

The fundamental normative questions in the evaluation of tax policy toward charitable giving are whether and to what extent such giving should be subsidized. If charitable giving were just another category of personal spending by consumers, there might be no reason to consider any form of subsidy whatsoever. A secondary question has to do with the proper form the subsidy should take, given that some subsidy is appropriate. In addressing questions such as these, it is useful to begin with the standard public finance criteria of efficiency and distributional equity. Other, more specific considerations may also be important. Before discussing these criteria, it is useful to note a fundamental distinction relevant to one specific form of subsidy: the deduction.

Two Views of the Charitable Deduction

Two quite different kinds of arguments have been offered to justify the present deductions for personal contributions and bequests. According to the first, the deductions are necessary adjustments in calculating the proper tax base. Andrews (1972) argues that contributions are properly excluded from the income tax base because they constitute neither accumulation nor consumption, the two components of income under the accretion concept.[4] Although contributions emanate from personal expenditures, he argues, they are not consumption in the usual sense because they effect a transfer of resources to others. Similarly, Wagner (1977) argues that a deduction is the correct mechanism for calculating the proper base for estate taxation, on the basis that funds set aside for charitable purposes are funds that cannot be enjoyed by the heirs of an estate. By this reasoning, horizontal equity thus requires that contributions be deducted in calculating the tax base.[5]

An alternative justification for the current charitable deduction is to view the deduction as an incentive by which the tax law encourages desirable behavior. According to this view charitable giving is an item of discretionary spending that warrants an incentive. A deduction is only one of several forms such an incentive might take; a tax credit or some matching arrangements might be as good or better. Since contributions are seen as discretionary expenditures by this view,[6] there is no necessity to provide the incentive in the form of a deduction from income. In contrast, the first view plainly requires the use of a deduction.

The implications of these views for the normative analysis of the tax treatment of charitable giving should be clear. If the deduction is seen as an absolutely necessary adjustment to income, it becomes "a matter of principle" (Break 1977, p. 1530), and there remains little to discuss concerning the proper tax treatment of charitable giving. If it is an incentive, however, alternative subsidies are fair game for consideration. The tax policy debate over the last two decades suggests that the first view is by no means universally accepted. That debate has focused on the form as well as scope of incentives for charitable giving. And, due to the existence of the standard deduction, the charitable deduction itself (along with the other itemized deductions) has been effectively limited to a minority of taxpayers. Accordingly, the remainder of this chapter is predicated on the assumption that the form of tax subsidy is not determined a priori, but rather is a question subject to normative policy analysis.

Efficient tax incentives for contributions
The concept of economic efficiency is important in any full assessment of tax provisions related to charitable contributions. Indeed, efficiency criteria are necessary for answering the primary question of whether charitable gifts should be subsidized at all. In order to give more concreteness to the application of economic efficiency to charitable contributions, it is useful to begin by presenting a stylized illustration of a tax policy decision involving incentives for charitable giving. Consider the choice between an increase of $1 million in government expenditures and an increase of the same amount in tax subsidies for charitable giving, both being financed by an increase in tax rates. Further suppose the new incentive leads to an increase in charitable giving of $X million. Obviously, government expenditures under the first option will be higher by $1 million. By the same token, charitable giving will be higher under the second; it will be $X million higher if increased government expenditures do not crowd out private charity. Assuming no crowding out, the income available to households after taxes and charitable giving will be $(X−1) million less in the second case.

One definition of efficiency used in connection with tax policy for charitable giving focuses on the size of the incentive's effect. As stressed in a number of empirical studies of tax effects on charitable giving, if the price elasticity of charitable giving is greater than 1 in absolute value, a tax incentive producing a marginal change in the rate of subsidy to contributions will increase giving by more than the associated revenue loss. Accordingly, some writers have defined the *efficiency* of the charitable deduction in terms of the ratio of increased contributions to foregone revenue.[7] By such a definition, the incentive described in the present example would be "efficient" if the elasticity is greater than 1 in absolute value because the rise in contributions ($X million) would exceed the revenue cost ($1 million). Clearly, this is quite a specialized definition of efficiency. This concept takes no account of the comparative social benefit derived from private contributions compared to public expenditures. Nor does it give any weight to the change in income after taxes and contributions.

In order to consider the implications of a more complete definition of efficiency, two kinds of theoretical models of incentives for contributions will be discussed. The first focuses on the presumed external benefits that result from contributions. The second includes more general optimal tax models that rest on an explicit maximization of welfare.

External benefits

It appears to be widely agreed that in contrast to most other types of expenditures, charitable contributions often contain a substantial element of external benefit. Although donors may reap some direct benefit from their contributions, much of that giving materially benefits others. It might also be argued that charitable organizations produce an external benefit for society to the extent that they offer alternatives to government services. One long-standing justification for public encouragement of charitable giving appeals to the value of diversity in a pluralistic society. It is a basic theorem in applied welfare economics that goods producing external benefits tend to be underprovided in private markets and that economic efficiency can be served by subsidizing such goods. In equilibrium the price faced by each individual ideally should equal his or her personal marginal valuation of the good, with the subsidy making up the difference between marginal cost and marginal valuation. Where the "good" is dollars of charitable contributions, (with a marginal cost of \$1), v is marginal valuation per dollar, and s is the subsidy per dollar, the relevant private optimality condition for individual i is simply

$$v_i + s_i = 1$$

Assuming the individual in equilibrium equates his or her marginal valuation with the price to be paid, $(1 - s_i)$, the social optimum will be achieved when s_i is set at the marginal external benefit. The greater the external benefit, the larger the optimal subsidy.[8]

Hochman and Rodgers (1977) and Posnett (1979) analyze the tax treatment of charitable contributions using similar normative models in which contributions are assumed to be pure public goods. Hochman and Rodgers show that a set of tax subsidies based on a Lindahl solution achieves the optimal allocation.[9] They argue further that, for a wide class of cases, a constant subsidy rate such as a tax credit satisfies the optimality condition. Posnett demonstrates, however, that the general superiority of a constant rate of subsidy cannot be shown. About the most that can be gleaned from these theoretical studies is that tax subsidies of some kind for contributions can be justified on grounds of efficiency.

Practically speaking, it is quite inconceivable that any subsidy scheme could be devised to meet the conditions of a Lindahl solution. Both the characteristics of gifts and the tastes of individuals differ too much. A more modest objective would be to set subsidy rates according to the average amount of external benefit from contributions of different kinds. Hochman and Rodgers (1977, pp. 13–15) recommend tax credits for contributions as a way of approximating the Lindahl solution and imply that subsidy rates might well differ by category of giving. They argue that religious giving may have a more important external component than gifts to organizations that have some government counterpart (p. 13). On the other hand, Schaefer (1968, p. 30) maintains that nonreligious giving involves much more redistribution than religious giving, the latter being used largely "to preserve houses of worship and to maintain the activities of congregations." Discrimination among donees on the basis of external benefits would be difficult, both analytically and politically, but there are precedents. Contributions to private foundations are accorded less favorable treatment in the lower percentage limitation of gifts, lack of carryover, and limitations on the deductibility of gifts of appreciated assets. And contributions to schools practicing racial discrimination are not deductible at all.[10]

Although the present deduction does not provide for any discrimination in subsidy rates by type of charitable donee (except for the nondeductibility of some gifts), subsidy rates definitely differ by income level. The rate of subsidy tends to rise with income because the marginal tax rate rises with income. For example, the average taxpayer in the $10,000–15,000 class in 1980 faced a marginal tax rate of 0.16, compared to a rate of 0.49 for a taxpayer in the $50,000–100,000 class. Distributional issued aside, this variation in subsidy rates may be judged in the light of the welfare economics of subsidizing goods with external benefits. If the charitable activities supported by high-income taxpayers—such as higher education, cultural institutions, and private foundations—have a higher component of external benefits than activities supported by lower-income households—primarily religion and community-welfare agencies—this structure of subsidies may be justified. However, if these activities cannot be distinguished on the basis of their external benefits in this way, differing rates of subsidy would not be efficient.[11] In any case, it is important to identify the structure of subsidy rates as primarily a question of efficiency, although distributional equity is relevant to the resulting pattern of tax burdens and the distribution of the benefits of charitable activities.

Optimal tax models

A more general treatment of the efficiency of tax incentives for charitable giving can be obtained with an optimal taxation model, as developed by Atkinson (1976) and Feldstein (1980). Atkinson's model incorporates an additive social-welfare function in which individual utilities depend on their contributions. The well-being of a needy group in society can be affected either by contributions or government expenditures. The effectiveness of private giving in aiding this group can be more or less than that of government. Atkinson (p. 21) shows that the optimal tax-credit rate for contributions is higher, among other things, the more effective private giving is.

Atkinson also spells out the special assumptions under which the narrow "efficiency" concept noted earlier is an appropriate rule for determining whether the introduction of a charitable deduction improves social welfare. Two conditions are necessary: contribution dollars must be as effective as public expenditures in helping the needy group, and the social-welfare function must be Rawlsian, with all weight being given to the utility of recipients. In terms of the example given earlier, the first assumption allows dollars of giving to be compared directly to dollars of government revenue; the second makes it unnecessary to be concerned with donors' incomes after taxes and contributions. The deduction is a social improvement if the rise in contributions exceeds the revenue cost (if $X > 1$); that is, if the elasticity is greater than 1 in absolute value. In general, however, the desirability of a deduction depends not only on the effectiveness of contributions but also on the weight given to the preferences of donors and the equity effects of a deduction compared to a credit.[12] An elasticity of -1 has no general efficiency connotations.

Feldstein's (1980) model compares the cost, measured by a representative individual's willingness to pay, of increasing the consumption of some preferred good through government expenditure versus private giving. The effectiveness of the two types of expenditure is allowed to differ. His model, like Atkinson's, implies that a subsidy for charitable contributions is desirable under certain conditions, particularly when the government is less efficient in provision, when labor supply is more sensitive to the marginal tax rate, and when there is no preexisting subsidy. Feldstein points out that these findings conflict with the view that all "tax expenditures" should be eliminated. Significantly, Feldstein's model implies that the optimal subsidy does not depend on the price elasticity of giving.

Other efficiency-related considerations

More generally, issues related to administrability or neutrality are proper considerations in the design of tax incentives for contributions. Administrability covers such issues as the compliance and administrative costs of tax provisions. As an illustration, proposals that would specify differing rates of subsidy for different types of charitable organizations might well entail higher enforcement costs. Alternatively, the extension of a tax subsidy for charitable gifts to low-income households might require significant increases in record keeping by taxpayers.

Neutrality arises as an issue particularly in the treatment of different types of charitable gifts. Long (1977) notes, for example, that the charitable deduction in the income tax provides neutral treatment as between gifts of time and money since the value of either kind of gift is excluded in the calculation of taxable income.[13] Thus, any important change in the tax incentive for contributions in any major tax could distort taxpayers' choices among lifetime gifts, volunteering, bequests, and even gifts made through a corporation. Another way of putting this point is that such tax incentives may affect the various tax bases. The elimination of the charitable deduction in the estate tax might well increase the amount of wealth given away during life, thus reducing the size of estates.[14]

Distributional Aspects of Tax Incentives

The charitable deduction has come in for sustained and vigorous criticism for its alleged favoritism toward high-income taxpayers. Because the tax savings per dollar obtained from the deduction rises with one's marginal tax rate, high-income taxpayers enjoy a bigger proportional tax reduction in their giving than taxpayers at lower income levels. One critic (Neilsen 1979, p. 16) states:

> The so-called "tax incentives" for charitable giving which are now embodied in the Internal Revenue Code are so extravagantly discriminatory as between poor and rich donors that for the social-action movements they are effectively meaningless as a help in soliciting individual gifts.
>
> The tax system as a whole is of no assistance in enabling them to be self-supporting through the contributions of their own members. Rather, it condemns them to dependence on baronial benefactors.

Others point out, however, that the differing rates of subsidy are merely an inevitable by-product of the progressive rate structure itself. If successive amounts are taxed at higher and higher rates, then a reduction of a dollar of taxable income must produce bigger tax reduction at higher incomes.[15] Clearly, this would not be the case with a tax credit, a fact that has led some critics of the deduction to favor a credit over the deduction on distributional grounds.[16]

It is important to ask whether this differential subsidy effect has any relevance for distributional equity. In doing so, it is useful to distinguish two kinds of effects resulting from the deduction: effects on the tax liabilities of taxpayers; and effects of changes in giving patterns. On the "tax side," the deduction affects taxpayers in much the same way as price reduction: there is both an income effect and a substitution effect. The income effect is associated with the improvement in utility for a taxpayer who makes donations. The substitution effect is the change in the relative price of giving. This substitution effect has no importance for distributional equity per se; it is important primarily for its efficiency implications. Its only distributional importance is in its effect on the pattern of support for charitable organizations, discussed later.

By contrast, there are clear distributional consequences in the deduction's income effect. These are reflected in the effect of the deduction on tax liabilities. If the charitable deduction were eliminated, the distribution of taxable income would change and along with it the measured progressivity of the income tax. As between any two taxpayers, the elimination of the deduction would raise average rate progression[17] if the product of the marginal tax rate and the proportion of income contributed rises with income. In order to see the likely effects on progressivity, Table 6.6 shows how tax liabilities would change based on the simulation model presented in my book, *Federal Tax Policy and Charitable Giving*. Revenues are held constant in each simulation by means of proportional changes in tax rates. The results show that it is quite likely that eliminating the deduction would

Table 6.6 Ratio of Taxes under Two Proposals to
Actual Taxes in 1983, by Income

Income ($1,000)[a]		Elimination of Deduction	20 Percent Tax Credit
$6.1 under	12.2	0.97	0.91
$12.2 under	18.3	0.98	0.96
$18.3 under	24.3	0.98	0.97
$24.3 under	30.4	0.99	0.99
$30.4 under	36.5	1.00	0.99
$36.5 under	60.9	1.01	1.02
$60.9 under	121.7	1.01	1.03
$121.7 under	243.4	1.00	1.03
$243.4 under	608.5	1.01	1.03
$608.5 under	1217	1.03	1.05
$1217 or more		1.04	1.06
All Classes		1.00	1.00

NOTE: Simulations use constant income and price elasticities.
[a]Taxpayers under $6,100 have no tax liability under any on the simulated taxes.

in fact increase the progressivity of the income tax. Accounting for the anticipated fall in contributions, those simulations imply that tax liabilities for taxpayers with incomes under about $30,400 would decline due to the overall reduction in tax rates made possible by the expansion of the tax base. For taxpayers with incomes over $36,500, taxes would rise. Conversion to a tax credit would increase tax progressivity even more. It is clear, therefore, that the existence and form of the incentive accorded to charitable contributions has effects on tax progressivity. Although it would certainly be possible to neutralize the impact of any change in the charitable deduction on tax progressivity by an appropriate restructuring in the tax schedule, it remains that the form of the incentive is a factor in determining the progressivity of the tax.

Tax incentives for giving may also have distributional consequences in their effect on giving patterns. Although the structure of net prices resulting from a tax incentive has no direct distributional effect on donors, the pattern of prices can affect the distribution of charitable support to various groups of charitable organizations. Because the present deduction results in net prices that fall with income, charities and charitable activities favored by the wealthy receive disproportionate encouragement. The result, in Vickrey's (1947, p. 131) words, is "a serious plutocratic bias to the activities of privately supported philanthropic, educational, and religious institutions."

To identify this bias is not to determine its ultimate distributional effect, however. A complete assessment of the distributional impact requires an examination of who ultimately benefits from the programs of charitable organizations. It is quite possible, as Schaefer (1968, p. 27) in fact suggests, that the charitable activities favored by the wealthy are more redistributive than organizations supported by lower-income taxpayers.[18] Unfortunately, little research into the distribution of benefits from charitable programs is available. It is quite conceivable that charities favored by the wealthy have no larger redistributive component.

Finally, it is possible that the examination of the distributional impact of the charitable deductions should go beyond conventionally measured economic benefits to include the distribution of economic power. Some criticisms of the current deduction clearly imply that the present tax incentives for contributions have the effect of concentrating power at upper-income levels.[19] This possibility is most evident in the private foundation. Simon writes (1978, p. 5): "We have to acknowledge that the fact that private economic power is being deployed, often dynastically, through the device of the charitable foundation and the power it gives the founder and the founder's family to select the objects of their charitable bounty and to manage the charitable assets." He concludes that, although the legal form and tax treatment of private foundations make it easier to achieve power, the "specter of privilege" applies to some degree in all tax subsidies for giving available to wealthy taxpayers (Simon 1978, pp. 17, 27). It is interesting to note that one likely, though unanticipated, effect of the relentless reduction in top marginal tax rates is to diminish this tendency to concentrate giving among the wealthy.

CONCLUSION

Federal tax policy has a substantial impact on the level and distribution of charitable giving in the United States. Empirical analysis suggests that support for charitable organizations responds both to explicit tax incentives for charitable contributions and to general changes in effective tax schedules.

In the normative evaluation of tax policy from the viewpoint of society as a whole, such a behavioral response is only one of a number of considerations. Efficiency and distributional equity are the two principal criteria for judging the desirability of tax incentives for charitable giving. The present deductions in the income and estate taxes have effects on the overall progressivity of those taxes, and the degree of behavioral response to tax incentives is relevant in measuring this effect. The differential pricing of contributions arising from the deduction is not itself an equity issue, but this price structure has distributional implications due to the particular pattern of contributions encouraged and the benefits enjoyed as a result. Again, the degree of behavioral response determines the importance of this distributional effect. In judging the efficiency of tax incentives for contributions, the magnitude of the price elasticity of charitable contributions is only one of several important factors that need to be considered; others include the external benefits derived from charitable giving, the value of diversity in the provision of services, the effectiveness of such giving com-

pared to government expenditures, and the distributional impact on donors and recipients. Except under very special assumptions, it is impossible
to state any simple relationship between the price elasticity and the efficiency of tax incentives for charitable giving.

The distinctive reliance of the United States on nonprofit institutions to
perform major social functions is fostered by federal tax provisions for
charitable giving. Changes in tax policy, effected through legislation or
inflation, can have a significant impact on the level and composition of
giving. As long as the nonprofit sector retains its important role in the
United States, understanding the effect of the tax structure on charitable
giving will be an essential part of the study of public policy in education,
health, and many areas of social welfare. Whether or not taxes are an
explicit part of policy in any of these areas, taxes are certainly an important implicit component.

NOTES

1. Total receipts by charitable organizations were some $28 billion in 1975 (see Table 6.2,
netting out foundation grants). Adding the $11 billion yields 89 percent for contributions to
501(c)3 organizations.

2. See U.S. Internal Revenue Service, *Statistics of Income—1974, Fiduciary Income Tax
Returns 1977* for the most recent published data on charitable deductions by fiduciaries.
Whether fiduciary contributions are reflected in the *Giving U.S.A.* estimates is unclear.

3. To avoid double counting, these figures omit foundation grants to other organizations
(Program on Non-profit Organizations 1981, p. 3).

4. See also Musgrave and Musgrave (1980, pp. 343–47) for a definition of the accretion
concept of income.

5. See also Posnett 1979 for a description of this view. Similar reasoning underlies the
justification for the deduction given in a 1938 congressional report: "The exemption from
taxation on money or property devoted to charitable and other purposes is based upon the
theory that the Government is compensated for the loss of revenue by its relief from financial
burden which would otherwise have to be met by appropriations from public funds, and by
the benefits resulting from the promotion of the general welfare" (U.S. Congress, House of
Representatives 1938, p. 19). The statement makes no explicit reference to the proper income
tax base.

6. Wagner (1977, p. 2342) notes, disapprovingly, that the "conceptualization of charity
as an act of personal consumption is conformable to the proclivities of many economists."

7. Feldstein testified: "a higher elasticity implies a greater efficiency; that is, more additional giving per dollar of lost tax revenues" (U.S. Congress, Senate 1980, p. 219). Also see
Boskin (1976, p. 55) and Donee Group (1977, p. 73) for similar references.

8. For a general treatment, see Musgrave and Musgrave (1980, pp. 78–80).

9. A Lindahl solution to the public-good allocation problem is one in which each individual pays a price equal to his marginal evaluation and the sum of marginal valuations equals
the marginal cost of the good. See Hochman and Rodgers (1977, p. 4).

10. Private schools in North and South Carolina whose practices were found to be dis-
criminatory were denied the right to receive deductible contributions in 1982. See *New York
Times* (16 October 1982), pp. 1, 7.

11. See Culyer, Wiseman, and Posnett (1976, pp. 44–46) for a proposal to replace the
British deduction by a matching grant with rates determined according to the externality
criterion. Posnett 1979 also endorses such a policy in general terms.

12. See especially Atkinson (1976, p. 25).

13. Boskin (1976, p. 50) makes a similar point.

14. See Boskin (1976) for a discussion of this point.

15. See, for example, Wagner (1977, p. 2344).

16. See, for example, Vickrey (1947, pp. 130–131) and Donee Group (1977, p. 72).

17. Using the average rate progression measure, a tax is progressive if the average tax rises with income. See Musgrave and Musgrave (1980, p. 376).

18. Boskin (1976, p. 50) also emphasizes the importance of identifying the beneficiaries of charitable programs, in the context of his discussion of the estate tax.

19. See, for example, Schaefer (1968, p. 25); Donee Group (1977); and Nielsen (1979).

REFERENCES

Andrews, William D. 1972. "Personal Deductions in an Ideal Income Tax." *Harvard Law Review* 86:309–85.

Atkinson, Anthony B. 1976. "The Income Tax Treatment of Charitable Contributions." In R. E. Grieson, ed. *Public and Urban Economics: Essays in Honor of William S. Vickrey.* New York: D. C. Heath.

Boskin, Michael J. 1976 "Estate Taxation and Charitable Bequests." *Journal of Public Economics* 5:27–56.

Break, George F. 1977. "Charitable Contributions under the Federal Individual Income Tax: Alternative Policy Options." In Commission on Private Philanthropy and Public Needs, *Research Papers,* vol. 3. Washington, D.C.: Treasury Department, pp. 1521–39.

Commission on Private Philanthropy and Public Needs. 1977 "Commentary on Commission Recommendations." In Commission on Private Philanthropy and Public Needs, *Research Papers,* vol. 1. Washington, D.C.: Treasury Department, pp. 3–48.

Culyer, A. J., J. Wiseman, and J. W. Posnett. 1976. "Charity and Public Policy in the UK: The Law and the Economics." *Social and Economic Administration* 10:32–50.

Donee Group. 1977. "Private Philanthropy: Vital and Innovative or Passive and Irrelevant." In Commission on Private Philanthropy and Public Needs, *Research Papers,* vol 1. Washington, D.C.: Treasury Department, pp. 49–85.

Feldstein, Martin. 1980. "A Contribution to the 'Theory of Tax Expenditures': The Case of Charitable Giving." In Henry J. Aaron and Michael J. Boskin, eds., *The Economics of Taxation.* Washington, D.C.: Brookings Institution, pp. 99–122.

Giving U.S.A. 1981–83. New York: American Association of Fund-Raising Counsel, Inc.

Hochman, Harold M. 1977. "The Optimal Treatment of Charitable Contributions." *National Tax Journal* 30:1–18.

Long, Stephen H. 1977. "Income Tax Effects on Donor Choice of Money and Time Contributions." *National Tax Journal* 30:207–11.

Musgrave, Richard A., and Peggy B. Musgrave. 1980. *Public Finance in Theory and Practice.* New York: McGraw-Hill.

Nielsen, Waldemar A. 1979. *The Endangered Sector.* New York: Columbia University Press.

Posnett, John. 1979. "The Optimal Fiscal Treatment of Charitable Activity," Discussion Paper 552–79, Institute for Research on Poverty, University of Wisconsin, Madison.

Program on Non-profit Organizations. 1981. *Research Reports* 1:3.

Schaefer, Jeffrey M. 1968. "Philanthropic Contributions: Their Equity and Efficiency." *Quarterly Review of Economics* 8:25–35.

Simon, John G. 1978. "Charity and Dynasty under the Federal Tax System." *Probate Lawyer* 5:1–92.

Sullivan, John, and Michael Coleman. 1981. Nonprofit Organizations, 1975–1978." *SOI Bulletin* 1:6–38.

Sumariwalla, Russy D. 1983. "Preliminary Observations on Scope, Size, and Classification of the Sector." In *Working Papers for Spring Research Forum: Since the Filer Commission.* Washington, D.C.: Independent Sector, pp. 181–228.

U.S. Congress. 1938. Committee on Ways and Means. *The Revenue Bill of 1938.* 75th Cong., 2d sess. Report No. 1860.

———. 1980. *Charitable Contribution Deductions: Hearing on S. 219.* 96th Cong., 2d sess.

———. 1977. *Statistics of Income—1974, Fiduciary Income Tax Returns.* Washington, D.C.: Government Printing Office.

Vickrey, William. 1947. *Agenda for Progressive Taxation.* New York: Ronald Press Co.

Wagner, Richard E. 1977. "Death, Taxes, and Charitable Bequests: A Survey of Issues and Options." In Commission on Private Philanthropy and Public Needs, *Research Papers,* vol. 4. Washington, D.C.: Treasury Department, pp. 2337–52.

7

Tax Policy, Charitable Giving, and the Nonprofit Sector: What Do We Really Know?

JERALD SCHIFF

The recent reform of the federal income tax has renewed concern among researchers, leaders of nonprofit organizations, and policy makers over the impact of taxation on charitable giving and, thus, on the nonprofit sector. A number of economists have predicted drastic reductions in money donations and, implicitly, in the quantity and quality of services provided by the nonprofit sector as a result of the reform.

In this chapter, the relationship between the tax system and charitable giving is reexamined. However, rather than adding to the rather large and growing number of statistical estimates of this relationship, an attempt will be made to analyze what we know and what still remains to be learned. It is argued that previous estimates have only limited relevance for policy decisions, for several reasons. First, important determinants of charitable giving may have been ignored in these analyses, causing estimates to be biased. More important, however, is that these estimates, even if not biased, may not be easily interpreted in ways that assist policy makers. In addition, even if previous estimates provide valuable information about the likely response of money donations by individuals to tax reform, they still leave unanswered many questions about the *overall* impact of such reform on the nonprofit sector. Past work has focused on the effect of taxation on donations of money by individuals, largely ignoring corporate and foundation giving as well as volunteering by individuals. Finally, potential responses to tax reform by the recipient organizations themselves—for instance, in the form of increased sales or solicitations—have not been adequately accounted for.

The chapter begins by describing the ways in which taxation and, in

Useful comments were provided by Michael McPherson and Richard Steinberg. I thank the Twentieth Century Fund for financial assistance.

particular, the recent tax reform influence charitable contributions of money. Next, the econometric literature with respect to this relationship between taxation and giving is reviewed briefly. Following that is a discussion, of some length, on the "holes" in our knowledge and how accounting for these shortcomings affects our conclusions with respect to the impact of tax reform on the nonprofit sector. The discussion examines the possibility that previous estimates are biased, problems in interpreting these estimates in a useful way, and the potential effects of tax reform on the nonprofit sector that have not generally been considered. Finally, the chapter concludes by analyzing how we might use our incomplete information to estimate the impact of tax reform on the nonprofit sector.

HOW DO TAXES AFFECT CONTRIBUTIONS OF MONEY?

Before we can understand the debate surrounding tax reform and money contributions, we need to consider how taxes, in general, influence giving. They do so in two basic ways. First, taxes affect the disposable, or after-tax, income of individuals. The more income a person has, the more he or she will tend to give. Thus, lower tax rates would be associated with greater levels of giving.

On the other hand, because contributions to charity are tax-deductible, individuals' donations are, in effect, subsidized by government, with the size of the subsidy increasing with the donor's tax rate. For instance, a donor in the 20 percent tax bracket who itemizes saves 20 cents in taxes for each dollar contributed. The cost, or price, of a dollar donation, then, is 80 cents. On the other hand, for the donor who faces a 30 percent tax rate, a dollar donation costs just 70 cents. The "price effect" of reduced tax rates works in the opposite direction from the "income effect" just described—the lower the tax rate, the less, other things equal, people will give. In addition, any tax law change that removes or reduces the value of tax deductibility or reduces the number of tax itemizers will increase the price of giving for the typical individual and so reduce money donations.

The 1986 Tax Reform

Several changes in the Tax Reform Act of 1986 have had, and will continue to have, a negative impact on charitable giving. First, the charitable deduction for nonitemizers has been eliminated. In addition, the elimination or reduction in value of a number of tax deductions will sharply reduce the total number of tax itemizers. These two changes will lead to a significant decline in the number of taxpayers eligible for a tax incentive for giving. In addition, marginal tax rates will decline for most individuals, so that even those taxpayers still itemizing will face a higher price of giving than previously. Finally, the inclusion of appreciated property in the minimum tax base increases the price of in-kind giving.[1]

The tax law changes will also affect the incomes of individual donors, further influencing contributions. However, because the tax reform is intended to be revenue-neutral, income effects are likely to be unimportant in the aggregate.[2]

THE RESPONSIVENESS OF GIVING: WHAT DO WE KNOW?

It is one thing to pinpoint those effects taxation should, in theory, have on charitable donations of money, and quite another to discover the extent to which those effects operate in reality. Economic theory allows the prediction that charitable contributions will respond to changes in income and prices but tells little about how large those responses will be. It is the magnitude of those responses that determines the actual impact of tax reform on money donations.

Economists have expended a great deal of effort attempting to estimate the responsiveness of money donations to changes in price and income. Responsiveness is generally measured in terms of an "elasticity," which reports the percentage change in some activity, such as charitable giving, associated with a 1 percent change in some other variable, such as price. A price elasticity of giving equal to one 1 would, then, imply that for each 1 percent increase in the price of giving, other things equal, giving will fall by 1 percent. An income elasticity of giving equal to, say, 0.5 would, similarly, reflect the fact that a 1 percent increase in income will be associated with a 0.5 percent rise in contributions.

A consensus has emerged among economists that the price elasticity of giving is slightly greater than 1, whereas the income elasticity is between 0.5 and 0.8. Table 7.1 presents a summary of elasticity estimates from some past studies.[3] Many of the estimates are quite similar—a comforting fact for those who use such information for policy decisions. We will see, however, that at least part of the confidence in these estimates may be unfounded.

Table 7.1   Estimates of Price and Income Elasticities of Giving, from Selected Studies[a]

Study	Price	Income
Feldstein (1975)	−1.26	+0.20
Feldstein and Taylor (1976)	−1.42	+0.77
Feldstein and Clotfelter (1976)	−1.55	+0.80
Boskin and Feldstein (1977)[a]	−2.54	+0.69
Reece (1979)	−1.40	+0.55
Clotfelter (1980)	−1.41	+0.53
Schiff (1984)	−2.45	+0.69
Reece and Zeischang (1985)	−.85	+1.43

[a] For low-income individuals only.

The result that the price elasticity of giving appears to exceed 1 has drawn a great deal of attention, because it is claimed to indicate that allowing contributions to be tax-deductible is an efficient, and so desirable, way for government to finance the goods and services produced by the nonprofit sector. In particular, it implies that the donations stimulated by deductibility exceed the tax revenue lost by government due to that provision, so that to transfer the same number of dollars *directly* to nonprofit organizations, say via grants, would be more costly to the government, in terms of revenue foregone, than allowing deductibility.[4]

While the elasticity is clearly of interest, other issues are involved in comparing tax-deductibility with other finance schemes. For instance, it is unclear whether it is more desirable to allow legislators or individual taxpayers to determine the mix of collective goods provided. Utilizing deductibility, in effect, allows individuals to direct foregone tax revenue into their favored uses. This allows the largest donors—typically, the wealthy—a disproportionate degree of influence; and, this is compounded by the fact that deductibility causes lower prices of giving for the wealthy, who face high tax rates and itemize.[5] On the other hand, allowing all decisions regarding revenue allocation to be made by legislators may not be desirable either. In addition, allocating revenues via deductibility may be more efficient than either tax credits or direct government spending, offsetting any inequities that might exist.[6]

Thus, the elasticity of giving will not, in and of itself, settle the normative issue of the desirability of the tax deduction for charitable donations; however, it clearly is relevant to that issue. In addition, it is of central importance to the positive issue presented by tax reform: what will happen to charitable giving? Therefore, the methodology by which estimates of these elasticities have been made will be examined briefly.

The Methodology: How Do We Know It?

To evaluate the usefulness of the estimates noted earlier requires some understanding of how they were generated. In estimating the determinants of charitable giving, and of consumer behavior in general, economists use what must seem to many outsiders a needlessly complex methodology. Most people, faced with the task of discovering how taxes influence contributions, would likely begin by taking a survey; that is, by simply asking. Economists, however, are generally distrustful of such surveys, believing that respondents are unable or unwilling to provide truthful and accurate answers.

Table 7.2 provides an excellent illustration of the potential pitfalls of surveys. People were asked, in a series of questions, whether the tax deductibility of contributions influenced giving by themselves, by others in a financial situation similar to theirs, and by others in general. As we can see, a typical response was that *I* do not allow my charitable giving to be influenced by taxation, but *others* do. Such information is not of much use to policy analysts.

Table 7.2 Survey Response: Does Deductibility Make a Difference in Giving Decisions?

	Yes	No	Don't Know
1. For Yourself	15.3%	81.5%	3.2%
2. For Others in your			
Financial Situation	30.5	50.7	18.8
3. For People in General	56.4	36.5	7.1

SOURCE: Survey Research Center, *The National Survey of Philanthropy* (Ann Arbor: University of Michigan, 1974).

In order to avoid such problems, economists generally attempt to infer the determinants of certain behavior, such as charitable giving, either by examining differences in behavior of individuals with varying characteristics or by looking at changes in behavior over a period of time in which key variables are changing. Using regression analysis, a number of such characteristics are held constant, and the impact of a change in one factor on the behavior in question is estimated. So, for example, an estimated price elasticity of 1.2 means that if we compare two individuals (or one individual at two points in time) who are identical except that they face prices of giving that differ by 1 percent, their contributions would be expected to differ by 1.2 percent.

In the real world, of course, many things are changing at once: prices, incomes, attitudes, and so forth. Tax reform, for instance, influences both the price of giving and the income of each taxpayer. It becomes necessary to simulate the effect on giving of tax law changes by first estimating the impact of the tax changes on prices and income and then adding the separate price and income effects on giving to reach an estimate of the *overall* impact. If, as a simple example, a tax change would increase each donor's income by 10 percent and raise the price of giving by 20 percent, and if our estimated income and price elasticities were 0.7 and 1.2, then the overall effect of the change would be to reduce giving per donor by 17 percent ($10 \times .7 - 20 \times 1.2$).

This sort of analysis (with, of course, a good deal of added complexity) lies behind recent estimates of the effect of tax reform on giving. Lindsey (1987), for instance, estimates that the 1986 Tax Reform Act will reduce donations of money by between 14.2 and 17.7 percent, and Clotfelter (1987) estimates a similar reduction of between 15 and 16 percent.

Although the economist's methodology avoids the shortcomings of surveys, it is not without its own problems. First, key variables, such as price and income, may be mismeasured or some relevant characteristics may not be controlled for. In either case, estimates of price and income elasticities may be *biased*. Obviously, using such estimates to answer policy questions is not satisfactory.

Second, estimates may be unbiased, but it may nevertheless be difficult to interpret them in terms useful to policy makers. That is, the estimates may provide a good, unbiased answer to some question but not necessarily the one asked by policy makers.

Finally, unbiased answers may be provided to just the right questions, and so give valuable information but tell only part of the story. For instance, the estimates may predict precisely what will happen to money donations by individuals as a result of tax reform yet offer no information about the influence on volunteering or corporate giving. This point is not so much a criticism of a particular methodology as a warning against framing policy questions in too narrow a manner.

ARE OUR ESTIMATES RELIABLE?

Recall that economists make predictions by observing differences in behavior across people and inferring the causes of those differences. We estimate the impact of, say, the price of giving on contributions by holding other relevant characteristics constant. The characteristics controlled for are those we believe to be important determinants of giving. If all the relevant characteristics are not held constant, however, the estimates will be biased.

Suppose, for instance, that we hypothesize that contributions depend *only* on price and ignore the role of income. The regression analysis, then, will cause all observed differences in giving to be attributed to differences in prices faced by various donors. Note, however, that people facing *low* prices will tend to be those people with *high* incomes, since high-income individuals generally face high tax rates. At least part of the difference in giving attributed to price, then, would actually be due to income differences. Thus the effect of price on giving, and so of a tax reform that raises that price, would be overestimated.

Economists do *not* make this obvious mistake in estimating the determinants of giving. However, we do generally hypothesize that giving depends only on a small number of factors: typically, price, income, and perhaps a few other characteristics such as age and marital status. Therefore, when estimates of price or income elasticity are made, only a small number of other characteristics are controlled for. Differences in giving that are imputed to price and income variation may result from some other, unobserved, differences, just as income and price effects were confused in the preceeding example. Suppose, for instance, that wealthier people are also more highly educated and that differences in education, rather than income per se, cause at least part of the observed differences in giving. Failure to hold education levels constant will tend to overestimate the impact of income changes on giving. A tax change that influences income, but not educational attainment, will have a smaller effect on giving than we would have predicted. The expectations will not hold true.

We all, economists included, recognize that giving depends on a great number of factors—whether, for example, a person's parents give regularly to charity. Why, then, do economists proceed as if only price and income (and a small number of other factors) matter? There are several explanations.

First, economists generally believe that much behavior can, in fact, be explained by concentrating on prices and income. Other factors, such as attitudes, may play a role, some economists may argue, but only a *small* role. We can ignore these secondary influences with little damage to our ability to predict. Economists acknowlege that preferences and attitudes differ across people and time but we do not attempt—as for example, psychologists do—to *explain* those differences and do not, in general, like to explain differences in behavior by variations in tastes, considering such explanations ad hoc.

A more sophisticated defense of the preoccupation of economists with prices and income would argue that, although other differences among people, such as in attitudes, may well be important, these differences are not correlated with price and income. If this were the case, ignoring them would still produce unbiased estimates of price and income effects. So, for instance, if altruisitic tendencies were an important determinant of giving, but low- and high-income individuals were equally likely to have these tendencies, ignoring them would not necessarily cause a misestimate of the effect of income on giving. In addition, it is believed that while prices and income can be influenced by government policy, preferences cannot. Thus, differences in tastes or attitudes would merely be "background noise."

Such an assumption about attitudes may not be justified when dealing with charitable, or altruistic, behavior. It seems at least plausible that high-income donors have different attitudes toward charitable giving than low-income individuals *apart* from differences in behavior induced by higher income (and the accompanying lower prices). Tax reform may well affect incomes and prices without influencing, at least directly or in the short run, those attitudes.[7]

A final, more practical defense pleads that, yes, other factors are important and, yes, ignoring them may bias our results, but data about these factors are simply not available. This is at least partly true—data on contributions, income, and prices are readily available from the IRS whereas those measuring attitudes, preferences, and other individual characteristics are scarce and, in some cases, difficult to conceptualize. Yet, there are data sources that provide a more complete description of the individual donors: telling whether one's parents gave to charity, attended religious services, and so forth. It is striking that economists employing such data bases still often ignore information other than price and income.[8] Finally, if a particular characteristic is believed to be an important determinant, or predictor, of charitable giving, the paucity of available data should not be allowed to limit investigation. It is always possible, after all, to go into the world and collect the necessary information. Economists, however, usually loathe to follow this route.

Are past estimates, in fact, biased? It is impossible to answer definitively: we can simply suggest what may have been left out and consider how results may have been affected. It may be instructive to note, however, that the one study in Table 7.1 that utilized background and attitudinal data (Schiff 1984) found a number of these variables to be important determinants of giving[9] and resulted in an estimated price elasticity much different from the consensus.

In addition to potential omitted variable bias, all of the studies noted in Table 7.1 with the exception of Reece and Zieschang (1985), suffer from a mismeasured price variable. Unlike the price of an ordinary good, which a single consumer must take as given, the price of a donation of money is, in part, dependent on a donor's behavior. An individual's gifts may be large enough to make itemization worthwhile, thus causing his or her price of giving to fall. In addition, extensive giving may push a donor into a lower tax bracket, thereby increasing the price of giving. Thus, price does not simply determine donations; they affect each other simultaneously. Reece and Zieschang adjust for this and find, as reported in Table 7.1, lower price and higher income elasticities, for the typical household, than the consensus. It is uncertain, however, whether qualitatively similar results would be obtained using the methodology of Reece and Zieschang on different data sets.

INTERPRETATIONAL PROBLEMS: WHAT DO THE RESULTS REALLY MEAN?

Suppose that all the relevant characteristics in the estimation of price and income elasticities have been accounted for, so that the estimates are unbiased. It may still not be a simple matter to estimate the impact of tax reform on giving. We must take great care interpreting and using our elasticity results—they may provide unbiased answers but to different questions than those asked by policy makers. Recall that elasticities tell how giving by an individual will change if his or her price or income, and nothing else, changes. These estimates need not shed much light on policy issues for several reasons. We turn to those now.

"Interactive Effects" Among Donors

My response to a change in my income or price of giving due, say, to an exogenous improvement in my personal fortunes may be quite different from my response to the *same* change in price or income brought about by tax reform, which changes everyone else's prices or incomes as well. It is the former response about which we generally know, but the latter that is more interesting from a policy standpoint.

Why would these responses differ? One reason is that my giving may depend, in part, on how much others around me give, or how much I expect them to give. When tax policy changes, not only is my giving affected directly, via changes in prices and income, but also indirectly, due to the change in my expectation of others' behavior.

In what direction does this "interactive effect" among donors operate?

Economists and noneconomists would, I believe, tend to disagree on this. Most economists would predict that as others around an individual increased their giving, for whatever reason, that individual would tend to cut back on contributions, preferring a "free ride" on the giving of others. Noneconomists would likely argue the opposite—donors may look to others for an indication of the socially appropriate level of contributions. As others give more each donor will adjust his or her giving by contributing more as well.[10]

Each of those views has very different implications for the effect of tax policy on giving.[11] Take the free-rider view; eliminating the deductibility of contributions will reduce each donor's expectation of giving by others and so tend to increase donations. This offsets, at least in part, the impact of the price effect of the tax change. A change in tax policy, then, would have a *smaller* impact than the estimates might predict. The alternative "social conformity" view leads to the opposite conclusion—tax policy changes will lead to *larger* changes in donative behavior than would otherwise be expected.

Little research has been done on this issue. However, available evidence (Schiff 1984) indicates that the free-rider effect dominates that of social conformity and that this reduces the impact of tax policy changes by about 10 percent.[12] It must be stressed, however, that the estimate of this effect is very imperfect and by no means should be the last word on this subject.

"Balanced-Budget" Impact

A second problem with interpreting tax policy effects in terms of estimated elasticities arises if that policy is not "revenue neutral"; that is, if it changes the total tax revenue collected. If we believe that such changes in revenue will bring about changes in government spending, we should calculate the net impact on contributions of the tax change plus the accompanying spending change: the "balanced budget" impact.

The relationship between government spending and charitable giving has attracted a great deal of attention recently from economists as well as public officials. Many, including President Reagan, believe—or at least hope—that recent social welfare budget cuts will stimulate giving, as private individuals attempt to compensate for lost government programs; and there is evidence that such a relationship does exist, to some extent.[13] Any tax change, then, that induced a cut in the overall budget would tend to increase contributions, apart from its direct impact on price and income.

Short- and Long-Run Effects

Consumers generally take time to respond to changes in economic variables such as prices and income—behavior changes only gradually. This is likely to be particularly true for charitable giving, which is closely tied to ethical considerations. Most estimates of price and income elasticities, and so of the effect of tax reform on giving, should be interpreted as *long-run* estimates. That is, they provide an answer to the question, How would giving by an individual change if his or her price (or income) changed, allowing enough time to pass for complete adjustment of the donor's behavior? However, Clotfelter (1980) estimates that only about *half* of the long-run change in giving will be realized over the first two years.

Predictions of short-run effects are made more difficult by the possibility of "bunching" of contributions in response to an announced change in tax policy. Individuals anticipating the recent increase in the price of giving may have given more in 1986 than they would have otherwise and so may reduce future giving more than predicted.

Long-run effects are, of course, very important. However, government decision makers and nonprofit sector leaders may also be concerned with changes immediately following—say, by a year or two—tax reform. These distinctions should be kept in mind when viewing results.

"SYSTEMWIDE" EFFECTS

Lags in behavior, then, imply that the more time passes following tax reform, the greater its impact on the nonprofit sector should be, *other things equal*. Other things, though, are generally *not* equal—the entire system within which donors and nonprofit organizations operate may respond to a policy change in ways that lessen or exacerbate the impact of the policy on the nonprofit sector. Even, then, if we have an excellent idea of how individual money donations will change, we may be unaware of *other* changes that the reform will bring about, perhaps indirectly, and that may have a strong effect on nonprofit organizations. The issue of tax reform and its impact on the nonprofit sector must be addressed in as broad a framework as possible to encompass all important avenues of influence.

Volunteering and Tax Policy

Volunteer labor is, for many nonprofit groups, as important a resource as money donations. In 1984, for instance, $57 billion and 16 billion hours were donated by individuals.[14] Even if these hours are valued at the minimum wage—a conservative estimate—volunteers are worth some $54 billion per year to nonprofit organizations. So, any impact tax reform has on volunteering should be of great interest to nonprofit sector leaders. Despite the obvious importance of volunteering, however, relatively little economic research has been directed toward examining such behavior.

Why might tax reform influence volunteering? There are, first, income and price effects analogous to those for money donations. Someone made richer by a new tax policy may volunteer more—it seems reasonable to expect wealthier donors to give more money *and* time, other things equal.

It may be less clear, at first thought, how taxation can influence the price of volunteering, or how volunteering can even be said to *have* a price. Economists view the use of any resource—and time is clearly a resource—as costing the user the value of the best opportunity foregone. This "opportunity cost" of donating an hour of time, then, is the value of the next-best use to which the volunteer could have put that hour. For many volunteers, that next-best use is a paying job, so that the cost of volunteering is the after-tax wage foregone. Lower tax rates, brought about by the recent reform, increase the attractiveness of working for pay relative to volunteering, and so tend to reduce volunteering.

In addition, tax reform may influence volunteering via its impact on the price of money donations, the so-called cross-price effect. If, for instance, donors view money and time donations as substitutes, as two means of reaching the same end, then the reduction in money donations caused by tax reform may be offset by an *increase* in volunteering. Donors may simply switch the form their giving takes. On the other hand, volunteering and money donations may tend to encourage each other. Donors may develop an "altruism syndrome," so that reductions in money donations may be accompanied by similar cutbacks in hours contributed. In this case, tax reform would hit the nonprofit sector particularly hard.

There has, as noted, been relatively little research done on the economic determinants of volunteering, due perhaps to the difficulty of obtaining necessary data. However, work that has been done[15] indicates that the cross-price effect may be important—increasing the price of money donations appears to reduce both money and time donated, for at least some types of nonprofits. To the extent this is true, our estimates that concentrate solely on money donations may underestimate the impact of tax reform on the nonprofit sector—nonprofit organizations will be losing money and volunteers. Schiff (1987), for instance, estimates that the recent tax reform will reduce volunteering by between 6 and 20 percent. The wide range of potential responses reflects the lack of consensus regarding the values of key elasticities.

Corporations and Foundations

Individuals are but one source of contributions for nonprofit organizations. Over 10 percent of all money donations come from corporations and foundations.[16] Again, relatively little is known about the determinants of such giving; however, it is likely that the tax reform, by reducing the top marginal corporate tax rate from 46 to 34 percent while increasing the overall corporate tax burden, will lead to small decline in giving. Clotfelter (1987) estimates that this decline will be on the order of 5 percent of corporate giving.

On the other hand, it is plausible that corporations could respond to reductions in individual contributions by raising their giving. This view is supported by anecdotal evidence that corporations increased and redirected their donations to aid those types of organizations hardest hit by recent federal budget cuts.[17] However, whether they would respond similarly to reductions in private giving is uncertain.

Foundations serve largely as intermediaries for charitable giving by individuals. Nevertheless, foundations may respond in several ways to the changes in individual giving brought about by the Tax Reform Act of 1986. They may, for instance, increase their payout rate—if they view current needs as particularly pressing. In addition, they may redirect their contributions to areas especially hard hit by cuts in individual giving or government support. To the extent that these responses occur, the impact of the recent tax law change will be muted. Again, however, very little is known about the likelihood or magnitude of such responses.

Responses by Nonprofit Organizations

Until now, nonprofit organizations have been viewed in this chapter as passive, simply accepting losses in donations without a whimper. In reality, they may respond to such losses in ways that lessen the impact of tax reform on their ability to serve their constituencies.

They may, for instance, increase their spending on solicitations. For instance, a recent New York Times article[18] reports that "nonprofit theatres have developed sophisticated fund-raising strategies in response to escalating operating costs, changing tax laws and uncertain economic conditions." If a similar increase occured following tax reform, and if the solicitations were successful in raising donations, then the tax policy change would have a less adverse effect on contributions.

We cannot, however, simply count the additional revenue raised by these extra solicitations as a net gain to the nonprofit sector. The gain in resources would only equal those extra contributions less the increase in expenditures to raise them. Any money spent on solicitations cannot, of course, be spent on providing charity to the needy, health care, or the other goods and services of the nonprofit sector.

In addition to increasing solicitations, nonprofit groups may search for other sources of revenue. They may, for instance, increase sales. This would tend to offset any loss of donations but would do so at a price. The beneficiaries of such activities would be forced to bear a larger share of the costs, and in many instances—such as services directed toward the needy—this would prove unacceptable. Schiff and Weisbrod (1986) found that nonprofit social welfare agencies respond to reductions in government support by increasing sales of output. Again, however, whether they will respond to losses of private contributions in a similar fashion is unclear.

Finally, it is possible that, facing reduced revenues, nonprofit organizations would begin to operate more efficiently. A predominant view among economists holds that because nonprofit managers cannot appropriate profits, they have little incentive to operate in an efficient or cost-effective manner. If there is any truth to this view, then there is room for potential cost-reduction without reductions in service levels. Thus, losses in contributions may—and it is only a possibility—not induce drastic reductions in nonprofit service levels, if productive responses can be found.

CONCLUSION: WHAT DO WE KNOW?

A good deal of time has been expended here discussing what is not known about charitable giving, and relatively little about what is. Before losing all hope of providing useful input into policy decisions, however, it should be recognized that it would be foolish to ignore all the information accumulated to date simply because it is less than complete. The intended lesson of this chapter is not that economists' estimates should be ignored but simply that they should be used with care. A corollary of this point is that questions about public policy to the nonprofit sector should be posed in a broad framework, so that all potential effects can be considered.

How Big an Impact?

It would be a useful exercise to begin with the consensus estimates of the responsiveness of giving to taxes and ask how the considerations discussed in this paper might affect the conclusions based on these estimates. In Table 7.3, we present the results of this exercise.

Focusing on the implications for individuals' donations of money (rows 1–3) we can very tentatively conclude that tax reform will have a significantly smaller impact than estimates imply in the short-run (say, two years) and a slightly smaller than predicted effect in the long-run, due to interactive effects. If results are biased because important variables are omitted, then another source of divergence from expectations is responsible, although the direction of this effect is uncertain. Given the current state of

Table 7.3 Implications for Impact of Tax Reform on Nonprofit Sector of Considerations Presented in This Chapter

Consideration	Implications for Impact of Tax Reform
1. Interactive Effects among Donors	Reduces impact on money donations; best guess is 10%
2. Balanced-Budget Effect	None, if tax reform is revenue neutral. If reduction in revenue raised, likely effect is to reduce impact on money donations
3. Short Run versus Long Run	50% of estimated impact on money donations will occur in first two years
4. Volunteering	Likely reduction in hours volunteered of 6–20%; increased impact on nonprofits
5. Corporate Giving	Slight decrease likely, increasing impact on nonprofit organizations
6. Offsetting Actions by Nonprofit Organizations	Possible increased solicitations, sales, and efficiency; reduced impact on nonprofit organizations

knowledge, then, it appears reasonable to continue to use elasticity estimates to predict the response of money donations to tax law changes,
particularly in the long run, although predictions may have to be adjusted
to account for interactive effects.

There are, however, those systemwide effects simply not addressed by
previous estimates (rows 4–6). We believe that tax reform will lead to a
fall, perhaps a significant one, in hours volunteered, increasing the pain
felt by nonprofit organizations. Corporate giving will likely fall as well,
unless corporations respond to reductions in individual giving by stepping
up their efforts. On the other hand, responses by the nonprofit sector itself
may lessen the blow.

Little is known about the magnitude of these effects, and this is crucial.
The analysis does, at least, suggest potentially fruitful areas of research.
Until more is known about the relationship between taxation and the nonprofit sector, only very imperfect and partial answers can be given to the
questions of policy-makers.

NOTES

1. For a more detailed discussion of the impact of the Tax Reform Act on giving, see
Lindsey (1987).

2. This need not be the case if the tax reform shifts income between groups with very
different income elasticities of giving.

3. For a more detailed review of the econometric literature, see Clotfelter (1985).

4. This argument is made, for example, in Feldstein and Taylor (1976).

5. See Simon (1978) for a discussion of this issue.

6. This would be the case if, for example, giving by high-income individuals was more
price elastic than giving by poor and middle-income donors.

7. Clotfelter (1980) holds constant, through the use of panel data, all individual characteristics that do not vary over time, partially avoiding the potential problems caused by the
correlation of attitudes and economic variables.

8. See, for example, Boskin and Feldstein (1977). They employ the National Survey of
Philanthropy, a very detailed data set with information on attitudes and background, of
donors, but include only price, income, and age as explanatory variables.

9. Coefficients on educational attainment, religious background, and another proxy for
altruistic tendencies were significant.

10. Some economists might also argue this way. See, for example, the discussion of "demonstration effects" in Feldstein and Clotfelter (1976).

11. Steinberg (1986) discusses this issue.

12. Specifically, Schiff (1984) finds that an individual will reduce giving by approximately
1 percent if all other donors increase giving by 10 percent.

13. For studies of this relationship see, for example, Abrams and Schmitz (1978, 1984),
Steinberg (1984), and Schiff (1985).

14. Hodgkinson and Weitzman (1986), pp. 61, 71.

15. See Menchik and Weisbrod (1981, 1987) and Schiff (1984).

16. See Hodgkinson and Weitzman (1986), p. 11.

17. "Corporate Giving Fails to Offset Cuts by U.S.," New York Times, (February 15,
1985), p. 1.

18. "For Nonprofit Theatres, A Victory over Red Ink," New York Times, (March 10,
1987), p. 48.

REFERENCES

Abrams, Burton A., and Mark D. Schmitz. 1984. "The Crowding-Out Effect of Governmental Transfers on Private Charitable Contributions: Cross-Section Evidence." *National Tax Journal* 37 (December): 563–68.

———. 1978. "The 'Crowding-Out' Effect of Governmental Transfers on Private Contributions." *Public Choice* 33: 29–37.

Boskin, Michael D., and Martin Feldstein. 1977. "Effects of the Charitable Deduction on Contributions by Low Income and Middle Income Households: Evidence from the National Survey of Philanthropy." *Review of Economics and Statistics* 59 (August): 351–54.

Clotfelter, Charles T. 1985. *Federal Tax Policy and Charitable Giving.* Chicago: University of Chicago Press.

———. 1985. "Tax Reform and Charitable Giving in 1985." Unpublished manuscript.

———. 1980. "Tax Incentives and Charitable Giving: Evidence from a Panel of Taxpayers," *Journal of Public Economics* 13 (June): 319–40.

———. "Life after Tax Reform for Higher Education." Working paper, Center for the Study of Philanthropy and Voluntarism, Duke University, January 1987.

Feldstein, Martin. 1975. "The Income Tax and Charitable Contributions: Part 1—Aggregate and Distributional Effects." *National Tax Journal* 28 (March): 81–100.

Feldstein, Martin, and Charles Clotfelter. 1976. "Tax Incentives and Charitable Contributions in the U.S.: A Microeconometric Analysis." *Journal of Public Economics* 5 (January): 1–26.

Feldstein, Martin, and Amy Taylor, 1976. "The Income Tax and Charitable Contributions." *Econometrica* 44 (November): 1201–22.

Hodgkinson, Virginia, and Murray S. Weitzman. 1986. *Dimensions of the Independent Sector.* Washington, D.C.: Independent Sector.

Lindsey, Lawrence. 1987. "Individual Giving Under the Tax Reform Act of 1986." In *Spring Research Forum: Working Papers*. Washington, D.C.: Independent Sector.

Menchik, Paul, and Burton Weisbrod. 1987 "Volunteer Labor Supply." *Journal of Public Economics* 32(March): 159–89.

———. 1981. "Voluntary Factor Provision in the Supply of Collective Goods." In Michelle White, ed., *Nonprofit Firms in a Three Sector Economy.* Washington, D.C.: Urban Institute.

Reece, William. 1979. "Charitable Contributions: New Evidence on Household Behavior." *American Economic Review* 69 (March):142–51.

Reece, William S., and Kimberly Zieschang. 1985. "Consistent Estimation of the Impact of Tax Deductibility on the Level of Charitable Contributions." *Econometrica* 53 (March): 271–93.

Schiff, Jerald. 1987. "Tax Reform and Volunteering." *Spring Research Forum: Working Papers.* Washington, D.C.: Independent Sector, pp. 137–46.

———. 1985. "Does Government Spending Crowd out Charitable Contributions?" *National Tax Journal* 38 (December): 535–46.

———. 1984. "Charitable Contributions of Money and Time: The Role of Government Policies." Ph.D. dissertation, University of Wisconsin—Madison.

Schiff, Jerald, and Burton Weisbrod. 1986. "Government Social Welfare Spending and the Private Nonprofit Sector: Crowding Out, and More." Unpublished manuscript.

Simon, John. 1978. "Charity and Dynasty under the Federal Tax System," *Probate Lawyer* 5: 1–92.

Steinberg, Richard. 1986. "Charitable Giving as a Mixed Public/Private Good: Implications for Tax Policy," *Public Finance Quarterly* 14 (October): 415–31.

———. 1984. "Voluntary Donations and Public Expenditures." Unpublished manuscript.

8

The Theory of Crowding Out: Donations, Local Government Spending, and the "New Federalism"

RICHARD STEINBERG

The nonprofit sector has long been recognized as a decentralized and pluralistic alternative to government for the provision of vital social services. Indeed, only in the last few decades have we come to rely upon a government "safety net" to provide for the impoverished and disabled, and large-scale government support for the performing arts is of even more recent vintage. Proponents of the "new federalism" have viewed the substitution of government (and especially federal) solutions with alarm. They worry that centrally imposed solutions take insufficient account of local conditions and preferences, stifling experimentation and innovation. They worry about the potential for corruption and waste when power is centralized. Perhaps most important, they worry about the alienation of the citizenry when social problems are handled in a distant, centralized fashion.

The basic tenet of federalism (old and new) is that social problems should be handled by the lowest capable level of government. Thus, the federal government should concern itself only with interstate commerce and international relations, the states with problems spanning local governments, and local governments with local public goods such as education, recreation, and sanitation. Proponents of the "new federalism" seek a return of centralized power to lower levels of government, and a return of local government power to the private sector. The ideal of perfect decentraliza-

Quite a few people have made helpful comments on earlier papers of mine, from which this paper is derived. I wish to thank the participants of seminars at Yale University, the University of Pennsylvania, Virginia Polytechnic Institute and State University, the University of Wyoming, and the State University of New York at Albany. Special thanks go to Robert Inman, Thomas Reiner, Susan Rose-Ackerman, Julian Wolpert, Avner Ben-Ner, Steven Craig, and Russell Roberts. Paul Dimaggio and Richard Magat made a number of helpful comments on this version of the paper. Support for the underlying research has been provided by grants from the National Science Foundation and Yale University's Program on Non-Profit Organizations (PONPO), with the latter providing the sole support for this paper.

tion would be reached if competitive for- and nonprofit organizations entirely replaced government at all levels, though no one seems to think this ideal is practical. In addition, there are limits as to how much decentralization is desirable, but the new federalists feel we are currently far from those limits.

Thus, advocates of the new federalism tend to recommend a substantial reduction in federal expenditure (in the forms of direct expenditure, targetted grants, and revenue sharing) on social problems. They believe that expenditure on these problems will not necessarily decrease with federal support, for state and local governments and nonprofit organizations can step into the breach. Centralized services will be replaced in a pluralistic fashion, with an improvement in quality and a decrease in voter alienation.

Nonprofit organizations provide a pluralistic alternative to government. Indeed, Burton Weisbrod (1975) has argued that in many cases the existence of nonprofit organizations can be explained by dissatisfaction with majoritarian provision of public goods by local governments. He argues that local governments tend to provide a level of public goods that satisfies the median preference voter in the community, for any other level of provision would be opposed by more than 50 percent of the electorate. But this implies that just under 50 percent would like to see a lower level of expenditure, whereas just under 50 percent would like to see a higher level and one person is satisfied with the existing level. Those desiring lower expenditures are stuck—though in the long run, Wolpert (1977) has argued, they can move—whereas those desiring a higher level supplement government spending through their donations to local nonprofit organizations.

Despite the inherent desirability of this pluralistic alternative, the empirical underpinnings of the new federalist position are suspect. Will local government and nonprofit expenditures rise in response to federal cutbacks? Will they rise sufficiently to replace the federal spending? This chapter summarizes a longer and more technical paper[1] that analyzes these questions. This chapter presents a theoretical model of the relationships among the three kinds of spending: donations, local government spending, and state and federal spending. Not surprisingly, although theory provides many insights, it provides no definitive answers on the extent of local replacement.

DONATIONS

In order to understand why government spending affects private donations, one must understand the underlying motives of the donor. One reason people might give is because they are unhappy with the level of some "local public good" and wish to raise that level. This employs the economist's definition: a good is locally public if everyone in the community can enjoy its provision regardless of who does the purchasing (consumption of the good is nonrival in that it is not restricted to one person). An example of a public good is voluntary income redistribution, for many of the benefits of redistribution (such as a lower crime rate and a reduction in exposure to sorry scenes of poverty) accrue to each person in the community.

The public good motivation for giving probably underlies some portion of contributions that support income redistribution, civic improvements, medical research, and other such causes, but it is doubtful that giving can be explained by this factor alone. For one thing, the magnitude of giving is too great. In larger communities, the individual motivation to contribute is limited in part by each donor's fear that his or her contribution would enable others in the community to reduce or eliminate their contributions and take a "free ride." Economists have shown that the aggregate level of contributions resulting from rational behavior of citizens motivated by a desire to provide public goods is likely to be quite small (Atkinson and Stiglitz 1980). When government spending is introduced in such a model, the predictions become even more unrealistic. Warr (1982) and Roberts (1984) have shown that when the only motive for giving is to augment the provision of a public good, there will be dollar for dollar crowdout; that is, a one dollar increase in government spending is entirely neutralized by the response of other donors (as long as donations remain positive). If government spending is great enough, donations would be driven to zero. Since donations can fall no further, government spending would only have impact past this point.

Roberts notes a further implication of this sort of model, that when the government chooses to provide any positive level of the public good, government spending would increase to the point that private donations would cease. This is because government spending increases have no impact on total expenditure on the public good as long as donations remain positive but add dollar for dollar to total expenditures when donations are driven to zero. Free-riding implies that total expenditures in the range of positive donations are too small, in that donors would unanimously support a government tax and expenditure increase that increased total expenditures on the public good. They support such an increase because the increase in (coerced) contributions of others more than compensates them for the (coerced) increase in their own tax-mediated contributions. The only way these donors could achieve this total expenditure increase is to vote for government spending in excess of the level that would drive donations to zero, and recipients, of course, vote for the same thing.

A second reason why people might donate is that they receive some private good (for which consumption is rival; that is, if one person consumes it, another cannot) in return for their donation. This private good may be something tangible, such as a front-row seat at the opera or an alumni magazine, or intangible but externally observable, such as increased prestige, friendship, or respect or less pressure from your boss or a fund raiser. Finally, the benefit may be strictly internal—the donor may feel good about the act of personally contributing.

It seems reasonable to allow for both sorts of motivation in any theory of giving, and this is done here. Furthermore, the two motives are inextricably intertwined. The level of private benefits obtained by a donor depends on the level of provision of the related public good. A donor is unlikely to obtain large satisfaction from the act of giving if needs are already met by government spending or the donations of others. Nonprofit organizations are unlikely to offer large private-good rewards unless their needs are acute.

Rational donors compare the additional benefits forthcoming from an additional donation with the additional costs. Any time these additional benefits (economists typically refer to them as *marginal benefits*) exceed the additional (marginal) costs, the donor would be better off increasing his or her donation. Utility is maximized when marginal benefits equal marginal costs. This is true because marginal benefits typically decline as consumption of any good increases (at least past some point). The first dollar of contribution provides larger additional utility than the last dollar, for the same reason that the first ice cream cone adds more happiness than the fifth cone. With diminishing marginal utility, a donor who set marginal benefits equal to marginal costs would break even on the last dollar contributed and come out ahead on every earlier dollar.

Neglecting tax considerations and administrative waste, the cost of giving is simply the dollar value of the donation, so that the marginal cost is always $1 (tax deductibility would lower this cost, whereas waste would raise it). If donations were simply a private good purchase, then each donor would simply maximize his or her own utility and aggregate donations would be independent of government spending. But since public and private good considerations matter, the donor decision depends on the levels of donations by others and government spending. If spending by others is regarded as an imperfect substitute for one's own donations, then the marginal benefit would fall at each level of one's own donations, so that the optimal personal donation would fall (see Figure 8.1). This phenomenon is denoted *simple crowdout*.

On the other hand, there are a variety of cases where one's own donations would rise in response to a government increase, denoted as *negative simple crowdout*, though these cases are probably less common. Rose-Ackerman (1981) analyzed the case where government expenditures take the form of grants to nonprofit organizations (rather than direct government provision). In this case, donors would generally feel that the government grant reduced the need for their own donation and would cut back. However, donors might view the grant as a signal that the recipient organization is especially efficient or meritorious and respond positively to this "seal of approval." Government grants may be accompanied by strings that affect the ideology of the organization so as to make it more attractive to smaller donors; for example, if it engages in less controversial activities. Grants may fund different services than donations, and these services may be complementary to the donative services. For example, if government

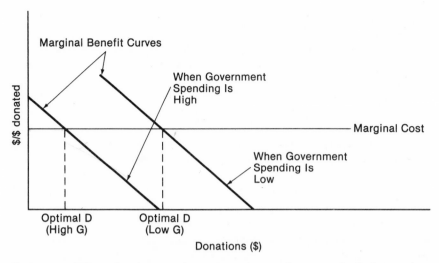

Figure 8.1 Effect of an increase in government spending on a single donor when donations and government spending are felt to be imperfect substitutes for each other by the donor.

funding built a symphony hall, donations for the symphony's operating budget will be encouraged. Finally, grants may be made on a matching basis, lowering the effective price of giving, thus stimulating more private donations.

When government expenditure takes the form of direct provision of a good that is similar to the good provided by nonprofits (a veteran's hospital, a public university, or Aid to Families with Dependent Children), there is a presumption that private donation will fall, but this is not necessarily so. The presumption stems from the so-called substitution effect, where the meeting of related needs by government spending reduces the marginal benefit of donations to the donor. In addition, government spending increases may be accompanied by tax increases, which would decrease after-tax income and hence donations. This will be denoted the *tax effect*.

This presumption could be false for some services for three reasons. First, the substitution effect would run in the opposite direction if donations were complementary to the governmentally provided service. Second, if tax increases are achieved by an increase in marginal income tax rates, then the value of the tax deduction for charitable donations will increase for itemizers. This would reduce or even reverse the tax effect.

A third reason the presumption could be false is the "income effect".[2] Often, some voter-donors would like to see government expenditures on the good increased. The reason is that taxes are compulsory, whereas donations are an individual decision. Taxpayers are assured that aggregate expenditures will increase immensely when their tax bill increases a bit. On the other hand, an individual donation would not typically be matched by anyone, so aggregate service expenditures would go up only by the amount of one's donation. Government spending on social services would thus appear to be a better buy than individual donations, and many voters are likely to feel undersupplied in the government service. Indeed, Weisbrod's theory suggests that exactly half the population would feel undersupplied in political-economic equilibrium (though there are complications applying that conclusion to the more elaborate model developed here). For these undersatisfied voters, an increase in government spending is analagous to an increase in income, for they are able to achieve a higher level of utility despite their fixed money income and decreased after-tax income. Since average donors increase the size of their donations when their income goes up, this "income" effect runs counter to the substitution and tax effects. On the other hand, donors who view government spending as excessive would feel poorer when government spending increased, and all else equal, these donors would cut back their donations. The direction of the income effect on overall donations is therefore indeterminate, depending on the relative sizes and reactions of the two groups.

Combining the substitution, tax, and income effects, we expect partial crowdout—total donations will fall, but by less than the increase in government spending. Donations will fall as long as the perverse income effect on donors who would like to see increased government spending is numerically smaller than the combined impact of the substitution and tax effects and the income effect on those wishing to see government spending decreased. Within this model, crowdout is perfectly symmetrical, so we expect that government decreases are only partly made up by donative increases. The prediction that crowdout will be partial represents my judgment as to the likely size of the effects and cannot be established in the general case on purely theoretical grounds. If the perverse income effect were sufficiently large, super crowdout would occur; that is, a $1 government spending increase could cause a $2 decrease in private donations. If this income effect were smaller and (coincidentally) of the exact right size, total (dollar-for-dollar) crowdout would occur. My judgment that partial crowdout is most likely is strengthened, however, by consideration of a fourth and final effect: the feedback effect.

Donors respond not only to government spending but to other donors. The three effects just described consider only the direct reaction to government spending increases by each donor. But, if each donor cut back in direct response to the government increase, then each donor would have to consider this response by other donors and readjust his or her own donation. Donations of others have a substitution effect on one's own donations just as government spending does, and I refer to this effect as the feedback effect. When donations of others are regarded as a substitute for own donations by each donor, the change in aggregate donations (accounting for this feedback) is smaller than what one would otherwise predict.[3] The feedback effect could work in the opposite direction, however, if donors wished to copy each other's behavior (that is, if donations of

Table 8.1 Effects of a Direct Government Expenditure Increase on
Related Donations

Name of Effect	Possible Impact on Donations	Probable Impact
1. Substitution	Reduction (if government expenditures are a substitute) Increase (if government expenditures are a complement)	Large reduction
2. Tax[a]	Reduction (if marginal rates are unchanged) Unclear (if marginal rates are increased)	Small reduction
3. Income[a]	Increase (if government expenditures are not excessive) Decrease (if government expenditures are excessive)	Small; direction unclear
4. Feedback	Decrease aggregate impact (if donations of others are a substitute) Increase aggregate impact (if donations of others are a complement)	Decrease aggregate impact
5. Total		Moderate reduction (partial crowdout)

[a] In both these cases, we assume that desired donations increase with income, all else constant. Exceptions to this assumption are quite rare.

others were complementary to one's own donations). Although we cannot rule out this sort of herd behavior on theoretical grounds, I think it is likely to be less common than the alternative. Table 8.1 summarizes the four effects.

Several studies have been done of simple crowdout. None of them is perfect, for there are serious problems with existing data sets and unresolved problems of statistical methodology. Nonetheless, they are collectively persuasive that crowdout does occur, that the extent of crowdout differs with the type of service provided, and that crowdout is generally partial. Table 8.2 summarizes these studies.

LOCAL GOVERNMENT SPENDING

Local government expenditures must generally be in accord with the wishes of the electorate,[4] so it is safe to predict the average behavior of governments from the preferences of median voters. These voters compare the marginal costs and benefits of local government spending. If marginal benefits exceed marginal costs at some initial level of government spending, then voters would ask for a government increase. Conversely, if marginal costs exceeded marginal benefits, a decrease would be warranted. In equilibrium, voters wish a level of government spending that equates marginal costs and marginal benefits. In this case, however, calculation of the marginal cost and benefit schedules is more complicated. The marginal cost of government expenditures is chiefly an increment to taxes. This cost is re-

Table 8.2 Empirical Studies of Simple Crowdout

Study	Crowdout Parameter[a]	Measure of Government Spending	Measure of Donations
1. Abrams and Schmitz (1978)	−0.236[b,d]	Federal expenditure on health, education, and welfare	Deductions on U.S. federal income tax returns
2. Reece (1979)	−0.011 to −0.100	AFDC + old age assistance + aid for disabled per recipient	Various, from national survey of philanthropy (U.S.)
3. Paquè (1982)	−0.06 to −0.35[c,d]	Social service public expenditures	Deductions on German (F.R.G.) federal income tax returns
	0.118[c]	Health and recreation public spending	
	0.11 to 0.31[c,d]	Cultural affairs public spending	
4. Amos (1982)	−0.002 to −0.462[d]	Total transfer income; AFDC; public welfare payments	Deductions on U.S. federal income tax returns
5. Steinberg (1983)	−0.001 to −0.003	Intergovernment grants for recreation	Local United Way allocations to specified service (U.S.)
	0.004 to 0.009[d]	Intergovernment grants for hospitals	
6. Jones (1983)	−0.015 to −0.016[b,d]	Central and local government spending on social services and housing	Family donations from family expenditure survey (U.K.)
7. Schiff (1985)	−0.66[d]	Local government expenditures	Various from national survey of philanthropy (U.S.)
	0.046[d]	State noncash welfare spending	
	−0.058[d]	State cash assistance	
8. Abrams and Schmitz (1984)	−0.30[d]	State and local social welfare payments per $1000 personal income	Deductions on U.S. federal income tax returns
9. Steinberg (1985)	−0.005[d]	Central and local government spending on social services and housing	Family donations from family expenditure survey (U.K.)

[a] Change in donations caused by a $1 increase in government spending.
[b] Calculated from reported elasticities and available data.
[c] Elasticity (relates percentage change in government spending to percentage change in donations); conversion to parameter not possible.
[d] Reported coefficients different from zero at .05 or better. In some cases, the crowdout parameter is a nonlinear function of reported coefficients and significance tests were not performed for the former.

duced for itemizers under current tax law because local taxes, like donations, are deductible from federal taxes. The cost is increased if local government is an inefficient provider of services.

Two factors affect the marginal benefits of local government expenditure. First is the leverage effect mentioned earlier: a vote to increase government expenditures is a vote to increase the (mandatory) contributions
of all other households in the community. Second is the crowdout effect:
when government spending goes up, private donations for related purposes
respond. We have argued that crowdout is generally partial. This means
that government spending would still have marginal benefits, but they would
be smaller since decreases in donations partly negate increases in government spending.

In general, communities with a greater crowdout effect will have smaller
local government expenditures.[5] Similarly, within any community, government spending will be larger on those services with smaller crowdout effects, all else equal. The level of private donations that would prevail absent government spending is also relevant. All else equal, communities with
a higher total propensity to give will have smaller government spending
on related services, and the size of government spending for any particular
service will vary inversely with the community's total propensity to give
for that service.

INTERGOVERNMENT AID TO LOCAL GOVERNMENTS

Local governments receive transfer payments from other levels of government (referred to here generically as *grants*) in the forms of categorical
assistance (a grant or reimbursement targeted to a specific purpose) and
general assistance (such as revenue sharing). Some grants are unconditional, whereas others have a matching requirement. Grants may be expected to alter both private donations (via simple crowdout) and local
government spending (via simple government crowdout, explained later).
In turn, induced changes in donations have feedback effects on local government spending, whereas induced changes in local government expenditures have feedback effects on donations. The ultimate impact on total
expenditures for the public good is denoted *joint crowdout*.

Simple government crowdout has been analyzed by a number of authors
(see Inman 1979 for a review of much of this literature). The type and
extent of such crowdout depends crucially on the details of grant administration. The simplest case to analyze is general nonmatching assistance,
such as revenue sharing. This type of grant induces only income effects,
for it is analytically equivalent to an increase in the average income of the
community (assuming the grant is financed in large part by taxpayers who
do not reside in the grant-receiving community). Unlike the case of donative crowdout, where the sign of the income effect is indeterminate, the
income effect here is almost certainly positive, so that some portion of the

grant will increase local public provision of each service and some other portion is devoted to local tax cuts. For example, if a community received $10,000 in revenue sharing, it might decide to cut local taxes by $6,000 (relative to what they would otherwise be) and allocate the $10,000 proportionately among local services. The net result is partial simple government crowdout, with local expenditures rising by a total of $4,000.

When there is a matching requirement, federal grants induce both price and income effects. Since local expenditures are matched, the community can purchase a greater quantity of services with each level of local tax effort. This decreases the marginal costs of government expenditure, inducing the community to vote for a greater quantity of services. The net result is that simple government crowdout is much smaller with matching than nonmatching grants. Indeed, it is possible that crowdout is negative if the price effect is especially large.

In theory, targeted and general grants should have much the same effects. The reason is that the grantor cannot be certain how the local government would have chosen to allocate its money across services in the absence of the grant. Thus, if a community received a $10,000 grant for welfare, it could reduce its intended locally raised expenditures for welfare by, say, $8,000 and allocate this $8,000 proportionately to other local services and to local tax relief. Although the grantor might suspect this was happening, the local books would indicate that the entire grant was devoted to welfare and the grantor would have no legal basis to challenge the allocation.

There is a difference between targeted and general grants, because government expenditures cannot be negative. Suppose a community received a grant targeted to the provision of swimming pools, a service that the local government would not otherwise provide. Since the community has no locally raised expenditures to reallocate, the grant alone would finance the targeted service. In summary, theory predicts that targeted grants will have only income effects unless postgrant local expenditures are exactly equal to the grant.

Theory does not seem quite adequate here, however; for a large body of evidence indicates that the expenditure increments induced by targeted grants are several times greater than the increments induced by increases in community income. This effect, dubbed the *flypaper effect* (money sticks where it hits), has been explained using a richer and more complex political model than the one adopted here (Craig and Inman 1986).

When intergovernment transfers go up or down, both private donations and local government respond. In turn, donations and local government expenditure respond to the induced changes in each other—joint crowdout, as noted earlier. Because the web of causalities is so complex and interdependent, I will skip the proofs of the following propositions here. For nonmatching grants, one can show:

1. When there is partial simple crowdout, joint crowdout is (generally) also partial.[6] Thus, total spending will rise and fall with intergovernment grants, but by a smaller amount than the grant.

2. When there is negative simple crowdout, joint crowdout is likely to be partial, but zero or negative joint crowdout are also possible.
3. When there is super simple crowdout, local government spending on the service will cease.
4. The portion of local government spending financed by local taxes will fall in response to a grant, though total local government spending (including the grant) will rise. Donations may rise or fall in political-economic equilibrium regardless of whether simple crowdout is partial or negative.
5. When the nature of the service provided by a categorical grant is different from the nature of the service when financed by local taxes (say, because the grant is accompanied by strings), a categorical grant will act as a quantity constraint on voters and will have differing effects from a general one.

Some of these conclusions are altered when grants are of the matching variety. An unconditional grant adds to the effective income of the median voter but does not change the marginal costs of local government spending. A matching grant lowers these marginal costs, for a given local tax increment will be matched and will purchase a larger service increment. Thus, with matching grants, the tendency toward partial joint crowdout is reduced or reversed and total spending may go up by an amount that exceeds the grant.

FURTHER COMPLICATIONS

The model is already quite complicated, encompassing local donations, local government expenditures, and intergovernment grants, but it remains too simple to encompass the range of relevant effects in the real world. Two additional factors are relevant but too cumbersome to model explicitly: government user charges and provision of related services by for-profit firms.

The decisive voter can separate the level of taxes and the level of public provision of services by setting the level of user fees. User fees are important in some areas, such as an admission charge to a public recreation facility, where governments and nonprofit organizations compete. In theory, since the voter is deciding between government tax-financed provision and fee-financed provision, any factor that affects one should affect the other. Thus, the level and response of donations and the level and type of intergovernment grant should affect user fee collections, and user fees should affect donations and tax-financed local government expenditures. Any empirical work should account for this interrelationship.

Provision of related goods by for-profit organizations may affect government spending and donations. Thus, empirical studies ought to include the revenues of for-profit firms as a control variable. Since we do not know (and have not modeled) the effect of intergovernment grants on the size of the for-profit sector, empirical analysts may wish to make the simplifying assumption that this variable is predetermined. The overall causal structure recommended is detailed in Figure 8.2.

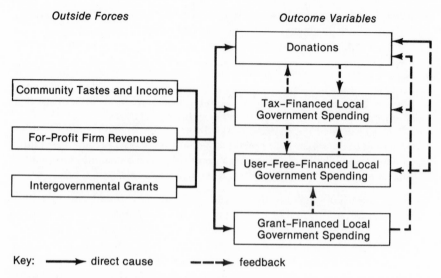

Figure 8.2   Assumed causal structure.

It is exceedingly difficult to conduct an empirical study of joint crowdout with currently available data sets. The reason is simple: in order to preserve confidentiality, most surveys of giving do not record the location of the respondent. The few that record this information survey too few individuals in each community to reliably estimate overall giving, so that this variable cannot be correlated with government spending at the local level. In three of my previous papers, I tried (without notable success) to circumvent this problem by using local United Way allocations as a proxy for local giving within a variety of assumed causal structures.[7] Evidently, this approximation is not very good as my results were not overly plausible. I hope that future researchers will develop data sets suitable for determining the extent of joint crowdout.

SUMMARY

One of the main tenets of the new federalism is that local governments and nonprofit organizations will increase their expenditures on social services in response to federal cutbacks. Although the case for federal cutbacks does not rest entirely on this empirical assertion, this relevant issue is examined on the theoretical level.

The relationship between government spending and private donations is not simple, but theory suggests that there will usually be partial simple crowdout; that is, that government expenditure increases (decreases) are partly neutralized by donative decreases (increases). A number of studies, each flawed but cumulatively persuasive, are cited to show that simple crowdout is usually partial, typically running between 1/2 and 30 cents on the dollar. This suggests that when a local government reduces its provision of social services, it cannot count on the nonprofit sector to replace a very large part of these services, although some replacement will occur.

When federal and state grants for social service provision are cut back, local donations, local government spending, and user charges can be expected to respond. If total spending goes up by less than an increment in intergovernment aid, joint crowdout is said to be partial. Partial joint crowdout is most commonly expected on theoretical grounds, but theory does not rule out other outcomes. Currently available data sets do not allow easy confirmation of this conclusion, but future data sets may allow estimation of the magnitude of crowdout.

NOTES

1. Richard Steinberg, "Voluntary Donations and Public Expenditures in a Federalist System," *The American Economic Review* 77 (March 1987).

2. Although I have used the terms *income* and *substitution effects*, this usage is not equivalent to the traditional usage of these terms in economics. Traditionally, these terms are reserved for the effects of a price change on desired purchases of a good. In this case, I am regarding the effects of a change in a quantity-constrained good (that is, a good whose purchase is not under the control of the donor) on desired purchases of a related good (donations) that is under donor control. Mackay and Whitney (1980) showed that the impact of a change in a quantity constraint can be broken down into effects that are analogous to those resulting from a price change, which they called *income- and substitution-like effects*.

3. The feedback is accounted for in the companion piece by calculating the change in the Nash equilibrium allocation. See any textbook on game theory such as Friedman (1977) for a discussion of the Nash equilibrium concept.

4. There are a number of reasons why local government expenditures would depart from the wishes of the electorate. Government bureaucrats provide information to voters, and this information can be manipulated to sway elections. Preferences of voters may be such that there is no stable voting outcome, just continuing political cycles. Voters may not be motivated to gather information necessary for their decisions, voting on whims or images. For a review of these and other problems, see any standard political economy text such as Atkinson and Stiglitz (1980). Nonetheless, the median voter model seems to have good predictive power (Inman 1978) and may serve as an adequate simplification of reality for our purposes.

5. There is a perverse income effect here as well, which could (in theory) reverse the relation between the crowdout parameter and local government spending. In practice, it is exceedingly unlikely that this income effect would be large enough to reverse the conclusion of the text.

6. Exceptions are possible, but sufficient conditions to ensure this result are that all goods are normal, the choice set is convex for the decisive voter, political-economic equilibrium is interior, and donations are locally linear.

7. These include my Ph.D. thesis, "Two Essays on the Nonprofit Sector," University of Pennsylvania, 1983; "Empirical Relations between Government Spending and Charitable Donations," V.P.I. & S.U. Working Paper #E84-09-02, September 1984 (a version of this paper, excluding the joint crowdout estimates, was published under the same title); and "Voluntary Donations and Public Expenditures," V.P.I. & S.U. Working Paper #E84-07-01, revised June 1985 (a version of this paper without the joint crowdout estimates was also published under a longer title as Steinberg 1987).

REFERENCES

Abrams, Burton, and Mark D. Schmitz. 1984. "The Crowding-Out Effect of Governmental Transfers on Private Charitable Contributions: Cross-Section Evidence." *National Tax Journal* 37 (December): 563–68.

———. 1978. "The 'Crowding-Out' Effect of Governmental Transfers on Private Charitable Contributions." *Public Choice* 33: 29–37.

Amos, Orley M. Jr. 1982. "Empirical Analysis of Motives Underlying Individual Contributions to Charity," *Atlantic Economic Journal* 10, no. 4 (December): 45–52.

Atkinson, Anthony, and Joseph Stiglitz. 1980. *Lectures on Public Economics*. New York: McGraw-Hill Book Co.

Craig, Steven G., and Robert P. Inman. 1986. "Education, Welfare, and the New Federalism: State Budgeting in a Federalist Public Economy." In Harvey Rosen, ed., *Studies in State and Local Public Finance*. Chicago: University of Chicago Press.

Friedman, James. 1977. *Oligopoly and the Theory of Games*. Amsterdam: North Holland.

Inman, Robert P. 1979. "The Fiscal Performance of Local Governments: An Interpretive Review." In Peter Mieszkowski and Mahlon Straszheim, eds., *Current Issues in Urban Economics*. Baltimore: Johns Hopkins University Press, pp. 270–321.

———. 1978. "Testing Political Economy's 'As If' Proposition: Is the Median Voter Really Decisive?" *Public Choice* 33: 45–65.

Jons, P. R. 1983. "Aid to Charities." *International Journal of Social Economics* 10, no. 2.

Mackay, Robert J., and Gerald A. Whitney. 1980. "The Comparative Statics of Quantity Constraints and Conditional Demands: Theory and Applications." *Econometrica* 48, no. 7 (November).

Paqué, Karl-Heinz. 1982. "Do Public Transfers "Crowd out" Private Charitable Giving? Some Econometric Evidence for the Federal Republic of Germany." Kiel Working Paper No. 152 (August).

Reece, W. W. 1979. "Charitable Contributions: New Evidence on Household Behavior." *American Economic Review* 69: 142–151.

Roberts, Russell D. 1984. "A Positive Model of Private Charity and Public Transfers." *Journal of Political Economy* 92, no. 1.

Rose-Ackerman, Susan. 1981. "Do Government Grants to Charity Reduce Private Donations." in Michelle White, ed., *Nonprofit Firms in a Three Sector Economy*. Washington, D.C.: Urban Institute.

———. 1980. "United Charities: An Economic Analysis." *Public Policy* 28 (Summer): 323:50.

Schiff, Jerald, 1985. "Does Government Spending Crowd Out Charitable Contributions?" *National Tax Journal* 38 (December).

Steinberg, Richard. 1987. "Voluntary Donations and Public Expenditures in a Federalist System." *American Economic Review* 77 (March).

———. 1985. "Empirical Relations between Government Spending and Charitable Donations." *Journal of Voluntary Action Research* 14 (Spring–Summer).

———. 1984. "Voluntary Donations and Public Expenditures. Virginia Polytechnic Institute and State University (V.P.I.S.U.) Dept. of Economics Working Paper #E84-07-01 (July, rev. June 1985).

———. 1984. "Empirical Relations between Government Spending and Charitable Donations," V.P.I. S.U. Working Paper #E84-09-02 (September).

———. 1983. *"Two Essays on the Nonprofit Sector."* Ph.D. dissertation, University of Pennsylvania.

Warr, Peter G. 1982. "Pareto Optimal Redistribution and Private Charity." *Journal of Public Economics* 19 (October): 131–38.

Weisbrod, Burton. 1975. "Toward a Theory of the Voluntary Non-Profit Sector in a Three Sector Economy." In Edmund Phelps, ed., *Altruism, Morality, and Economic Theory*. New York: Russell Sage Foundation.

Wolpert, Julian. 1977. "Social Income and the Voluntary Sector." Regional Science Association, *Papers* 39: 217–29.

IV

FOUNDATIONS: DONORS, TYPES, AND MANAGEMENT

9

Independent Foundations and Wealthy Donors

TERESA ODENDAHL

Foundations are funds of private wealth established for charitable purposes. Such institutions are usually intended to last in perpetuity. Most foundations are small and controlled by donors. Over 5,000 of the 24,000 independent foundations in the United States have more than $1 million in endowments. Today, these large private grant-making organizations hold over $90 billion in assets. They distribute $3 billion annually to a variety of nonprofit causes. However, with the exception of the early 1980s, the rates of establishment and asset growth of independent foundations have declined since mid-century.

The project summarized here focused on independent foundations, which constitute the vast majority of the grant-making field, rather than community, company-sponsored, or operating foundations. To determine and analyze trends in the creation and growth rates of independent foundations, seven interrelated studies were conducted, using a host of historical and social science methodologies.

The research was designed to be of use to policy makers and others interested in foundations and charity. It went beyond existing published materials, examining historical trends in the formation and growth of foundations of all sizes. In addition, it enlarged upon studies of the Internal Revenue Service covering one- and two-year periods, and surveyed organizational and donor characteristics that had not been examined systematically before. Extensive interviews were conducted with wealthy individuals

This chapter, originally published as the overview to *America's Wealthy and the Future of Foundations* (1987), is reprinted in a slightly revised form with the permission of the Foundation Center. It grew out of the Foundation Formation, Growth, and Termination Project (Foundation Project) undertaken in 1983 by the Council on Foundations and the Program on Non-Profit Organizations at Yale University. Except where indicated, statistical data beyond 1983 were not available for this study and are incomplete. The work presented here was carried out by a team of researchers, although the interpretation is my own. I am especially grateful to Elizabeth Boris, Kathleen McCarthy, and John Simon for their extensive comments on earlier versions.

and their advisors on attitudes, motivations for giving to charity, and incentives and disincentives to foundation creation. These add a new dimension to the discussion of foundation births and deaths.

The interdisciplinary approach of the study permitted examination of
the same phenomenon from a variety of perspectives: anthropology, economics, law, political science, and sociology. A straightforward single-dis-
cipline approach might have produced a neater explanation of the phenomenon under investigation; however, it would have provided only one
view of the independent foundation field.

This chapter deals with matters of subjective intention on the part of
hundreds of persons in a wide variety of settings. Some of the results are
only suggestive, but clues that emerged in one study led to findings in
another. Certain trends are clear; others are modestly corroborated by more
than one team member. As a result, the explanations for foundation formation and growth that emerge from this study have to be stated in ten-
tative language.

CREATION AND GROWTH TRENDS

After the Civil War and through the turn of the century, rapid industrialization enabled some individuals to amass vast personal fortunes. Some of
the wealthy funded benevolent activities, especially churches, hospitals,
schools, and universities, and a few formed trusts to carry their philanthropic goals forward after they died; but their charitable giving was usually not systematic nor devoted to broad social purposes. In the early years
of the twentieth century, the habit of establishing general-purpose foundations to address some of the root causes of social problems took hold
(Hall 1987; Karl and Katz 1981). Early foundation donors, such as Andrew Carnegie, John D. Rockefeller, and Margaret Olivia Sage, had visions of the society they wanted to promote through their philanthropy.

The number of new independent foundations grew at a fairly steady
rate through mid-century. Then the rate of foundation establishment began to decline, especially for the largest independent foundations (see Table 9.1). The pattern differs somewhat for the smallest foundations. The
growth of new small foundations also peaked in the 1950s and declined
in the 1960s; but unlike the large foundations, there was a recovery in the
1970s. However, as with the largest foundations, the rate of formation has
declined since mid-century.

Only six new very large foundations (those with more than $100 million
in assets) have been formed since 1970. And, those created are more likely
to have been established at the death of the donor. This trend may nega-
tively affect the future of the field since a major source of overall foundation growth had been gifts from donors during their lifetimes.

Foundations with $100 million or more in assets account for a smaller
proportion of all foundations asset than did comparably sized foundations

Table 9.1 Creation Rate of Independent Foundations

Decade Created	Number Created	Cumulative Total	Percent of Increase
Before 1910	51	—	—
1910–1919	44	95	86
1920–1929	108	203	114
1930–1939	164	367	81
1940–1949	571	938	156
1950–1959	1120	2058	119
1960–1969	766	2824	37
1970–1979	401	3225	14
1980–1983	135	3360	—

SOURCE: Foundation Center data, for foundations with $1 million or more in assets or $100,000 in grants, calculated from a special run.

(in constant dollars) in the early 1960s. Total foundation assets grew from 1970 to 1983 through the creation of new foundations, contributions or gifts from donors into existing foundations, and returns on investment. However, foundation assets declined as a share of the general economy, and the real (inflation-adjusted) value of their holdings showed no growth during this period (see Figure 9.1). At least part of this lack of asset growth may be attributed to poor stock market performance, federal regulations that required foundations to pay out most of their income during this period, as well as the more than 10,000 foundations that terminated in the last decade (see Table 9.2). After 1981, a combination of good stock market performance and less stringent regulations favorably affected foundations assets.

Table 9.2 Terminated Foundations of All Sizes

Year	Number	Percent
1970	833	7.9
1971	1506	14.3
1972	1733	16.5
1973	946	9.0
1974	952	9.0
1975	792	7.5
1976	784	7.4
1977	618	5.9
1978	585	5.6
1979	616	5.9
1980	555	5.2
1981	455	4.3
1982	151	1.4
Total	10,526	100%

SOURCE: The Foundation Center, special runs. These numbers are approximate, but it is clear that many foundations were terminated in the early 1970s.

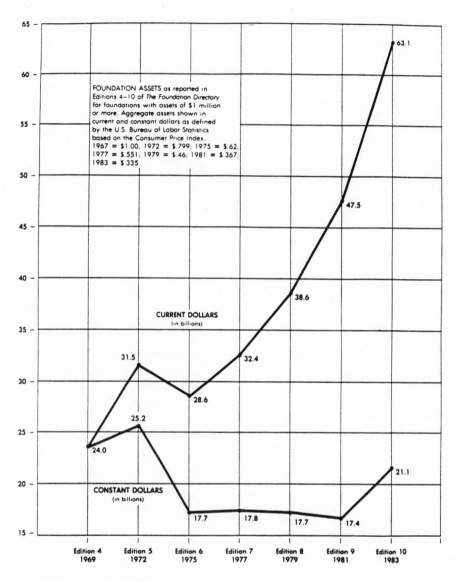

FOUNDATION ASSETS as reported in
Editions 4–10 of *The Foundation Directory*
for foundations with assets of $1 million
or more. Aggregate assets shown in
current and constant dollars as defined
by the U.S. Bureau of Labor Statistics
based on the Consumer Price Index.
1967 = $1.00; 1972 = $.799; 1975 = $.62;
1977 = $.551; 1979 = $.46; 1981 = $.367;
1983 = $.335

CURRENT DOLLARS
(in billions)

CONSTANT DOLLARS
(in billions)

SOURCE: The Foundation Center © 1985

Figure 9.1 The effect of inflation on foundation assets 1969–1983. Foundation assets are reported in Editions 4–10 of *The Foundation Directory* for foundations with assets of $1 million or more. Aggregate assets shown in current and constant dollars as defined by the U.S. Bureau of Labor Statistics based on the Consumer Price Index. 1967 = $1.00; 1972 = $.799; 1975 = $.62; 1977 = $.551; 1979 = $.46; 1981 = $.367; 1983 = $.335. (*Source:* New York: The Foundation Center © 1985.)

The eleventh edition of *The Foundation Directory*, reporting 1985 data, reveals "the largest two-year increase in foundation assets and total giving in the last ten years." This is attributed to stock market gains and low inflation during the period as well as favorable legislation. Although this growth rate appears greater than in the 1970s, it does not match that of the 1950s. The increase seems to be the results of business acquisitions and mergers and the use of proceeds from the sale of nonprofit hospitals to form foundations, and it may also reflect a changed climate for foundation philanthropy. Detailed analysis of these findings have not been undertaken, but it is possible that economic conditions allowed more accumulation of wealth, the basis of foundation formation.

What explains these trends? The forces and events that appear to have contributed to the decline in foundation formation over the past three decades may be grouped under four major categories (see Table 9.3):

1. Legal changes that directly affect independent foundations;
2. Tax law changes affecting charitable giving by wealthy persons;
3. Attitudinal changes regarding charitable giving;
4. Shifts in wealth structure.

Economic history trend data formed the initial view of the foundation field. Ralph Nelson, an economist at Queens College, selected 1962 as his base year, comparing it with data from 1981. He compiled detailed information on the assets, contributions, grants, and expenditures of independent foundations with more than $250,000 in assets or grants of at least $25,000.

Nelson found that, although the growth of private foundations during this period had been substantial, it had not been great enough to maintain their position in the larger economy. There were almost twice as many foundations of comparable size in 1981 (some newly created; some increased in size), holding 3.3 times the value of assets and granting 4.25 times as many dollars; thus, they had not kept pace with inflation.

In a separate study, Nelson focused specifically on the largest independent foundations, comparing those in 1960 with those in 1982. Although sixty-four foundations fell into the category in 1960, holding assets of $25 million or more, eighty-eight qualified in 1982, with assets of $75 million or more to account for inflation. However, this increase masks a high turnover among such foundations. Only forty-two of the sixty-four largest in 1960 were among the largest in 1982. And, only six large foundations were established after 1970. Nelson uncovered a pattern of fluidity among the largest foundations, with movement in and out of the largest asset category.

The number of big foundations grew less than the foundation field as a whole, which almost doubled in number. The relative share of the largest foundations also declined dramatically, from 69 percent of all foundation assets in 1962 to 48 percent two decades later (Nelson 1987).

Gabriel Rudney (1987), economist and senior research scientist at Yale University, investigated a sample of 367 out of the 937 independent foun-

Table 9.3 Clues to Decline of Independent Foundations

Legal Changes that Directly Affect Independent Foundations:

Increased regulation since the 1950s, particularly the Tax Reform Act of 1969, has reduced tax incentives and imposed a complex web of regulations on foundations, producing a negative impact on attorneys and through them on donors.

Decreased deductibility of gifts to foundations over the last thirty years also militates against formation of new foundations and growth of established ones. Attorneys regard it as their responsibility to recommend charitable vehicles with the greatest tax deductibility.

Increased reliance on attorneys for advice on charitable giving focuses more sharply on the incentives and disincentives of the tax code. Almost all of the attorneys interviewed reported that they rarely recommend that their clients form private foundations, for reasons directly traceable to regulations of the Tax Reform Act of 1969.

A host of other charitable alternatives, such as community foundations, direct giving, split interest trusts, and support organizations, provide greater deductibility than independent foundations and more freedom from regulation.

Tax Law Changes Affecting Charitable Giving by Wealthy Persons:

Top individual income tax rates in the last thirty years have declined from 92 percent to 50 percent. The higher the rate, the greater the incentives for charitable giving of all kinds.[a]

The unlimited marital deduction, passed in 1982, allows an individual to bequeath all of his or her property to the surviving spouse without paying estate tax. This will probably encourage people to pass their entire estates on to their spouses, rather than to give to charity, and may leave charitable decisions to the surviving spouses.

Attitudinal Changes Regarding Charitable Giving:

The desire for personal involvement, especially on the part of younger wealthy people, has led to some rejection of the bureaucracy of formal grant making.

The declining popularity of foundations as a result of congressional hearings, media and public attacks, and rebellion of recent succeeding generations against the values of their parents has had a negative impact on the creation of new foundations.

Shifts in the Structure of Wealth:

These factors have some bearing, although they do not directly cause a decline in foundation formation and growth.

The age of a wealthy person frequently affects giving patterns. Older people are more likely to give larger sums of money.

The gender of wealthy individuals also affects giving patterns. With few exceptions, men were the major philanthropists in the past. Today, as women gain greater control over their wealth, they are increasingly establishing foundations, but this does not compensate for the overall decline.

The type of assets held by the wealthy may also be relevant. In the last fifteen years, new foundations were more likely to be formed with assets derived from real estate and less likely to include stock in family-held corporations.

NOTE: This list does not include the many positive reasons reported throughout the study for the use of foundations. These will be reviewed later. The factors just outlined in skeletal form have overcome the positive factors in enough cases to bring about the thirty-year decline in independent foundation formation and asset growth.

[a]The 1986 Tax Reform Act was being debated just as the Foundation project was being completed. The new top individual income tax rate has been lowered to 28 percent, which will probably lead to a decline in charitable giving by the wealthy.

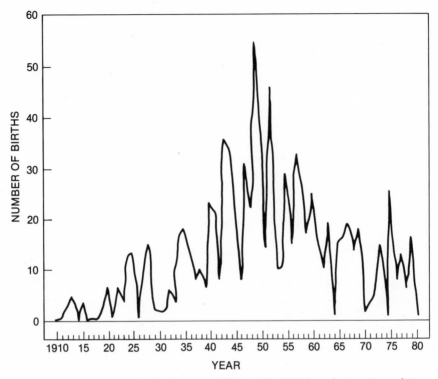

Figure 9.2   Foundation births by year, 1910–1980 (937 foundations in study).

dations with 1982 assets of $5 million or more. His research revealed that there was a period of moderate foundation creation up to the 1940s; followed by increased foundation formation then decline between 1940 and 1965; and a leveling-off of foundation establishment in the ensuing years (see Figure 9.2). Like Nelson, Rudney found that foundations with $5 million or more showed little real asset growth. Although the total assets held by these foundations more than tripled between 1962 and 1982, when inflation is taken into account, their real value remained approximately the same in both years.

A survey questionnaire was distributed by the Roper Center at Yale University to 1075 foundations, the same that Nelson and Rudney had investigated as well as a random sample of foundations with from $100,000 to $5 million in assets. These small foundations make up the bulk of the field. The survey solicited information on the foundations' creation, characteristics and motives of their donors, and past as well as anticipated gifts.

Elizabeth Boris (1987), a political scientist and vice-president for research and planning at the Council on Foundations, analyzed these data. She found that the typical foundation was formed in 1959 by a living

donor with a small endowment of $100,000. Half have subsequently re-
ceived gifts, and 24 percent expect to in the future.

Established with a small initial gift, most foundations grew through additional contributions of the donor, bequests, and asset appreciation. More
than half of those with holdings in excess of $100 million were originally
endowed with less than $1 million. Smaller foundations were most likely
to have received gifts from living donors, while the large ones were more
likely to have received bequests.

During the course of investigation, members of the project coined the
phrase *the acorn theory of growth*. All three of the studies cited support
the notion that many foundations, even large ones, start out small and
grow to their full size through a combination of appreciation and contributions. However, some foundations never mature beyond the acorn stage,
others stop growing when they are saplings, and a few are created as giant
oaks.

Many die, for various reasons. Some donors arranged a limited lifespan
for their foundations. In other cases, boards of trustees decided to liqui-
date foundations by making grants out of capital. The data on foundation
terminations were the least satisfactory of all used, and the analysis was
therefore limited (see Table 9.2).

In-depth interviews were conducted with 135 wealthy individuals and
more than 100 advisers—attorneys, accountants, bank trusts officers,
foundation employees, and personal advisers—to help illuminate the decision-making process preceding the establishment of foundations. Donors
were asked how they became involved in the nonprofit sector, why they
give to it, and why they had formed a foundation or decided against it in
favor of another charitable vehicle (Odendahl 1987a).

The wealthy people we interviewed presented a wide range of somewhat
paradoxical views about foundations. The desire to exercise more control
over their giving led many to create foundations. However, other philanthropists felt they could be more effective by giving directly to specific
charities or by using charitable vehicles other than foundations. (Motives
for giving are discussed in greater detail later.)

Advisers were asked to describe their personal backgrounds, the nature
of their work, the types of interactions they had with wealthy clients, and
their views on foundations and other charitable or legal instruments. Many
advisers no longer view the private foundation as a desirable vehicle for
distributing charitable funds (Ostrower 1987).

Over the years, a variety of other charitable options have become available to the wealthy donor. Giving directly to a nonprofit organization is
often favored by both advisers and their clients. A host of generation-
skipping and split-interest charitable trusts are also alternatives. Many establish donor-advised funds in community foundations or set up support
organizations that provide grants to one or more nonprofit agencies on a
regular basis.

THE TAX CODE AND FOUNDATIONS

In a society suffused with taxes and reliant on them as engines of social and economic policy, the union of charity and taxes is in reality indissoluble—and controversy therefore inevitable. Charity seems destined to be enmeshed in tax policy debate not only because so is everything else in our society, but also because, over the years, we have come to entrust to the tax system a central role in the nourishment and regulation of the nonprofit sector (Simon 1987).

For more than seventy years, federal income, estate, and gift tax laws have served to encourage or discourage charitable giving. Policy makers have attempted to use tax legislation to create incentives and disincentives to various forms of private philanthropy. Tax laws have also made some types of gifts or donees more attractive than others.

The prevailing view is that taxes have at least two effects. They reduce discretionary income and the amounts of wealth that can be transferred. And, changes in rates of taxation that otherwise reduce income, estate, and gift taxes make it more "costly" to contribute to charity, especially for wealthy donors. For example, a 50 percent tax rate costs the taxpayer 50 cents on every dollar, while a 25 percent rate has a price of 75 cents. Tax benefits derived from charitable giving are often greatest when tax rates are highest (Clotfelter and Steuerle 1981).

Most of the millionaires we interviewed said that they give up to the maximum for which they receive a deduction. Fewer than 10 of the 135 wealthy people claimed that taxes had no importance. According to one interviewee: "I try to maximize my philanthropy and minimize my taxes."

The first American foundations appeared before the enactment of an income tax. However, we found that the greatest number of foundation creations coincided with the highest income and inheritance tax rates.

The Revenue Acts of 1950, 1954, and 1964

The earliest tax codes included deductions for charitable donations, but congressional regulation of foundations did not begin until passage of the Revenue Act of 1950 (Edie 1987). The 1950 law required that business and charity operate at an "arm's length" and denied tax exemption to certain accumulations of assets. Private foundations were not specifically defined as such within the tax code at this time, yet the new regulations differentiated them from other charities such as churches or schools.

Gabriel Rudney (1987) argues that regulatory constraints have contributed to the decline in the formation of new foundations, beginning with the 1950 Revenue Act. According to Rudney, lower income and wealth-transfer taxes in the 1960s also diminished incentives for foundation development, although he notes that the full impact of government intervention is "difficult to isolate and measure."

The Revenue Acts of 1954 and 1964 provided for higher tax deductions (up to 30 percent of gross adjusted income) on contributions to certain religious, educational, and hospital organizations, the so-called public charities. Foundations retained a lower level of deductibility (up to 20 percent of adjusted gross income), probably because they had come under fire in a series of congressional hearings, which resulted in a request for a special investigation of private foundations by the Treasury Department. The report and the recommendations that resulted from that study became the model for the next legislation to affect foundations.

The Tax Reform Act of 1969

Numerous regulations on private foundations were imposed by the Tax Reform Act of 1969. These included a 4 percent tax on investment, stronger provisions against self-dealing, a minimum pay-out requirement, and rules against "excess business holdings" and jeopardy investments. Excise taxes were imposed for noncompliance. Furthermore, the discrepancy in tax deductibility between foundations and public charities was increased. Donations to foundations continued to be deductible up to 20 percent of adjusted gross income, with gifts of appreciated property included in that ceiling on the basis of the donor's cost plus 60 percent of appreciation. In contrast, gifts to public charities became deductible up to 50 percent of the donor's income, including full appreciated value of most property.

The 1969 act weakened tax incentives for foundation creation and had a negative impact on asset growth. Interviews with attorneys revealed that they began advising against foundation creation as a result of the 1969 act (Ostrower 1987). Boris and Rudney found no real growth in the foundation field after 1969. In addition, Nelson documented higher foundation expenditures among the largest foundations in the 1970s and attributed this to the pay-out requirement.

The "excess business holdings" provision of the 1969 act prohibited a foundation and its "disqualified persons" together from holding more than 20 percent of the stock of any one company. Only 14 percent of the independent foundations that Boris studied had been endowed originally in such a way that donors, managers, and other related persons represented a controlling interest in a company. But 50 percent of the largest foundations were created with enough stock to constitute control.

As might be expected, 81 percent of foundations established with excess business holdings had been formed before the 1969 act. Most of these foundations disposed of the control stock after passage of the act. While additional foundations continue to be formed with control stock, the small number suggests that the restriction against excess business holding may be one of the reasons that so few major foundations have been formed in recent years.

Nelson analyzed external and internal factors in the growth of the largest foundations (those with over $100 million in assets). The overall reduced growth rate reflects the effects of a significantly higher ratio of expenditures to assets mandated by the pay-out requirement and a significantly lower total return on investments. In the period between 1970 and 1982, large gifts to foundations accounted for three-fifths of the overall growth rate of the largest foundations, whereas the combined effects of investment return and expenditure policy accounted for only two-fifths of the growth. Although it is likely that many wills that provided bequests to foundations may have been written prior to 1969, this was a reversal in the roles that the two components had played in the 1961–1969 period.

Nelson found that there was an average eighteen-year time lag between establishment and major gifts to big foundations. This would lead us to expect that foundations formed in the 1960s should receive most of their endowment by the end of the 1980s. The decade of the 1990s, then, if the pattern continues, would be the period when foundations created in the 1970s would be fully endowed.

According to Boris, since 1970, there has been an increase in the number of testamentary creations, which accounted for over half of all new foundations. Donors have increasingly specified that foundations would be formed only after their deaths. Both the survey and personal interviews indicate that with few exceptions it is not likely that any other family members would supplement the endowments provided by the donor. Again, many wills providing bequests to foundations may have been written prior to 1969, but Boris warns that a potential impact of this trend toward bequests is that fewer foundations and the field as a whole will experience major growth.

Recent Legislation

Since 1969, there have been several changes in the regulations: lowering the investment tax; modifying the pay-out requirement to a flat 5 percent of assets; increasing deductibility of foundation gifts to 30 percent of income; and permitting deductibility of the full value of some types of appreciated property, notably publicly traded stock. Foundation assets increased in 1983, the first real growth in a decade, which may partially reflect changes in the pay-out requirement. Technical adjustments simplified some of the other requirements (Edie 1987).

The Marital Deduction Act of 1982 may reinforce the trend toward bequest giving. It allows a decedent to pass all property on to a spouse without paying estate taxes. Charitable decisions may be left to the surviving spouse.[1]

Wealth Accumulation

By matching estate and income tax returns, economist Eugene Steuerle (1987), director of finance and taxation projects at the American Enterprise Institute, showed that the wealthy tend to retain their assets during their lifetimes, making their largest contributions as bequests. Steuerle's comparison of estate tax returns filed in 1977 with the tax returns of decedents in the years prior to death demonstrated that top wealth holders gave only a slightly higher percentage of their income to charity than did all itemizers. The elderly were an exception; Steuerle's research supports the finding that giving increases with age (Clotfelter and Steuerle 1981; Yankelovich, Skelly, and White 1986).

This pattern of wealth retention is remarkable in light of the greater tax benefits of lifetime giving over bequests. Lifetime contributors enjoy a savings in both income and estate taxes, which can then be used to increase either consumption or total giving. In effect, the rich hold on to wealth that they will most likely never consume, and they pay a greater price for it. Lack of familiarity with tax laws, according to Steuerle, may be one reason the wealthy do not take better advantage of existing charitable deductions, but there may be other reasons. The small incentive to give means that much income is not recognized in the first place. The wealthy tend to show a preference for holding wealth regardless of the tax consequences. Steuerle's findings reinforce the idea that tax considerations are not paramount in charitable giving. His results also complement those in the foundation survey (Boris 1987), which document a movement toward the establishment of foundations by bequest.

Tax Incentives and Disincentives

In the survey, tax incentives ranked sixth to other motivations for establishing a foundation (Boris 1987). These results reinforce the findings of the personal interviews. Taxes are important considerations in charitable giving generally, but not the impetus. Yet, taxes may be particularly important in causing a donor to choose a charitable option other than a foundation.

According to Ostrower (1987), tax issues structure attorney and client consultation. Attorneys provide technical expertise about taxes. One wealthy person told us: "They advised me on how much of my income I had better contribute to charity. Otherwise Uncle Sam was going to take it."

Not only are there so many alternative charitable vehicles, but we might also speculate that the availability of tax planning devices such as non-charitable shelters compete with charitable dispositions (Steuerle 1987). However, we did not attempt to measure this in the research.

Positive tax incentives are significant for encouraging charitable giving. Both the personal interviews with wealthy individuals and their advisers and the survey results indicate that foundations were often used as vehicles to protect "windfall" profits or other assets where capital gains taxes would be very high. By using a foundation, the donors did not have to decide immediately how to give the money away, and by giving the assets they received a deduction for the property, they avoided the capital gains tax.

Foundations are still created because they are convenient holding spots for charitable funds and they can exist in pertetuity. Many donors and advisers (Odendahl 1987a; Ostrower 1987) indicated that they had more control when they made their charitable contributions through a foundation and applauded the systematic disposition of grants. But the lower deductibility and the complex regulatory framework act as disincentives to foundation development. Tax policies, on balance, provide more disincentives than incentives for foundation establishment.

WEALTHY DONORS AND THEIR CHARITABLE ATTITUDES

In personal interviews and in the survey, the project sought information about the general charitable attitudes and motives of the wealthy as well as their views on foundations.

Philanthropic Ideology

Many wealthy people believe that the private nonprofit sector is essential to the American way of life and contrasted it favorably with European welfare states. The people we interviewed thought that their funding decisions were better than those of the government. A woman from New York expressed the sentiments of a number of others when she said:

If there were not a nonprofit community out there, two things would have happened. I do not think we would still be a democratic country. This may be exaggerated, but the public demand for the government to do this, that, and the other. . . . We would probably be a huge big government doing all these things, socialistic.

Our respondents said that a free-market system fosters a giving environment. Whatever their political perspectives, interviewees had strong anti-welfare state ideologies. They seemed to endorse individual social responsibility,[2] rather than electoral decision making. A member of a family of multigenerational wealth told us: "I would rather decide where that money goes than have the government do so."

From the interviews, we also found that a key motive for philanthropy is the desire to maintain control over the disposition of money, rather than paying it in taxes (Odendahl 1987a; Ostrower 1987). Although the issue of control was not always linked to taxation, most wealthy study participants spoke of them together, declaring that they give up to the maximum for which they receive a deduction.

About 75 percent of the attorneys mentioned donor control as an important incentive for creating a foundation rather than giving to charity through another legal vehicle (Ostrower 1987). Attorneys seemed to think foundations are appropriate when an overriding concern of the donor is to exert influence over the future uses of charitable gifts. Donor control distinguishes the foundation from other philanthropic options.

Motives for Establishing Foundations and for Charitable Giving

The idea of starting a foundation, rather than to give through some other means, is almost always the donor's, rather than an adviser's. Attorneys who are consulted usually present a wide range of charitable options to their clients (Ostrower 1987). Personal philosophy was the strongest variable accounting for the establishment of a foundation, according to respondents to the survey. The second-ranking motivation was the desire to establish a vehicle for the systematic conduct of charitable giving. Survey results indicated that foundation donors were motivated by a cluster of "altruistic" variables including a concern for the welfare of others, religious heritage, a belief in social responsibility, and a family tradition of charitable giving.

Most of the rich people we interviewed cited family tradition as a reason to give to charity. For some, the family had a history of responding to needs of communities where they had been "leading families." Others, who had acquired their fortunes more recently, hoped to establish a family tradition.

Some of the wealthy people we interviewed admitted that their philanthropic motives are mixed. One man said:

Some people really want to get out there and do a lot of work and make some kind of contribution. I did that with the ———— a lot, but I do not do it with a lot of this other stuff. . . . There is some of just writing checks to friends. And there is some sort of feathering your nest in the hospital in your areas or whatever. And there is just giving because you have the money to give away, giving to organizations that you think are doing the kind of things that you want to see done or you think are important.

Personal interest, involvement, and satisfaction are important motivations for charitable giving. One interviewee commented: "I think my funding comes out of both helping others as well as how it affects me. I feel like the organizations I have funded are important toward creating change and I want to see that continue after I die as well." In addition, many rich people find their charitable activities pleasurable. A woman with inherited wealth told us: "there is a part of me that really enjoys giving." Although prestige and status are widely recognized motives for philanthropy, most of the wealthy people we interviewed did not offer them among their own reasons for giving.

Donor Characteristics

Foundation donors, over the course of a century, have tended to be Protestant men over the age of fifty (Boris 1987). But in recent years, more women have created grant-making institutions. These women were mostly the recipients of inherited wealth, in comparison to male foundation donors, who were entrepreneurs.

We interviewed fifty-six women and seventy-nine men. Almost all were involved in charitable activities in their communities. They served on the boards of nonprofit organizations, engaged in fund raising, and volunteered in other ways.

Gender

Some intriguing trends related to gender appeared. Among married people we interviewed, the husband was usually the entrepreneur and the wife the philanthropist. We also found that men who inherited wealth tended to invest their fortunes in business endeavors, whereas women in the same position set up charitable funds and foundations that they administered themselves. The ideology of "doing good" corresponds with traditional expectations about women's aspirations and interests. Philanthropic work is considered appropriate for rich women (Daniels 1988; Ostrander 1984).

Although attorneys and accountants work with both upper-class men and women, it is commonly held that women have less control over their wealth (Tickamyer 1981). A middle-aged woman interviewed explained:

The problem is the financial advisers and the lawyers, particularly with women. . . . I am talking about friends whose situations are controlled by lawyers and financial advisers, friends who have a considerable amount of money. . . . The people in my generation have depended on their male advisers. . . . The women in the younger generation are well-educated and career professionals themselves. So I think that is going to change.

A number of these young, wealthy women to whom our informant referred have become directly involved in the disposition of their fortunes. Many spend much of their time working on issues related to their charitable giving. They are among several hundred women around the country who have established "women's funds," grant-making organizations that focus on the concerns of women and girls.

Ethnicity and Religion

Like the survey donors, the interview group was primarily white and Protestant, although over a third were Jewish. Jews across the country and fundamentalist Protestants in the South mentioned religious motives for philanthropy most frequently.

More foundations were created by Protestants and Jews than their proportion in the general population: Protestants, representing 55 percent of Americans, created 69 percent of the foundations surveyed; Jews comprise only 4 percent of the total population, but they formed 19 percent of all foundations; and Catholics, who are 37 percent of the population, established only 8 percent of foundations.

The strong Jewish tradition of giving was revealed over and over in the personal interviews, in every part of the United States.

I guess you have to go into Jewish history and law to understand how deeply felt it is and how very basic. The Talmud, which is the written law, makes one feel very strongly about [charity]. . . . The fact that this is a guiding principle of our religion means that it is something that every Jewish child understands. They start out in their first Jewish educational experiences giving for some Jewish purpose. The reason that it is for Jewish purposes and why the Jewish community is so highly organized is because throughout our history we have always lived a little bit outside of the general society.

In addition, Jews give to all types of philanthropies, including non-Jewish causes. Synagogues were not the main beneficiaries. A wide range of service and educational institutions were usually supported, often through the United Jewish Appeal or the Federation of Jewish Philanthropies.

Age, Generation, and Source of Wealth

A majority of the philanthropists we interviewed and surveyed were older, married, and had children. A southern philanthropist indicated: "I characterize people as going through, first an acquisitive period where they want more money . . . then they reach either a maturity or a recognition of their own mortality, and then, they become dispositive. Some people never do—just up until the day they are put in the box." Several other studies confirm that people contribute more to charity when they get older (Clotfelter and Steuerle 1981; Jencks 1987; Steuerle 1987; Yankelovich, Skelly, and White 1986). It is not only the sense of one's mortality but also that people usually have more resources later in life.

Manufacturing was the prevalent source of wealth for both the donors in the foundation survey and those interviewed. The survey reveals that nearly 75 percent of the foundation donors were "self-made," having earned rather than inherited their fortunes. Most people who establish foundations have been philanthropists for some time, giving primarily to educational institutions. The families of 10 percent of the survey respondents had more than one foundation. When there are several heirs, family wealth may be broken up, and few assets may be available in succeeding generations for forming foundations.

One of the interesting developments in philanthropy over the last decade involves young people with inherited wealth who have been contributing generously to charity, although not necessarily through a foundation. In an attempt to aid in this process and urge the rich to use their fortunes for social change, a book was published under the auspices of the Vanguard Public Foundation (1977) by an informal group of wealthy funders. As the authors note: "The subject [of money] is laden with embarrassment, guilt, secret pleasure, and the fear of other people's envy. At family gatherings, it is discussed rarely, if at all." And they continue: "So some of us who have inherited money lie to our friends, and we hope new acquaintances won't find out. Yet we can if we wish, learn to see our money as a useful tool, and put it to work to help produce the kind of world we'd like to live in." Younger donors shared the ethic of social responsibility voiced by earlier generations, but specifically referred to their desire to combine political beliefs with funding priorities.

Wealthy entrepreneurs may be more likely to establish foundations than their offspring, but today they are no more disposed toward foundations than to other institutionalized forms of charitable giving. Many wealthy people prefer to contribute directly to charitable organizations. A young woman said that she had not started a private foundation because:

I think that individual funding is important. If it is done intelligently, it can be very effective. You have a lot more flexibility if you are not a foundation. You have much less administrative burden. And, I think you can play a very creative role, because your timetable is your own. You have complete independence. You can respond to requests almost as a kind of 911 number.

FINAL THOUGHTS

Big foundations may be a phenomenon of the twentieth century, treated in history books as the charitable by-products of capitalist accumulation and dynasty formation. A few of the largest independent grant-making organizations were created in the era of the robber barons. Foundations' popularity rose steeply over the course of the next four decades. Large foundations established by wealthy people during this period, along with more recent ones, continue to capture the public's eye today.

Foundations of all sizes proliferated in the 1940s. The postwar prosperity of the late 1940s and early 1950s saw the endowment of a number of big foundations, some of which had been formed earlier. During the 1950s, the number of independent foundations with more than $1 million in assets doubled. In addition, there was an increase in new small foundations. By the 1960s, however, the large foundation had started to wane in popularity among potential wealthy donors.

There were exceptions. In places like California and Texas, where new money was being made from the computer and oil industries, a few foundations formed in the 1960s grew into large ones. Some small foundations created during this period may have been victims of the stagnating economy and failed to grow.

Between 1960 and 1982, the composition of the foundation field changed. Small grant-making organizations doubled in number, but only six giant grant makers emerged. The relative asset share of the smaller foundations increased and exceeded the assets of the largest foundations. Of course, smaller foundations have the potential to grow into large ones. More than half of the largest foundations in the survey were formed with less than $1 million in assets. They grew through gifts during the donors' lifetimes and bequests. If the pattern of former years were to continue, foundations created in the 1960s should receive bequests in the 1980s. Yet since the 1970s, this trend seems to have shifted. More foundations are being formed by bequest, which lessens the likelihood of future gifts. Foundations established in the 1970s would have been endowed in the 1990s, but not if they were initially formed as testamentary bequests.

Tax Law and Attorneys as Catalysts

One interpretation of this history is that increasing regulation triggered a convergence of developments that had a bearing on the grant-making field. Attorneys began to develop specialties in nonprofit law. The more complicated the regulations became, the more the wealthy depended on attorneys to help them make decisions about their methods of charitable disposition. New charitable options were developed that competed with foundations. These alternatives combined greater tax deductibility with less demanding rules than those that applied to foundations. Since attorneys consulted in connection with charitable options are usually tax oriented, they advised their clients to use the newer alternative vehicles, such as split-interest trusts and support organizations, to enhance their tax benefits. Attorneys became primary mediators of technical knowledge regarding charitable donations, though many may never have mastered the foundation rules.

In New York City and other places where attorneys were more sophisticated, alternative charitable vehicles were used more widely (Cooper 1977, p. 210). In the big cities, where the greatest wealth was concentrated, the most practiced attorneys and larger law firms quickly responded to the 1969 Tax Reform Act. Knowledge of the changes spread to small towns relatively slowly, especially among attorneys in general practices.

Charitable Ideology

People give to charity because of altruism, personal interests, and political beliefs. They choose the foundation form as a systematic, flexible philanthropic vehicle with tax deductibility. Altruistic motivations may stem from the donor's personal philosophy or a family tradition of charitable work. Altruistic motivations may also reflect long-standing cultural factors in American society such as concern for the welfare of others, a religious heritage, and a belief in social responsibility.

Most major donors believe that the nonprofit sector is able to accomplish goals better with their gifts than the government can with their taxes. In addition, charitable gifts can be directed to the organizations and causes that the donor favors. There is an implicit belief in the power and responsibility of individuals and a corresponding distrust of big government bureaucracy.

Many wealthy people endorse an ideology that leads them to try to save as much in taxes as possible while exercising control over their wealth. Independent foundations offer certain advantages to donors that other charitable vehicles do not. Wealthy people attach value to the accumulation of assets. Holding wealth provides benefits such as prestige and influence. Foundations may have been particularly popular when company assets could be placed in their portfolios, and families could run the companies. But foundations continue to combine the value of accumulation with systematic giving and control over wealth, even if not for personal use and with no pecuniary benefit.

Speculations and Questions

Smaller foundations may provide status to the moderately wealthy, and they are usually lasting memorials to the donor. Foundations seem to be particularly popular among those who may be trying to increase their status; for example, men and women with new wealth and women who have inherited estates. The unanswered question is whether these smaller foundations will grow through additions to endowment.

We know that older people are more likely both to give to charity and create foundations than the young. There may be many reasons for this tendency. Recognition of one's mortality was mentioned as a possible motive. Wealthy young business people today may also be more involved with establishing their companies and making money than giving it to charity. It may also be the case that their wealth is not liquid—heavily leveraged or tied up in business ventures. The young entrepeneurs of today may be creating foundations in the decades to come, or they may follow their attorneys' advice and establish other kinds of charitable vehicles.

What are the implications of changing sources of wealth in an increasingly complicated and fluid world economy? The major foundations of the past were established from fortunes won in manufacturing. Yet, as a growing proportion of gross domestic product, service and trade sectors may create the millionaires of the future. Will they behave like their predecessors?

What are the implications of generational differences? Current trends suggest that people of inherited wealth may be rebelling against the course their parents chose. It seems the foundation as a symbol of great wealth holding has become socially unacceptable among some young elites. Consequently, many young people with the means to form foundations have decided to combine their philanthropy with that of other wealthy people or to establish grant-making organizations that are not foundations. Some inherited wealth is still tied up in trusts, and when these mature, the recipients may form foundations. Or, the wealth may have been dispersed over the generations.

Government policy makers of the late 1960s may have accomplished their goal of discouraging foundation formation. Curiously, however, small foundations that are largely charitable checkbooks for donors have persisted. A partial explanation is that many of these are "pass-through" foundations used to make yearly gifts not to build an endowment. Contributions that flow through such a foundation enjoy the maximum charitable deduction.

Individuals who value a systematic, perpetual vehicle that offers strong donor control in accomplishing charitable goals still form foundations. Directly and indirectly, however, increased regulation and less favorable tax deductibility have continued to limit the formation of new large independent foundations.

NOTES

1. Although it is too early to tell, this law may have particular ramifications for the foundation field. This supposition arises from two converging trends: increasing numbers of foundations are being established by women, and at the same time, more foundations are created by donors at death. As most surviving spouses will be women, they may be making charitable decisions that favor foundations.

2. Some might call this *civic stewardship* or *noblesse oblige*. See Robert Bremner, *American Philanthropy* (1988) and Kathleen McCarthy, *Noblesse Oblige* (1982).

REFERENCES

Bittker, Boris K. 1972. "Should Foundations Be Third-Class Charities?" In Fritz F. Heimann, ed., *The Future of Foundations*. Englewood Cliffs, N.J.: Prentice-Hall.

Bremner, Robert H. 1988. *American Philanthropy*. Chicago: University of Chicago Press.

Boris, Elizabeth T., 1987. "Creation and Growth: A Survey of Independent Foundations." In Teresa Odendahl, ed. *America's Wealthy and the future of Foundations*. New York: The Foundation Center, pp. 65–126.

Clotfelter, Charles T., and Eugene Steuerle. 1981. "Charitable Contributions." In Henry J. Aaron and Joseph Pechman, eds., *How Taxes Affect Economic Behavior*. Washington, D.C.: Brookings Institution.

Cooper, George. 1977. "A Voluntary Tax? New Perspectives on Sophisticated Estate Tax Avoidance." *Columbia Law Review* 77, no. 2: pp. 161–247.

Daniels, Arlene Kaplan. 1988. *Invisible Careers: Women Civic Leaders From the Volunteer World*. Chicago: University of Chicago Press.

Edie, John A. 1987. "Congress and Foundations: An Historical Summary." In Teresa Odendahl, ed., *America's Wealthy and the Future of Foundations*. New York: The Foundation Center, pp. 43–64.

Hall, Peter Dobkin. 1987. "An Historical Overview of the Private Non-Profit Sector." In Walter W. Powell, ed., *The Nonprofit Sector: A Research Handbook*. New Haven: Yale University Press, pp. 3–26.

Jencks, Christopher. 1987. "Who Gives to What?" In Walter W. Powell, ed., *The Nonprofit Sector: A Research Handbook*. New Haven: Yale University Press, pp. 321–339.

Karl, Barry D., and Stanley N. Katz. 1981. "The American Private Philanthropic Foundation and the Public Sphere, 1890–1930." *Minerva*, pp. 236–70.

McCarthy, Kathleen D. 1982. *Noblesse Oblige: Charity and Cultural Philanthropy in Chicago, 1849–1929*. Chicago: University of Chicago Press.

Nelson, Ralph L. 1987. "An Economic History of Large Foundations." In Teresa Odendahl, ed., *America's Wealthy and the Future of Foundations*. New York: The Foundation Center, pp. 127–79.

Odendahl, Teresa. 1987a. "Wealthy Donors and Their Charitable Attitudes." In Teresa Odendahl, ed., *Independent Foundations and Wealthy Donors*. New York: The Foundation Center, pp. 223–46.

Odendahl, Teresa, ed. 1987b. *America's Wealthy and the Future of Foundations* New York: The Foundation Center.

Ostrander, Susan A. 1984. *Women of the Upper Class*. Philadelphia: Temple University Press.

Ostrower, Francie. 1987. "The Role of Advisors to the Wealthy." In Teresa Odendahl, ed., *America's Wealthy and the Future of Foundations*. New York: The Foundation Center, pp. 247–66.

Rudney, Gabriel. 1987. "Creation of Foundations and their Wealth." In Teresa Odendahl, ed., *America's Wealthy and the Future of Foundations*. New York: The Foundation Center, pp. 179–202.

Simon, John G. 1987. "The Tax Treatment of Nonprofit Organizations." In Walter W. Powell, ed., *The Nonprofit Sector: A Research Handbook*. New Haven: Yale University Press, pp. 67–98.

Steuerle, Eugene. 1987. "The Charitable Giving Patterns of the Wealthy." In Teresa Odendahl, ed., *America's Wealthy and the Future of Foundations*. New York: The Foundation Center, pp. 203–21.

Tickamyer, Ann R. 1981. "Wealth and Power: A Comparison of Men and Women in the Property Elite." *Social Forces* 60, no. 2.

Yankelovich, Skelly, and White, Inc. 1986. *The Charitable Behavior of Americans: A National Survey*. Washington, D.C.: Independent Sector.

Vanguard Public Foundation. 1977. *Robin Hood Was Right: A Guide to Giving Your Money for Social Change*. San Francisco.

10

The Community Foundation in America, 1914–1987

PETER DOBKIN HALL

Community foundations are among the most interesting and least understood creations of American philanthropy. They differ from the majority of foundations in important ways:

- Rather than being based on endowments created by a single wealthy donor or family, they bring together endowment funds—large and small, restricted and unrestricted—to benefit charitable endeavors in particular cities or regions.
- Their boards are not self-perpetuating: many members are public officials serving ex-officio, others are designated by community organizations for the purpose of assuring a broadly representative character to the foundations' grant-making decisions.
- Their grant-making function is usually separated from the management of their assets, which, in many cases, is delegated to private for-profit firms.

Because of their peculiarly public and representative character, community foundations have avoided the taint of self-aggrandizement that has shadowed foundation philanthropy.[1]

With representative, often publicly appointed, governing boards and decentralized financial management, community foundations have not been suitable vehicles for serving the dynastic intentions of large charitable donors. Since 1914, when the first community trust was established, only 300 have been created—a fraction of the 23,000 or so private foundations set up in the United States during the same period. Of the $50 billion in assets held by American foundations in 1984, community foundations accounted for slightly more than $3 billion. Community foundations vary in size, ranging from giants like the New York Community Trust and the Cleveland Foundation, with assets exceeding $300 million, to tiny operations

The research on which this paper is based was funded by grants from the American Council of Learned Societies, AT&T Foundation, Ellis Phillips Foundation, the Equitable Life Assurance Society of the United States, Exxon Education Foundation, General Electric Foundation, Teagle Foundation, and the Program on Non-Profit Organizations, Yale University.

like the Branford (Connecticut) Community Foundation, with assets of less than $30,000. Of the 300, only 39 community foundations had assets exceeding $10 million in 1984; 97 had assets of less than $1 million.[2]

In spite of their organizational peculiarities and their relative rarity, community foundations are an important part of the overall philanthropic picture. Because the 1964 Tax Reform Act set them apart from private foundations and placed them in the special category of public charities, donors to community foundations enjoy unusually favorable tax treatment.[3] This, combined with Reagan administration's encouragement of community-based private voluntarism, has revived interest in community foundations among both large and small donors. Of the 300 community foundations established since 1914, 47 percent were set up after 1960. If federal regulation of foundations continues along the increasingly restrictive path it has followed since the early 1950s, community foundations and other kinds or public charities may well become the distinctive philanthropic vehicle of the next century.

THE ORIGINS OF THE COMMUNITY FOUNDATION

In eroding traditional forms of mutual support based on family and community, the industrial and political revolutions of the nineteenth century highlighted age-old problems of disease, dependency, ignorance, and poverty. Urban life, internal migration, and large-scale immigration aggravated these problems. But, the same ethos that led Americans to search for better and more efficient ways of making things also led them to seek social improvement. The wave of religious evangelism of the first quarter of the nineteenth century produced secular spin-offs—Abolitionism, Temperance, utopian socialism—as many Americans translated their search for spiritual perfection into efforts to reform society through political and legal changes.

Through the 1890s, political panaceas, ranging from Greenbackism and Free Silver through anarchism and socialism, were presented as solutions to the problems of progress and poverty. But frustration with the political process led many influential Americans to seek other ways of changing society. The emergence of science as the primary source of intellectual authority, led to a reevaluation of political economy. Reformers increasingly emphasized the usefulness of scientific approaches to society: they called for experts to study social problems, to devise technical solutions, and to implement policies administered by trained personnel. Beginning in such fields as public health, these reformers worked to create nonpartisan support for municipal and state regulatory action.[4] As the movement proceeded, it removed important issues from the arena of political debate and placed them under the authority of appointed boards of certified health professionals. Because the knowledge out of which these experts formed policy was generated for the most part in privately-supported universities and because the professional certification process itself was a function of those institutions, a set of private nonprofit corporations—universities, medical societies, and hospitals—began to assume unusual power in the forming of public policy.

The establishment of open-ended private foundations by Rockefeller, Carnegie, Sage, and other industrialists can only be adaquately understood in this institutional context. The beneficiaries of these large philanthropies were not to be the poor, sick, and dependent but rather the *institutions* that dealt with them, either in delivering needed services or in creating knowledge about them. The private foundations were to their nonprofit beneficiaries what the investment banks were to the railroads, manufacturers, and other capital-seeking industrial enterprises. Because the governing boards of foundations interlocked significantly with the boards of major recipients, certain philanthropies possessed the power to set and coordinate policy for such fields as higher education and public health.[5] And because these institutions were the centers for the study of science and society, the foundations were able to exert a profound influence on public policy.

The rise of the great industrial corporations and the philanthropic activities of their proprietors had their greatest impact on the national level. The creation of genuinely national economic, political, and cultural life in the United States required the development of institutions that were self-consciously national in orientation. But, in spite of its importance, this nationalization was highly selective. Its impact on the villages, towns, and middle-sized cities where most Americans continued to live varied considerably. A city like Bethlehem, Pennsylvania, for example, might, by 1910, be a virtual satillite of the national economy: its single largest industrial enterprise, Bethlehem Steel, owned by outsiders and its prosperity dependent on large contracts from government and other national firms, particularly the railroads.[6] Bethlehem's social structure was an expression of the corporation's bureaucratic hierarchy. Cleveland, Ohio, on the other hand, was far more diversified.[7] Although such national enterprises as the railroads and oil, coal, and steel companies were key elements in its economy, these were owned by a variety of groups, some based in the city, some based in New York or Philadelphia. In these diversified settings, the leadership structure tended to be tied more to local interests, particularly to families whose prominence preceeded the industrial revolution.

Leaders in cities like Buffalo, Chicago, Cleveland, and Detroit were ambivalent about the increasingly national character of American life. On the one hand, they admired and identified with the achievements of the Carnegies, Rockefellers, and Morgans: their success justified capitalism itself and the national prosperity they create brought benefited their communities. On the other hand, they could not help resenting the extent to which the growth of the national economy was transforming their communities. Towns that were once relatively isolated and self-determined were becoming cities afflicted by the same problems of poverty, dependency, disease, and disorder characteristic of the great metropolises.

The community leaders of the early twentieth century were historically rooted in the intimacy of small-town and village life. They yearned not only to recreate traditions of mutuality and collective responsibility but also to recapture the power once enjoyed by the wealthy, learned, and respectable in earlier times. This nostalgia meshed amazingly well with the popularized variants of Social Darwinism that became current in the United States in the decades following the Civil War.

Darwinism portrayed society—"the life of the race"—as an organic process in which "the struggle for existence" led to "the survival of the fittest." Though academic ideologues like William Graham Sumner might use this conception of society as an excuse to write off the poor, for many Americans, especially the reform-minded community leaders in mid-sized cities, it became the basis for reappraising the relation between social classes.[8] Carnegie's 1886 essays on the labor question made sense: the modern industrial society required both skilled managers and skilled laborers.[9] Clearly, the principle of "divided and subordinate responsibilities" that characterized both the army and large industry was also one that could be applied to the organization of community life.[10] For both national and local leaders, Social Darwinism reaffirmed the essential place of enlightened leadership in the social organism. And, the touchstone of enlightenment was scientific expertise.

This credo had important implications for local philanthropy. Throughout the nineteenth century, private charities had flourished in most American cities. Denominational and nonsectarian schools, churches, hospitals, colleges, mutual benefit societies, fraternal organizations, and funds established by charitable individuals and families had kept pace with the growing industrial economy. No effort was made to coordinate their activities, or even to ensure that the interests of their designated beneficiaries were served. In some cities, "charity organization societies" were set up in an effort to eliminate corruption, duplication, and waste in charitable organizations. Some states even created state boards of charities with limited regulatory powers. But by the mid-1880s it was clear that the problem was not the multiplicity of charitable objects or even maladministration of charitable institutions but the nature of charity itself.

The bellweather of this change was Andrew Carnegie's 1889 essay, "The Gospel of Wealth," in which the Great Ironmaster, who considered himself the leading American student of social Darwinist Herbert Spencer, applied Darwinism to charitable giving.[11] Mere almsgiving, Carnegie argued, did more harm than good, debilitating its recipients and doing nothing with the underlying forces that caused social problems. The time had come, he asserted, for a scientific philanthropy that would deploy charitable resources to maximum effect. The rich, he further suggested, had a particular not only to promote effective philanthropy, but to do so in ways that would economically empower future generations. If capitalism was to survive, it must be a self-renewing system. And only by putting back into the social organism what they had taken out could its future be assured. For those who did not share his view that "he who dies rich dies disgraced,"

Carnegie suggested that the government institute a sharply progressive income and confiscatory inheritance taxes.

Well-heeled reformers in mid-sized industrial cities found Carnegie's words particularly appealing because they transformed their desire to lead from an ineffectual yearning for lost power to a positive injunction to seek it. Further, they pointed to scientifically justified, nonpartisan activity as the most effective means of achieving it: it stood to reason that charitable resources should be administered in ways that would do the community the most good. Finally, like Carnegie, they saw no necessary opposition between charity and profit making: prosperity increased the sum of funds available for philanthropy; properly administered, such funds should not only alleviate suffering and eliminate its causes but should reaffirm the power of enlightened leaders.

By the beginning of the twentieth century, few Americans were willing to defend the status quo. Everyone agreed that something must be done to narrow the gap between the rich and the poor, the powerful and the powerless, but there were enormous differences of opinion about what ought to be done. At the extremes, some proposed doing away with democracy while others agitated to abolish capitalism. In between, the majority "hesitated, vacillated, swayed forward and back," between two visions.[12] One looked back to a past in which the town was the arena of political life, the family the building block of society, and the personally owned and managed firm or farm the basic unit of production. The other looked forward to a brave new world in which the mechanical and industrial forces of the nation would bring about a golden age of opportunity and fulfillment.

The term *Progressive,* as it came to be used in the second decade of the twentieth century, covered a multitude of possibilities.[13] In its broadest usage, it included all those who envisioned an urban and industrial nation. To most, the term connoted efficient and rational government run by experts but accountable to the public. Beyond that point, however, the various kinds of Progressives went in different directions. Business elements in the Progressive coalition remained firmly opposed to the growth of government power, preferring instead to deliver of basic social services through the private sector and self-policing over government regulation.[14] Intellectuals and a considerable number of social welfare professionals and academics increasingly favored direct government intervention in the economy and in society.

Although many intellectuals turned to the Left—a trend that would be accelerated by the World War, the Russian Revolution, and the Red Raids—progressively oriented businessmen searched with increasing urgency for a middle course between socialism and laissez-faire capitalism. They were convinced that capitalism could be a humane and just system if the principles of efficient organization toward which they strove in their enterprises were applied to efforts to solve social problems.[15] The community foundation and the federated giving campaign were the two most important products of this search.

FEDERATED GIVING

It was appropriate that the "middle course" for business Progressives should have originated in Cleveland, Ohio. A Cleveland businessman, John Hay, had written one of the first "social realist" novels to be published in America, *The Breadwinners*.[16] This political melodrama, published in 1886, depicted the situation of a well-intentioned but ineffectual patrician reformer in his efforts to combat an alliance of corrupt politicians and anarchists. Though the book was hardly sympathetic to the poor, it compellingly depicted both the obstacles encountered by immigrants when they tried to better themselves and the frustrations of elite reformers when they tried to intervene in the process.

The political culture that produced Hay also nurtured his contemporary, Marcus Alonzo Hanna.[17] Although Hanna was, by the 1870s, a millionaire in the coal and iron business, he was remarkably thoughtful on social questions. When the miners in the Ohio coalfields struck in the 1870s, he was the first operator to sign a contract with his workers. And, as the national struggle between capital and labor grew more intense in the 1880s, he became increasingly outspoken about the need for employers to deal fairly with their workers. He was a leading advocate of labor arbitration and, by the 1890s, became involved in the Civic Federation movement, which attempted to institutionalize dialogue between the major organized groups within communities.

In 1896, Hanna engineered the presidential candidacy of William McKinley in a campaign that pitted the populist radical, William Jennings Bryan, against the forces of corporate capitalism. The Republican slogan Four More Years of the Full Dinner Pail was more than a "you never had it so good" exercise in smugness. It encapsulated the Darwinist views of men like Hanna and Carnegie, who believed that, in an industrial economy, labor and capital were mutually dependent. As the national Republican "boss" during the presidencies of McKinley and Theodore Roosevelt, Hanna continued to promote conciliation between labor and management. When, during the great coal strike of 1902, coal operater George F. Baer asserted that "the interests of the laboring man will be protected and cared for—not by the labor agitators, but by the Christian men to whom God in his infinite wisdom has given the control of the property interests of the country," Hanna's response was "any man who won't meet his men halfway is a damned fool."[18]

Besides national figures like Hay (who became Secretary of State in the cabinets of McKinley and Roosevelt) and Hanna, Cleveland nurtured a remarkably active set of local reformers, many of whom became nationally prominent.[19] The reform impulse was not confined to politics, Cleveland's urban ills increasingly absorbed the energies of civic activists from the business community.

The center of concern about the effectiveness of private charity was the city's Chamber of Commerce, which, in 1900, formed a Committee on Benevolent Associations to investigate the increasing number of causes making appeals in the city, some of which were clearly fraudulent.[20] In that year, the committee began the practice of endorsing those charities found "worthy." Subsequently, it encountered increasing complaints even from established charities about the difficulties of fund raising, particularly those due to competition for the attention of a limited pool of donors. In 1909, the committee launched a full-scale investigation of the problem, inviting seventy-three charities to submit their lists of donors and donations for analysis. The results were astonishing:

Out of a city of over 600,000 people, it was found that the whole charitable enterprise, receiving current contributions of $500,000 was supported by only 5386 separate contributors of $5 or more—less than one per cent. of the population. Of these, furthermore, more than 800 were commercial firms and corporations. Moreover, of the 5386 contributors 54 were giving 55 per cent. of the total contributed, while 1066 individuals and firms were contributing 90 percent. of the total.[21]

The response of the committee to these findings was immediate. Not only did it express concern about the need to make fund raising more effective and efficient, it gave particular emphasis to the need to broaden the base for charity by educating small givers. To do this, the Chamber of Commerce created a new organization, the Cleveland Federation for Charity and Philanthropy, which began operation in March 1913.

The federation brought together fifty-three charitable organizations endorsed as legitimate by the Chamber of Commerce's Committee on Benevolent Organizations. These donees elected ten of their number to serve on the federation's thirty-member governing board. Of the twenty remaining members, ten were elected by "the city's larger givers" and the rest were "selected to represent the city at large by the president and directors of the Chamber of Commerce."[22] The federation planned to make a coordinated appeal that would permit donors either to designate particular organizations as beneficiaries or to place their gifts for distribution at the discretion of the governing board. The federation conducted its first campaign between October 1912 and February 1913. It was a conspicuously successful effort. Over this six-month period, the federation brought in over two-thirds of the city's annual "benevolent budget": donations totaled $300,000 from 4000 givers.[23]

The Cleveland federation became the model for the national movement that came to be known as the Community Chest. Within two years, Denver, New Orleans, Baltimore, South Bend, Indiana, Dayton, Ohio, and Elmira, New York, had set up similar organizations; by 1917, 13 cities were participating; by 1929, more than 129 cities were mounting "coordinated appeals," raising an annual total of $58.8 million.[24] The success of federated giving sparked interest in other ways of making philanthropy more efficient and responsive, while at the same time broadening its base. Once again, Cleveland set the pace.

THE COMMUNITY TRUST

For all of its obvious benefits, the Cleveland federation only dealt with one aspect of charity: the problem of meeting the current needs of participating organizations through annual gifts. It did not address the desire of some donors to create charitable trusts, which would yield income for benevolent purposes in perpetuity; nor did it deal with the "dead hand," the problem of assuring that future needs would be met by such perpetual trusts.[25] Open-ended grant-making foundations, which began to appear in the first decade of the new century, were one way of dealing with the "dead hand." Placing endowment trusts "for the benefit of mankind" at the discretion of a board of trustees seemed to assure flexible and responsive applications of these charitable endowments. But such vast enterprises seemed to have limited application to the situation of mid-sized cities or to the charitable impulses of the holders of mid-sized fortunes. Their national focus did not address the communitarian concerns of elites in provincial cities.[26] And, many Progressives found the self-perpetuating boards of the Carnegie and Rockefeller philanthropies to compromise their continuing responsiveness to public needs.

In 1913, Frederick H. Goff, president of the Cleveland Trust Company, developed a plan to accommodate the desires of donors to establish charitable trusts while, at the same time, ensuring their flexible and responsive application.[27] Under this plan a foundation would be set up to receive and manage charitable trust funds, which, as under the federation scheme, their donors could either designate for particular purposes or leave to the discretion of the foundation for distribution. Recipients of such discretionary funds would be determined by a five-member distribution committee consisting of two directors of the Cleveland Trust Company, the mayor of Cleveland, the senior probate judge of Cuyahoga County, and the senior presiding judge of the United States District Court for the Northern District of Ohio.

The virtues of this "community trust" are best described by its creator:

It is a fund created by the union of many gifts—many different estates or parts of estates—held in trust; contributed by the people of Cleveland and managed by them for the benefit of the city of Cleveland.

It provides a plan of organization sufficiently flexible to meet conditions that cannot be anticipated at the present time. The income from the fund will be available at all times for the most pressing civic needs—even a part of the principle may be used in great extremity.

Does the Cleveland Foundation interest only men of wealth? On the contrary, it appeals to men and women of moderate means whose surplus (after caring for children and relatives) would not be great enough to endow a chair or a charity or accomplish any other notable purpose. By the combining of many small funds a large income is provided with which work of real significance to the community may be accomplished. It makes appeal to the possessors of wealth, large or small.

Men of great wealth have in the past created private foundations, but no way has been provided by which even greater foundations may be created out of the contributions of many citizens.[28]

Doubtlessly benefiting from the success of the Cleveland federation, Goff drafted and promoted his plan in the closing months of 1913. In January 1914, the board of his bank passed resolution creating the Cleveland Foundation. The new organization was greated enthusiastically by the press and, evidently, by charitable donors as well. Within ten years of its creation, the foundation was administering more than $100 million in charitable funds.[29]

The community trust idea was widely but selectively emulated in the ten years following the establishment of the Cleveland Foundation (see Table 10.1). Of the twenty-six trusts set up between 1914 and 1924, thirteen (50 percent) were established in the Midwest, 9 (69 percent) of them in cities that were also the pioneers in organizing federated annual giving campaigns (Community Chests).[30] This was a striking contrast to the cities of the wealthier Northeast, which represented only 8 (31 percent) of the trusts formed, only one of which had set up a Community Chest. Of southern cities, only the most progressive, Atlanta and Louisville, had trusts—and both, significantly, also possessed Community Chests.

Overall, community foundations seem to have had their greatest appeal to inland cities with strong, elite-dominated reformist political cultures. These midwestern elites differed significantly from their metropolitan counterparts on the East Coast, particularly in their willingness to trust the public to use community charitable resources intelligently. One token of this is the pioneering role of midwestern cities in the Community Chest movement. Another is the predominance of public, rather than private, universities in the states that were early to adopt community trusts: of the fifteen states represented in Table 10.1, nine were noted for the high quality of their public universities.

Other evidence for a uniquely public philanthropic orientation can be found in the actions of certain wealthy midwestern Progressives. When William Rockhill Nelson, proprietor of the *Kansas City Star*, died in 1910, he directed that his newspaper be sold to its employees.[31] His estate became the corpus of the Nelson Trust, whose trustees were to be appointed by the curators of the University of Missouri. Although the designated beneficiary of the trust was to be an art museum, the people who benefited most turned out to be the people of Kansas City. Because the terms of the trust specified that its assets be invested in real estate within a hundred mile radius of the city, the trustees, led by urban planner Jesse Nichol, used this provision to create one of the most extensive and innovative planned developments in the United States, giving reality to the ideas of the City Beautiful movement, of which Nelson had been a major supporter.[32] A similarly progressive orientation was evident in the foundation set up by Chicago's Julius Rosenwald. Though managed by trustees of his

Table 10.1   Regional Distribution of Community Trusts and Federated Appeals, 1914–1924, by Year of Formation

Cities with Trusts, by Region	Year Trust Formed	Cities with Federated Appeals	Percent of Total
Midwest:			
Cleveland	1914	1912	
Chicago	1915		
Detroit	1915	1924	
Minneapolis	1915	by 1920	
St. Louis	1915		
Milwaukee	1915	1924	
Indianapolis	1916	1924	
Youngstown, Ohio	1918	by 1920	
Dayton, Ohio	1921	by 1920	
Grand Rapids, Mich.	1922	by 1920	
Ashtabula, Ohio	1920		
Toledo, Ohio	1924	by 1920	
Troy, Ohio	1924		
		9/13	(13) 50
Northeast:			
Boston	1915		
Providence, R.I.	1916		
Williamsport, Penn.	1916		
Philadelphia	1918	1924	
Buffalo	1919		
New York	1923		
Waterbury, Conn.	1923		
Lancaster, Penn.	1924		
		1/8	(8) 31
South:			
Louisville, Ky.	1916	by 1920	
Winston-Salem, N.C.	1919		
Atlanta	1921	1924	
		2/3	(3) 12
West:			
Los Angeles	1915		
Honolulu, Hi.	1916		(2) 8
Total	26	12 (46%)	26 100

SOURCES: Data from the *Foundation Directory, 1984–85*; Pierce Williams and F. E. Croxton, *Corporation Contributions to Organized Community Welfare Services* (1930).

selection, Rosenwald specified that his fund have a duration of only twenty-five years. This not only maximized its pay out, but also avoided the problems of "the dead hand" which so preoccupied many midwestern philanthropists.[33]

Of course, *trusting the public* is, in this context, a relative term. The board members of midwestern community trusts, federated giving campaigns, and state universities were hardly typical members of the middle and lower classes. Still, as the Lynds pointed out in *Middletown*, a study of Muncie, Indiana, in the early 1920s, the essence of civic life in midwestern cities was a pattern of loyalty and deference that, despite divisive forces of income, religion, and ethnicity, tended to draw the various groups in the population together.[34] Organizing a Community Chest in a city like New York would be difficult because the sheer scale of the place, and the diversity of its population—and its elites—stood in the way of broad commitments to common enterprises. But in a Cleveland, Buffalo, or Detroit, where civic leadership was clearly articulated and deeply rooted in patterns of well-established deference, it was easy for urban elites to conceive of themselves as representing the public.

SLOW GROWTH, 1930–1960

For the foundation world as a whole, the years between 1930 and 1960 were boom times: almost 60 percent of all foundations came into being during these three decades. This was a stark contrast to the slow growth of community trusts, which increased by only 33 percent. Trusts continued to be a largely midwestern phenomenon: of the 129 trusts in existence by 1959, 53 (41 percent) were concentrated in the central states with an additional 13 (10 percent) located in cities like Buffalo, Erie, and Pittsburgh, which were essentially midwestern in their economic and cultural orientation.[35]

Whereas the growth rate for all foundations averaged 118 percent between 1920 and 1959, the growth rate for community trusts averaged only 65 percent. Although almost two-thirds of the foundations in existence by 1983 had been created in the twenty years between 1950 and 1969, less than half of the community trusts were. In the 1950s, when foundation formation was ten times greater than it had been in the 1920s, the number of community trusts formed was less than twice what it had been. Nevertheless, the 1950s were a turning point for community foundations, heralding dramatic growth during 1960s and 1970s. This would coincide with a slowdown in the growth of other kinds of foundations.

How do we account for the slow growth of community foundations between 1930 and 1960? The most likely explanations appear to lie in three areas of American political and economic culture. The business-oriented progressivism that brought the trusts into being before the Great War remained intact through the boom of the 1920s. But the Great Crash shattered the Progressives' hopes of abolishing poverty through for-profit and nonprofit private sector activity. Although Community Chests raised more funds than ever before during the protracted crisis of 1931–1933, the effort drained the resources of local welfare agencies, private and public. Such agencies would continue to be important components of the health and welfare system after the coming of the New Deal, but the struggles of the early 1930s made it clear to everyone that they could serve, at best, as adjuncts to and conduits for the vast resources of the state and federal governments.[36]

This shift in the relation of localities to higher levels of government was paralleled by a wholesale abandonment of social responsibility by business leaders in American communities after 1935. Most business people had initially welcomed the New Deal's programs, inasmuch as the NRA, with its industry codes, appeared to be a formalized version of the "associative state" ideas of Herbert Hoover.[37] But in 1935, the New Deal turned sharply leftward, responding not only to the Supreme Court's nullification of the NRA, but also in response to the political threat posed by Huey Long and other populist agitators. Its encouragement of organized labor, the sharply progressive tax increases in the 1935 Tax Act, and an explicit "soak the rich" rhetoric made clear the administration's hostility to business.

The response of the wealthy, not surprisingly, was defensive. They shied away from forms of giving that would place their charitable funds under public control. Private foundations became a wonderfully efficient way of avoiding increased income and estate taxes. In addition, because they were virtually unregulated, private foundations enabled the wealthy to do things that were not possible through normal testamentary arrangements. As the great beneficiaries of the economic booms of the late nineteenth and early twentieth century began to die off in the 1930s and 1940s, private foundations became an important component in the institutionalization of dynastic wealth, and community trusts were viewed as curious artifacts of a vanished time.

Finally, the relation of local elites to their communities changed drastically after 1930. The privileged young in places like Scranton, Cleveland, and Buffalo were increasingly unlikely to take up the leadership roles their parents and grandparents had occupied in their communities. As early as the mid-1920s, the Lynd's *Middletown* had noted that the spread of high school education, the national print media, movies, and radio were pushing boys in Muncie to enter occupations other than those of their fathers.[38] This alienation from communities of origin was intensified by the Depression, which closed off many local opportunities, and by World War II, which opened vast realms of opportunity for the ambitious. Novels like J. P. Marquand's *Point of No Return* (1948) and Sloan Wilson's *Man in the Gray Flannel Suit* (1950) compellingly describe the exodus of potential leaders from small towns and cities during this period.[39]

Though declining, interest in community trusts did not entirely vanish (see Tables 10.2 and 10.3). Although only ten were set up during the 1930s, during the 1940s and 1950s they rebounded and seventy were established. Although thirty-one (44 percent) of these were in the central states, which had always been the center of community trust activity, twenty-two (31 percent) were in the southern and Pacific states, which had experienced enormous economic and population growth during the war. Community foundations resumed their growth during the 1940s and 1950s, and their number increased at a much slower rate than private foundations of the more conventional type (see Table 10.4): between 1940 and 1959, foundations overall increased in number from 367 to 2058 (a growth rate of

Table 10.2 Number of Regional Distribution of Community Foundations, by Decade, 1914–1979

Decade of Establishment	West	Central	South	North	Total	Percent of 1979 CFs
1910–1919	2 (10%)	9 (45%)	3 (15%)	6 (30%)	20	8%
1920–1929	4 (14%)	10 (35%)	2 (7%)	13 (45%)	29	12%
1930–1939	1 (4%)	3 (30%)	2 (20%)	4 (40%)	10	4%
1940–1949	4 (9%)	14 (52%)	1 (4%)	8 (30%)	27	11%
1950–1959	4 (9%)	17 (40%)	13 (30%)	8 (19%)	43	18%
1960–1969	3 (5%)	21 (37%)	16 (28%)	17 (30%)	57	24%
1970–1979	8 (15%)	20 (37%)	12 (22%)	14 (26%)	54	23%
Total	26 (11%)	94 (39%)	49 (20%)	70 (29%)	240	100%

SOURCE: Council on Foundations, *Status of Community Foundations in 1984* (Washington, D.C.: Author, n.d.)
NOTE: States are allocated by region as follows:
West: Arizona, Alaska, California, Colorado, Idaho, Montana, Nevada, New Mexico, Oregon, Utah, Washington, Wyoming.
Central: Illinois, Indiana, Iowa, Kansas, Michigan, Minnesota, Missouri, Nebraska, North Dakota, Ohio, South Dakota, Wisconsin.
South: Alabama, Arkansas, Florida, Kentucky, Louisiana, Mississippi, North Carolina, Oklahoma, South Carolina, Tennessee, Texas.
North: Connecticut, Maine, New Hampshire, New York, Pennsylvania, Rhode Island, Virginia, Vermont, West Virginia.

461 percent), while community foundations increased in number from 59 to 129 (a growth rate of 95 percent).

TAKEOFF, 1960–1979

Not until the 1960s did the establishment of community foundations begin to show real vigor. Not only did the number set up exceed previous levels, their geographical distribution began to shift. Although central and northern states continued to lead, with thirty-eight of the fifty-seven new trusts (59 percent), the revival of interest in creating community endowments continued to evince the national characteristics that had already begun to

Table 10.3 Regional Distribution of All Foundations and Community Foundations in 1984

	West	Central	South	North
All Foundations	14%	29%	21%	36%
Community Foundations	11%	39%	20%	29%

SOURCES: Data from Teresa Odendahl, ed., *America's Wealthy and the Future of Foundations* (New York: Foundation Center, 1987); Council on Foundations, *Status of Community Foundations in 1984* (Washington, D.C.: Author, n.d).
NOTE: States are allocated by region as follows:
West: Arizona, Alaska, California, Colorado, Idaho, Montana, Nevada, New Mexico, Oregon, Utah, Washington, Wyoming.
Central: Illinois, Indiana, Iowa, Kansas, Michigan, Minnesota, Missouri, Nebraska, North Dakota, Ohio, South Dakota, Wisconsin.
South: Alabama, Arkansas, Florida, Kentucky, Louisiana, Mississippi, North Carolina, Oklahoma, South Carolina, Tennessee, Texas.
North: Connecticut, Maine, New Hampshire, New York, Pennsylvania, Rhode Island, Virginia, Vermont, West Virginia.

Table 10.4 Number and Rate of Increase of All Foundations and Community Trusts, by Decade, 1920–1979

Decade of Establishment	Number of Foundations Established	Percent of Increase All Foundations	Community Trusts Established	Percent of Increase Community Trusts
1910–1919	44	86	20	—
1920–1929	108	114	29	145
1930–1939	164	81	10	20
1940–1949	571	156	27	46
1950–1959	1120	119	43	50
1960–1969	766	37	57	44
1970–1979	401	14	54	29
Total	3225		110	

SOURCES: Data from Teresa Odendahl, ed., *America's Wealthy* (New York: Foundation Center, 1987) and Council on Foundations, *Status of Community Foundations in 1984* (Washington, D.C.: Author, 1985).

emerge in the 1950s. Growth was most dramatic in the West and South, which had always lagged far behind other regions: 43 percent of the community foundations in the West and 50 percent of those in the south were created after 1960. There seem to be two major reasons for this shift. One involved changes in the federal tax code, the other stemmed from national demographic and economic trends.

In 1943, the federal government created a distinction between public and private charities based on the proportion of annual support derived from government and public appeals. Donors to the former could deduct contributions up to 30 percent of adjusted gross income, as opposed to the 20 percent deduction available to donors to conventional "private" charities.[40] Under the 1943 law, favored organizations included schools, churches, medical institutions, fund-raising organizations for state and local government-supported colleges, universities, and government agencies, and "publicly supported organizations meeting certain tests."[41] Although publicly supported, community foundations did not benefit from the 1943 law, because their heavy dependence on income from invested funds made it difficult for them to demonstrate that a third of their annual income was derived from public sources.

In 1964, however, IRS regulations were revised. A "facts and circumstances test" was introduced to supplement the older "mechanical test" that had been used to determine public charity status. The new test allowed community foundations to qualify as "public charities" if they possessed ongoing programs of public support, publicly representative governing bodies, and published annual reports. The new regulations also recognized community foundations as organizations, rather than treating each trust fund within the foundation as a separate entity. Including community foundations within the category of "public charities" was of enormous importance in an increasingly tax-conscious society.

By 1969, this provision of the tax code was further strengthened by the clarification of guidelines under which community foundations could qualify for the favorable "public charity" status. These included enlarging the amount of support such foundations could receive from individual donors from 1 to 2 percent, lowering the public support requirement from 33 to 10 percent, and increasing the deductibility of gifts 30 to 50 percent. "Public charities" were also exempted from the 4-percent excise tax on investment income imposed on private foundations by the 1969 act.

Not all community foundations could qualify as public charities under IRS guidelines. Some of the older foundations had been established as trusts and for that reason did not meet IRS operating requirements. Others found it difficult to demonstrate the requisite levels of public support.[42] Although the new regulations created some problems for older community trusts and foundations, they appear to have acted as a powerful inducement for the formation of new ones. In the decade of the 1970s, the number of new community foundations being established continued to increase: by 1979, this sector's growth rate finally surpassed that of private foundations.

Economic and demographic trends also appear to have played a role in the revival of interest in community foundations. During the 1950s and 1960s, an increasing number of industries and their employees moved to underpopulated regions of the West and South. These areas contributed disproportionately to the growth of community foundations. Of the 111 established between 1960 and 1979, 39 (35 percent) were set up in Pacific and "southern rim" states, like Texas, Florida, and California, which had more than doubled their populations since 1950.

Neither tax laws nor mere growth of population and business activity can fully account for the revival of interest in community foundations after 1960. Certainly, for many newly prosperous places, community foundations were viewed as emblems of urban achievement. They also seem to be tied to the rise of new local and regional elites, groups whose power was based on the rise of new businesses. Lacking deep community roots, these new leaders were willing to by-pass older charitable organizations to establish new ones more reflective of their personal interests and management styles. Because many of them were politically conservative, they regarded locally focused voluntary activity as an important alternative to government action while shying away from the private foundation form, which was tainted with an aura of liberalism.

COMMUNITY FOUNDATIONS AND THE FUTURE OF PHILANTHROPY

Before Congressman Wright Patman's hearings of the 1960s, American foundations grew almost unnoticed. No general reference work on foundations was available until 1960, when the first edition of the *Foundation Directory* appeared. In spite of their considerable assets, foundations enjoyed minimal government oversight. Scholars devoted no attention to the subject.

The Patman Committee hearings changed all this because its findings were widely publicized in the press and such best-sellers as Ferdinand Lundberg's *The Rich and the Super Rich*.[43] As the hearings revealed widespread abuses, it became apparent that, rather than serving redistributional purposes, many had been used to further enrich the very rich. Although some might do good works, a distressing number were shown to be no more than dynastic mechanisms, holding companies, and methods of tax avoidance. Community foundations were shown, however, to be exceptionally free of abuse.

Not surprisingly, as Congress drafted the 1969 Tax Act, the example of the community foundation proved to be particularly compelling. With its broadly representative distribution committee, its separation of financial management from grant-making activity, and its dependence on public support, it was not amenable to the kinds of abuse that characterized private foundations. For those who favored responsive and genuinely redistributional philanthropy, it appeared to be a model to be encouraged. The events of the 1970s and 1980s would help this model of responsiveness to fulfill its possibilities.

The 1970s proved to be a watershed in our political culture as great as the 1930s had been. If the Great Depression had smashed the business Progressives' belief that poverty could be abolished through private sector action, the war in Vietnam and the breakdown of public order in the 1970s shattered liberal faith in the capacity of big government to ensure social and economic justice. Disillusionment with government carried over to the foundations, the largest of which were closely identified in the public mind with the discredited liberal social programs and foreign policy. Americans did not lose faith in philanthropy, but their interests began to shift in the direction of more tangible forms of charity, particularly those with a local impact.

The election of the conservative Ronald Reagan in 1981 further increased interest in locally focused voluntarism and, as a result, the possibilities of community foundations. The new president expected to dramatically reduce federal spending. He hoped that private charity would be able to "take up the slack." Accordingly, he convened a Task Force on Private Sector Initiatives to explore what private voluntary activity could do to take the place of government. Although the task force generated neither substantial research nor concrete policy guidelines, it played an important role in promoting voluntarism, particularly on the local level and in regions, like the South and West, where political conservatism had previously impeded any substantial development of private philanthropy.

Philanthropy has generally been thought of as something that only the wealthy could afford. The community foundation, like the federated giving campaign, was created to combat this impression. The growth of big government, big business, and big philanthropy between 1930 and 1960 fostered the illusion that philanthropy was only for the rich, that community responsibility was a form of vulgar boosterism, and that most problems could not be effectively dealt with at the local level. Since the 1960s, we have gradually been rediscovering the satisfactions and benefits of entrepreneurship, civic pride, and "hands-on" solutions to problems.[44] The revival of community foundations is only one part of this rediscovery, the rise of local partnerships between business, government, and private philanthropy to improve public education, the Boston Compact being the most notable example, is another. These partnerships have opened new vistas for community foundations, many of which have become active in as brokers for public and private resources.[45]

While serving new kinds of needs, community foundations also serve the desires of more traditional philanthropic impulses. As increasingly stringent government regulations discourage the growth of private foundations, community foundations continue to permit donors to set up perpetual charitable endowments, directed to any legal purpose, on which donors can, if they wish, place their names. Such funds can serve many of the purposes of private foundations, as well as bringing conspicuous tax advantages and freedom from red tape.

At the same time, events similar to those that first brought the community foundation into being appear to be contributing to its revival. As at the turn of the century, we are faced with seemingly intractable problems of disease, ignorance, and poverty—all of which seem peculiarly resistant to conventional solutions. At the same time, the gap between the very rich and the middle and lower classes, which had been narrowing since the Depression, has again begun to widen alarmingly.

If present trends continue, it possible that the political climate will become increasingly hostile to the wealthy. Questions will doubtless be raised about the inequity of the tax code, particularly the failure of its charitable provisions to produce effective redistributional outcomes. Private foundations may be subjected to even more intense scrutiny than they experienced during the Patman Committee investigations. In such a grim scenario, community foundations would be seen in even more favorable a light than they were in the 1960s, with commensurate regulatory provisions in their favor.

NOTES

1. For excellent summaries of the structural aspects of community foundations, see Wilmer Shields Rich, *Community Foundations in the United States and Canada* (New York: National Council on Community Foundations, 1961); Norman A. Sugarman, "Community Foundations," in Commission on Private Philanthropy and Public Needs, *Research Papers* (Washington, D.C.: Department of the Treasury, 1977), p. 1689 ff, and *The Status of Community Foundations in 1984* (Washington, D.C.: Council on Foundations, 1985).

2. On the fiscal features of community foundations, see Sugarman.

3. On the tax status of community foundations, see Sugarman, "Community Founda-

tions," and John G. Simon, "The Tax Treatment of Nonprofit Organizations: A Review of Federal and State Policies," in Walter W. Powell, ed., *The Nonprofit Sector: A Research Handbook* (New Haven, Conn.: Yale University Press, 1987), pp. 67 ff.

4. On the development of a scientific approach to social problems, see Thomas Haskell, *The Emergence of Professional Social Science* (Urbana: University of Illinois Press, 1977); Roy Lubove, *The Professional Altruist: The Emergence of Social Work as a Career, 1880–1930* (New York: Atheneum, 1969); and Barbara G. Rosenkrantz, *Public Health in Massachusetts* (Cambridge, Mass.: Harvard University Press, 1974).

5. On the early development of this system of private power, see Burton Bledstein, *The Culture of Professionalism* (New York: W. W. Norton, 1976). On its later development, see Guy Alchon, *The Invisible Hand of Planning: Capitalism, Social Science, and the State in the 1920s* (Princeton, N.J.: Princeton University Press, 1985).

6. On Bethlehem and its social structure, see W. Ross Yates, *Bethlehem of Pennsylvania* (Bethlehem, Penn.: Bicentennical Commission, 1976).

7. Hoyt L. Warner, *Progressivism in Ohio, 1897–1917* (Columbus: Ohio Historical Society, 1964).

8. Peter Dobkin Hall, *The Business of America* (New York: Simon and Schuster, forthcoming).

9. Andrew Carnegie, "An Employers View of the Labor Question," *Forum* (1886), pp. 114 ff; and "Results of the Labor Struggle," *Forum* (1886), pp. 538 ff.

10. Charles W. Eliot, "Inaugural Address as President of Harvard University," in *Educational Reform: Essays and Addresses* (New York: Macmillan Co., 1898), pp. 1 ff.

11. Andrew Carnegie, "The Gospel of Wealth," *North American Review* 148 (1889): 653 ff, 149: 682 ff.

12. Henry Adams, *The Education of Henry Adams* (Boston: Massachusetts Historical Society, 1918), p. 343.

13. On the expanded meaning of *progressivism*, see Bledstein, *Culture of Professionalism*; and Robert Wiebe, *The Search for Order* (New York: Hall and Wang, 1967).

14. A wonderful example of the split between intellectuals and businesspeople in the reform movement can be found in a 1915 exchange between Norman Hapgood and Julius Rosenwald: Hapgood, "Modern Charity," *Harper's Weekly* 51 (March 20, 1915), p. 268; Rosenwald, "Charity," *Harper's Weekly* 51 (March 27, 1915), p. 522.

15. The best expression of business's vision of achieving social justice through the private sector is Herbert Hoover's *American Individualism* (Garden City, N.Y.: Doubleday, Doran and Co., 1922).

16. John Hay, *The Breadwinners: A Social Study* (New York: Harper and Brothers, 1883).

17. Though characterized by his enemies, and by many subsequent historians, as an arch-conservative, a measure of his achievements as a reformer can be found in the 1919 biography, *Marcus Alonzo Hanna: His Life and Work* (New York: Macmillan Co., 1919). The book was written by Herbert Croly, founder of *The New Republic* and the leading Progressive thinker of his time.

18. George F. Baer to William F. Clark (July 17, 1903), in William Cahn, *A Pictorial History of American Labor* (New York: Crown, Publishers, 1972), p. 189.

19. On progressive political culture in the Midwest, see Russel B. Nye, *Midwestern Progressive Politics, 1870–1958* (New York: Harper and Row, 1959) and Warner, *Progressivism in Ohio*.

20. On the origins of federated giving, see Scott M. Cutlip, *Fund Raising in the United States: Its Role in American Philanthropy* (New Brunswick, N.J.: Rutgers University Press, 1965), pp. 29–109. On its Cleveland origins, see Charles Whiting Williams, "Cleveland's Federated Givers," *Review of Reviews* 48 (1913): 172 ff; C. W. Williams, "Cleveland's Group Plan," *The Survey* 29 (1913), pp. 603–606; E. M. Williams, "Essentials of the Cleveland Experiment in Cooperative Benevolence," *Conference on Charities and Corrections* (1913): 111–15.

21. Williams, "Cleveland's Federated Givers," p. 472.

22. Ibid., p. 473.

23. Ibid., p. 473.

24. Pierce Williams and William R. Croxton, *Corporate Contributions to Organized Community Welfare Services* (New York: National Bureau of Economic Research, 1930) contains the most complete list of Community Chests established before 1930, see pp. 91–138, 247–253. See also Cutlip, *Fund Raising in the United States*, p. 73 ff. The Community Chest idea was spread through an aggressive education campaign under the leadership of the United States Chamber of Commerce.

25. On the background of the establishment of the Cleveland foundation, see "Cleveland Foundation Not Self-Perpetuating," *The Survey* 31 (January 31, 1914), p. 511; I. M. Tarbell, "He Helps Capitalists to Die Poor," *American Mercury* 128 (September 1914), pp. 56–57; "Survey of Cleveland by New Foundation," *The Survey* 31 (March 21, 1914), p. 763; "Private Wealth for Public Needs," *The Independent* 70 (January 12, 1914), p. 50; "Trust Company and the Community," *Outlook* 106 (January 17, 1914), pp. 108–109. The best early overview of the community trust movement in Walter Greenough, "The Dead Hand Harnessed: The Significance of Community Trusts," *Scribner's* 74 (December 1923), pp. 697 ff.

26. The tie between charity and boosterism is suggested in Cutlip, *Fund Raising in the United States*, p. 79.

27. On the Goff Plan, see Greenough, "The Dead Hand Harnessed."

28. Quoted in ibid., p. 698.

29. Ibid.

30. On the spread of community foundations, see "Four New Community Trusts Established," *The Survey* 34 (June 12, 1915), pp. 239–40; "Charity Federation and Its Fruits,"*The Survey* 36 (May 13, 1916), pp. 187–88; W. J. Norton, "Progress of Financial Federation," *National Conference on Social Work* (1917): 503–507; F. M. Hollingshead, "Community Foundation," *The Survey* 45 (January 29, 1921), pp. 639–40; F. J. Parsons, "New York Community Trust," *American City* 23 (August 1920), pp. 138–39; T. Devine, "Uniform Trust for Public Uses," *The Survey* 45 (February 12, 1921), p. 694; T. Devine, "Community Trusts in Chicago," *The Survey* 45 (March 26, 1921), pp. 318–19; F. D. Loomis, "Community Trusts," *The Survey* 46, p. 219; R. M. Yerkes, "Science and Community Trusts," *Science* 53 (June 10, 1921), pp. 527–29; J. Cowen, "Community Trust," *Canadian Magazine* 63 (July 1924), pp. 169–74; Raymond Moley, "Community Trust," *National Conference on Social Work* (1921): 427–32; Ralph Hayes, "To Help the Dead Hand by a Living Mind," *American City* 29 (December 1923), pp. 600–602; Pierce Williams, "Could Community Trusts Work?" *The Survey* 58 (May 15, 1927), pp. 223–24; Ralph Hayes, "Dead Hands and Frozen Funds," *North American Review* 227 (May 1929): 607–14; H. Wickenden, "What the Community Trust Is Doing for American Cities," *American City* 40 (January 6, 1929) pp. 124–26; "Community Trust Funds in the United States," *The Survey* 66 (April 15, 1931), p. 100. Eduard Lindeman's *Wealth and Culture: A Study of One Hundred Foundations and Community Trusts and Their Operations during the Decade 1921–1930* (New York: Harcourt Brace and Co., 1936) lists twenty community trusts with their dates of establishment.

31. On the Nelson Trust, see *William Rockhill Nelson* (Cambridge, Mass., Riverside Press, 1915) and I. F. Johnson, *William Rockhill Nelson and the Kansas City Star* (Kansas City: Burton Publishing, 1935).

32. On the role of Nelson, his trust, and his chief trustee, Jesse Nichols, in the transformation of Kansas City, see William H. Wilson, *The City Beautiful Movement in Kansas City* (Columbia: University of Missouri Press, 1964).

33. On Rosenwald's views of foundation philanthropy, particularly his anxiety about "the dead hand," see Julius Rosenwald, "Principles of Public Giving," *Atlantic Monthly* (May 1929), pp. 599–606.

34. Robert S. Lynd and Helen Merrell Lynd, *Middletown: A Study in Modern American Culture* (New York: Harcourt, Brace and World, 1929), pp. 478 ff.

35. For data on foundation establishment, see Teresa Odendahl, ed., *America's Wealthy and the Future of Foundations* (New York: The Foundation Center, 1986); and Thomas R.

Buckman and Loren Renz, "Introduction," *The Foundation Directory, Eleventh Edition* (New York: Foundation Center, 1987), pp. v–xxviii.

36. For a detailed study of the relation between public and private relief efforts in the Depression, see P. D. Hall and K. L. Hall, "Allentown, 1929–41," in Mahlon Hellerich, ed., *Allentown, 1762–1987* (Allentown, Penn.: Lehigh County Historical Society, 1987), pp. 81 ff.

37. On Hoover's ideas, see Ellis Hawley, "Herbert Hoover, the Commerce Secretariat, and the Vision of an Associative State," in Edwin J. Perkins, ed., *Men and Organizations* (New York: G. P. Putnam's Sons 1977), pp. 131–48; and Paul Johnson, *Modern Times* (New York: Harper and Row, 1983), pp. 234 ff.

38. Lynd and Lynd, *Middletown,* pp. 50 ff. On the philanthropic consequences of out-migration by the young, see P. D. Hall, "Philanthropy as Investment," *History of Education Quarterly* 22, no. 2 (Summer 1982): pp. 85–91.

39. John P. Marquand, *Point of No Return* (Boston: Little Brown, 1947) and *Sincerely, Willis Wayde* (Boston: Little Brown, 1954), and Sloan Wilson's *Man in the Gray Flannel Suit* (New York: Simon and Schuster, 1955) eloquently depict outmigration and alienation in the managerial class of the 1930s and 40s.

40. Sugarman, "Community Foundations," pp. 1700–1706.

41. Ibid., p. 1701.

42. The public support test poses major problems for those attempting to identify community foundations. Because the Foundation Center's National Data Base uses IRS files rather than foundation charters, the definition of a community foundation depends on whether or not it has met IRS public charity tests. For this reason, it is difficult to obtain either a simple time-series or community foundation formation or a fully comprehensive list of community foundations. Listings in *The Foundation Directory* compound this problem because they only include large community foundations. For a useful view of the consequences of the public support test, see *The Public Support Test: A Survey of 82 Community Foundations,* Final Report of the Subcommittee on Legislation and Regulation (Washington: Council on Foundations, 1982).

43. Ferdinand Lundberg, *The Rich and the Super Rich* (New York: Lyle Stuart, 1968). Waldemar Nielsen's *The Big Foundations* (New York: Columbia University Press, 1972) also drew extensively on the Patman Committee hearings. For a useful, if highly partisan, overview of the hearings, see F. Emerson Andrews, *Patman and the Foundations* (New York: Foundation Center, 1968).

44. On the renewal of interest in community foundations, see Richard Magat, "Out of the Shadows," *Foundation News* (July–August 1984): p. 24.

45. For a good summary of the new uses of community foundations, especially in brokering and leveraging, see Lois Roisman, "The Community Foundation Connection," *Foundation News* (March–April 1982).

11

Working in Philanthropic Foundations

ELIZABETH T. BORIS

Professional or paid staff are a rarity in American private foundations. Most foundations are small endowment funds created and managed by family members and their trusted friends. Fewer than 2,000 of the 24,800 philanthropic foundations in the United States employ paid staff, and most of the estimated 8,000 staff members work for the 500 largest foundations. Outside of a handful of the big foundations such as the Ford and Rockefeller Foundations, and several operating foundations with more than 100 employees, most staffed foundations have only a few employees. These positions are very desirable, however, and are sought by many who think they would like to make grants for worthy causes.

Little was known about foundation staff members until the first major study of foundation employees, *The Foundation Administrator*, based on survey responses, was published in 1972 (Zurcher and Dustan 1972). Since 1978 the Council on Foundations' biennial surveys have provided an increasingly rich source of information about foundation staffing and management. Among other findings, these surveys have revealed that an increasing number of women have entered foundation work in recent years. Questions about the current status of women within the grant-making field led to this study of foundation employees and their work.

The study examined the nature of foundation work: the daily grant-making tasks, the administrative hierarchy, and the relationships among staff, board, grantees, and the wider community. It also looked at the staff members: their job satisfaction, career patterns, recruitment, and educational backgrounds, as well as gender-related issues, including salary differentials and the impact of the work on the family.

Sixty in-depth personal interviews were conducted with a sample of chief executive officers (CEOs), program officers, and administrative assistants

Adapted with permission from *Working in Foundations: Career Patterns of Women and Men*, by Teresa Jean Odendahl, Elizabeth Boris, and Arlene Kaplan Daniels (New York: The Foundation Center, 1985). The study was sponsored by Women and Foundations—Corporate Philanthropy in cooperation with the Council on Foundations with funding from the Russell Sage Foundation and the John Hay Whitney Foundation.

Table 11.1   Foundations and Staff Members in Interview Sample

	New York City	Midwest	California	Total
Geographical Distribution:				
Foundations	14	16	12	42
Individual Participants	19	22	19	60

	Women	Men	Total
Position by Gender:			
Chief Executive Officer	11	16	27
	34%	67%	45%
Program Officer	15	8	23
	43%	33%	38%
Administrative Assistant	10	0	10
	28%	0%	17%
Total	36	24	60
	60%	40%	100%

from forty-two foundations in New York City, the Midwest, and California, between March and October 1982 (see Table 11.1). Although sixty interviews constitutes a small sample, it was selected randomly from among the 329 staffed foundations that responded to the 1980 Council on Foundations biennial management survey; these represent about one-fifth of staffed foundations in the country. They tend to be the larger, more active foundations, although foundations of all sizes are included in the sample.

Data on board and staff demographics and on salary levels are from the *1982 Compensation and Benefits Report* and *1986 Foundation Management Report* (Boris and Hooper 1982; 1986). The council's management surveys are sent to a relatively stable sample of 1000 council member foundations every two years.

THE IDEOLOGY OF PRIVATE GRANT MAKING

Foundation work takes place in a milieu where doing good, promoting beneficial activities, and trying to develop solutions to societal problems is the normal daily activity. In this world there is a dominant ideology that stresses the independence of private philanthropy from the constraints of business and government and celebrates its unique ability to take risks and creatively address society's needs. Private philanthropy is viewed as an important facet of democratic pluralism and civic involvement. Workers in the philanthropic sector take pride in the many important contributions that private foundations have made during a relatively short history. Foundations are viewed as a precious national resource: society's risk capital, a source of innovation, a symbol of altruism and individual responsibility.

This ideology may create tremendous pressure both for governing boards and for those who work in philanthropy. There is a luster of power and status attached to foundation work that makes it a desirable occupation. But the ability to live up to the ideals of fostering creativity, taking risks, and facilitating innovative solutions to social problems is a difficult charge. Success depends on judgment, timing, and good luck, as well as on skill. It is often difficult to measure success, at least in the short term. Failure, a significant number of unsuccessful grants, is implicit in the assumption that grant making embodies risks. Also, it can be frustrating for many staff members to work toward social and cultural goals indirectly, by providing funding for others, rather than by direct action.

National organizations, including the Council on Foundations and the Independent Sector, have begun to sponsor broad-ranging discussions of the role of philanthropy in society and the management and efficacy of philanthropic institutions. Several such discussions organized by these organizations have centered around questions of values, ethics, and attitudes of those who work in philanthropy. For foundations these issues include the following questions. Can grant making be neutral? Must foundations stand for something? What role does political orientation play? What role should foundations play in public policy, in promoting social or political values? What is ethical grant-making behavior? Does it include openness and accountability to the public? How should we define and avoid conflicts of interest? What obligations do grantors have toward grantees and toward the public? Are there reciprocal obligations of grantees and the public toward grant-making institutions?

The diversity within the grant-making field ensures that there is little agreement on answers to most of these questions. At most, there is a general, though not unanimous, consensus that a Statement of Principles and Practices for Effective Grant Making that the Council on Foundations adopted in 1980 after several years work by a committee of leading grant makers, represents minimum standards for responsible and responsive practices. But, continuing concern and self-consciousness among foundation staff members about appropriate norms is an indication of the increasing professionalism among those who work in this field.

PRIVATE GRANT-MAKING INSTITUTIONS

Individuals included in the study worked in all of the major types of foundations: independent, community, company-sponsored, and operating foundations. The majority, however, worked in private independent foundations, which make up the bulk of the foundations in this country. Community foundations, a small segment of the foundation field, numbering about 300 organizations, are the fastest growing part of the foundation field. Company-sponsored foundations that award grants from funds donated by the parent company are legally independent private foundations, but few are fully endowed. Staff members usually operate within the corporate culture and may have additional company responsibilities. The president and other corporate executives usually serve on the foundation board. There are 781 corporate foundations among the 5148 largest foundations ($100,000 in grants or $1 million in assets) reported in the *Foundation Directory* (Renz and Olson 1987). Operating foundations are private foundations that generally provide direct services. They may operate a museum or a home for the aged, or conduct research; grant making is not their primary activity. There are 675 operating foundations, but only 107 are large enough to be included in the *Foundation Directory*. The staff may include subject matter specialists, professionals who provide services, administrators, and those who manage a grant-making program, if there is one.

THE ADMINISTRATIVE STRUCTURE

Staffed foundations vary greatly in the style of management, ranging from extremely bureaucratic, and sometimes authoritarian, to nearly consensual and democratic. The board of directors (or trustees, or distribution committee) sets the policies and the tone for the foundation.

Members of the foundation's governing board may include the donor, family members or heirs, the family attorney, business associates and friends, corporate officers (in company-sponsored foundations), dignitaries, experts in specialized fields, and representatives of community groups (in community foundations). Although the number of women and minority group members who serve on foundation boards has increased in the past decade, white men still dominate in these positions, 75 percent of which are held by men and 95 percent by whites (Boris and Hooper 1986).

The highest ranking staff position in the foundation usually has the title of president, executive director, or director. In the smaller foundations this chief executive officer (CEO) is often the sole employee, a generalist responsible for all day-to-day operations of the grant-making organization. In larger foundations, the CEO oversees the activities of other staff members who have specialized functions.

Foundation employees, and particularly chief executive officers, often have personal attributes and backgrounds like those of the board members for whom they work. The majority of foundation CEOs are white men and many attended prestigious private colleges and universities. In recent years, however, an increasing number of women and minority group members are becoming foundation CEOs. In 1982, only 26 percent of CEOs in the survey sample were women; in 1986, 35 percent were women. The proportion of minority group CEOs rose from 1 percent to 5 percent. Still only a handful of women have been chosen to manage the largest foundations, those with more than $100 million in assets or $5 million in grants (see Table 11.2).

Table 11.2 Number and Gender of CEOs and Program Officers by Foundation Asset Size

Asset Group (millions)	Women		Men		Total	
	Number	Asset Group (%)	Number	Asset Group (%)	Number	Total (%)
			CEOs			
$100 and over	4	7%	51	91%	56	20%
$25–99.9	24	26%	69	74%	93	34%
$10–24.9	22	41%	32	59%	54	19%
Under $10	40	54%	34	46%	74	27%
Total	90	32%	186	67%	277	100%
			Program Officers			
$100 and over	132	56%	100	42%	236	71%
$25–99.8	48	70%	21	30%	69	21%
$10–24.9	11	73%	4	27%	15	5%
Under $10	8	73%	3	27%	11	3%
Total	199	60%	128	39%	331	100%

SOURCE: Elizabeth T. Boris and Carol Hooper, *1986 Foundation Management Report* (Washington, D.C.: Council on Foundations, 1986).
NOTE: Totals may differ from number in position because each respondent may not have supplied gender information.

Program officers are in the middle of the foundation administrative hierarchy. They review grant applications, investigate applicants' credentials and previous work, summarize proposals, and may develop funding recommendations for the CEO or for the board. Women and minority group members have tended to enter foundation professional ranks through program officer positions. From 1982 to 1986, the proportion of women in these position grew from 51 to 60 percent and the proportion of minority people grew from 13 to 16 percent.

Administrative assistants are responsible for a wide variety of tasks that may range from administering grants or finances to providing routine secretarial services for their bosses. Many serve as office managers, particularly in the smaller foundations where they are responsible for organizing and maintaining the office support systems.

Whether the staff is large or small, its fundamental role is to implement the policies of the board. Trustees are responsible for exercising financial oversight, developing grant-making criteria and overall policies, hiring a CEO, and ultimately, deciding which projects to fund. Some boards seem to rubber-stamp employee recommendations; others are active participants in the daily operations of the foundation. Boards of the smaller family foundations are usually more involved in the day-to-day work of the foundation than are the boards of the larger foundations, where there may be little, if any, donor or family involvement.

CEOs and other staff members serve at the pleasure of the board. It is clear why trustees may wish to choose executives who resemble themselves; they can respect the judgments and find it easier to trust the recommendations of those with whom they share some common perspectives. An increasing number of boards, however, recognize the virtue of drawing on diverse perspectives, and seek staff members who have experience in the nonprofit community.

The dynamics of the board-staff relationship is one of the critical determinants of a foundation's effectiveness. Successful foundation management requires good communication and a delicate balance of power between trustees and employees.

FOUNDATION MODELS

The balance between foundation board and staff roles ranges from situations where the donor makes all decisions to those where employees have almost complete control. On this continuum are what might be called the donor model, the administrator model, the director model, and the presidential model.

In the donor model, the foundation is run by the individual(s) who created it or by the family or trustees. They are volunteers who use only part-time legal or clerical help, if any, to accomplish their purposes. This is usually the pattern in the smaller or newer foundations.

For slightly larger foundations, the administrator model is more common. Here the trustees are clearly dominant in the grant decision process, and they may also be involved in the day-to-day operation of the foundation, but one or two staff members help to manage the paperwork and the finances. Assets are usually limited, and the grants directly reflect the wishes of the donor. The chief staff officer usually does not have much authority and rarely has the title of president or executive director. Many women serve in these positions. Trustees usually discuss all proposals and make all funding decisions.

The CEO has more authority in foundations that fall into the director model. Here, the operating style is generally collegial, with the director or executive director consulting often with trustees about the grant-making program and policies. This model primarily occurs in middle-sized to large foundations that may still have some donor or family participation. Trustees discuss grant proposals recommended by staff, vote on funding, and decide on fiscal and grant program policies. The director often has the authority to investigate potential grant-making opportunities and may play an important role in the community. Women are increasingly finding positions in foundations that operate in this mode.

The presidential model is usually found in large foundations that do not have significant donor or family involvement. Trustees delegate wide authority to the chief staff officer, who normally holds the title of president and is considered a member of the board. The CEO provides leadership to foundation employees and to the board. Trustees set fiscal and program policies, monitor progress, and make decisions only on very large grant proposals. The CEO is usually someone of national stature.

Obviously these are models, and actual foundations mix characteristics from each. The incentives for the board to delegate more authority to the executive of a large foundation where hundreds of grants and millions of dollars are involved each year are understandable. The progression from model to model generally follows a common life cycle of foundations based on maturity, increasing size, and distance from the original donor and family. For example, in family-oriented foundations, even if very large, the donor or trustees may be deeply involved in running the foundation and in maintaining the charitable directions established by the donor. Other similar foundations may be governed by a completely different philosophy and may have a minimum of family or donor involvement.

BOARD-STAFF RELATIONS

Interaction with the board of trustees is a central responsibility mentioned by all of the CEOs whom we interviewed. Most of these executives reported "good working relationships" and many stressed the informal power they have acquired as a result (Odendahl and Boris, 1983a). For example, one president focused on providing leadership for the trustees as follows: "My tasks are to help the board formulate its policies, both about the areas in which we make grants, and the ways in which the foundation will operate. The care and feeding of the board, keeping up its motivation, making the board work interesting and rewarding—satisfying—is an important responsibility." An executive director spends more time paving the way for board decisions:

Before taking proposals to the board I have an idea how each one of them is going to vote on each proposal. If I feel that it is a shaky vote, I will meet with the individual board members where I feel that there may be a problem. I may have them come to meetings of the organization in order to sell them so that when we have the board meeting I can get my vote.

All CEOs in the sample attend board meetings; those who are considered board members have a vote. Presidents of foundations with over $25 million in assets and CEOs of corporate foundations often serve as full members of the board or distribution committee. Few women CEOs have trustee status, in part because many head smaller foundations. Although the actual benefit of having a vote is not clear, CEOs who are board members are widely perceived as having greater authority than their nontrustee counterparts. One CEO explained: "I also have a vote on the board. There are twenty-five foundations that do it, and the best foundations do. There are times when you have to address your trustees not as an employee, but as a fellow trustee. There are all kinds of issues in which it just makes it easier."

Attendance at board meetings is not the rule for other staff members. Over half of the program officers that we interviewed do not attend board meetings unless they are asked to make a special presentation. The roles of those who do attend vary. One program officer reported that he explains the nature of the projects and the rationale for his recommendations but does not take part in any discussions. Another reported, "We do the actual presentation, and we defend, and we interact with the board as the staff responsible for that particular grant." Most program officers who are not permitted to attend board meetings wish that they could and view this restriction as a limitation on their authority.

FOUNDATION WORK

Foundation work involves a curious mix of altruism and arrogance. As one president explained: "The only thing that really counts is doing some good in the world. You're not trying to make a profit. You're not trying to win a championship. You're not trying to do anything but make a difference on the right side. I think that can be a very comforting circumstance, provided that you don't get smug and complacent." But, the danger is always there. As one employee remarked: "People sitting on a big pile of money get very pleased with themselves." Staff members, however, report feeling isolated. They feel constrained in developing friendships with grant seekers, who are often ingratiating because they are competing for scarce resources.

Grant making is the central activity of most of the foundations in our sample. The CEO usually determines who will review applications on any given topic, although in the larger foundations, administrative assistants may route proposals to program officers, who manage specific funding areas and screen out proposals that are clearly outside the foundation's funding areas. In ten of the forty-two organizations in the study, all of the program staff, including the CEO, review every proposal that falls within the foundation guidelines (Odendahl and Boris, 1983b).

Some proposals are rejected after deliberations at staff meetings, though if an idea seems worthwhile, staff may seek additional information from the organization. The next steps depend on the management style of the CEO, the nature of the proposal, and the tradition of the foundation. The proposal may be sent for peer review to specialists with experience in the research or program area. In these cases foundation employees recruit readers and sometimes organize meetings of reviewers to discuss their recommendations.

For some projects, foundation staff members may arrange to make a site visit, or key staff members from the applying organization may be invited for a discussion at the foundation offices. The staff might prepare for such meetings by checking with colleagues at other foundations or with individuals from organizations that work in the same program area.

At the end of the investigation, staff members usually prepare summaries of the proposals and make recommendations that are sent to the board in preparation for the discussion at a board meeting. Exceptions to this pattern are in the largest foundations, where the CEO and top program staff make the final decisions, and in some of the smallest foundations, where the board makes decisions without staff recommendations.

In addition to grant making, "brokering" is a significant activity in some foundations. Employees may provide technical assistance, help grantees raise additional revenues, or promote a particular program area among other public and private grant makers. One program officer at an independent foundation explained:

I see the responsibility to be accessible to the nonprofit community, to provide in-kind or technical assistance to the community-based organizations to the extent that I can, whether or not we can grant to them. And I have a responsibility to be as informed as I can of the characteristics of this community, and mediating be-tween community-based organizations and nonprofits, and foundation-based staff. I think of myself as straddling the institutional setting where I work and the com-munity.

Foundations such as his often deliberately seek to play an entrepreneurial role in identifying needs and seeking good people who need funding. Ac-tivism often leads to joint funding with public or private organization, an increasing phenomenon in the foundation field.

Networking is another aspect of foundation work. Employees are often expected to act as representatives of their grant-making organizations in social activities, meetings with other grant makers or community leaders, and in sessions with national policy makers and the media. Many CEOs and program officers consider networking essential to their work because it provides a necessary framework for brokering and joint funding efforts. A growing group of regional associations of grant makers helps to facili-tate links among grant makers and between grant makers and the wider community. In some settings, club and professional memberships may fa-cilitate networking. Such memberships are likely to be provided to CEOs and top staff of larger foundations.

JOB SATISFACTION

Foundation staff members report working long hours, with the majority of CEOs in the sample working more than fifty hours per week. Yet, most find their work thoroughly enjoyable. CEOs have high status and feel they have almost unlimited access to ideas, information, and people. They also believe they have the opportunity to make a real impact on their program areas. As one CEO said: "The flexibility is vast. The work is interesting. We are able to have some effect. I feel potent. I feel the institution is effective. I enjoy what I think is a generally well-deserved reputation for doing pretty good work."

Program officers also report high job satisfaction, but are not as overwhelmingly pleased as are CEOs. They value the opportunity to meet many different types of people and to deal with ideas. But having less authority and limited career options, they expressed more frustration. As one commented on what she likes least: "Having to say no. Doubling your values and priorities as the basis for saying no. It creates a lot of self-doubt".

Administrative assistants in the sample like their jobs but expect less of them. Most feel fortunate to be associated with people they admire and seem to derive their job satisfaction from the work of the foundation or their bosses rather than from their own activities.

COMPENSATION

One aspect of employee satisfaction is the level of compensation. On the average, salaries of foundation employees compare favorably with those in other occupations. Benefits, except in the smallest foundations, are adequate (see Table 11.3).

Salaries vary greatly within the field, however, depending primarily on the asset base of the organization. For example, the median salary of CEOs from private foundations with $100 million or more in assets is $115,300, compared to $41,500 for CEOs of foundations with less than $10 million (see Table 11.4). Salaries for staff members in community foundations are generally lower than those of private foundations.

Gender is also a major determinant of salary differentials within foundations (see Table 11.5). Statistical analysis of 1982 data shows that even when variables such as asset size, seniority, and education are controlled, salaries of women are significantly lower than those of men (Odendahl, Boris, and Daniels 1985, pp. 70–95). Foundations reflect the larger society in paying women less than men for equivalent work.

Table 11.3 Comparisons of Average Foundation Professional and Academic Salaries to Similar Groups of Workers, 1984–1985

Annual Earnings	Federal Government	Private Industry	Higher Education Faculty	Foundation Professional
$90,000				
		$87,570 (Attorneys)		
$80,000				
				$79,700 (CEOs)
	$70,790 (Engineers)			
$70,000				
$60,000				
	$59,050 (GS-15)			
		$54,340 (Chemists)		
				$53,195 (Other Professionals)

Table 11.3 *Continued*

Annual Earnings	Federal Government	Private Industry	Higher Education Faculty	Foundation Professional
$50,000				
				$45,300 (Program Officers)
	$41,960 (GS-13)			
$40,000				
		$37,840 (Buyers)	$37,400 (Full Profs.)	
		$35,720 (Accountants)		
		$34,880 (Job Analysts)		
$30,000				
			$29,130 (All Ranks)	
	$27,760 (GS-11)			
			$28,220 (Associate Profs.)	
			$23,210 (Assistant Profs.)	
$20,000				
			$18,660 (Instructors)	

SOURCES: Data from American Association of University Professors, 1984–85; Boris and Hooper 1986.

Table 11.4 Chief Executive Officers' Salaries

Assets ($ millions)	Number in Position	Number of Foundations	Salary ($ thousands)			
			Low	High	Mean	Median
Independent, Corporate, Private Operating Foundations:						
$100 and over	50	49	$60.0	$225.0	$128.2	$115.3
$25–99.9	67	66	$29.0	$165.0	$75.6	$70.0
$10–24.9	36	36	$20.0	$101.0	$50.3	$51.3
Under $10	27	27	$13.5	$65.0	$39.1	$41.5
All Asset Groups	180	178	$13.5	$225.0	$79.7	$66.8
Community and Other Foundations:						
$100 and over	6	6	$105.0	$153.6	$130.8	$129.3
$25–99.9	16	16	$42.5	$95.0	$59.9	$59.5
$10–24.9	7	7	$27.0	$55.0	$41.7	$40.0
Under $10	47	47	$11.0	$100.0	$40.8	$37.4
All Asset Groups	76	76	$11.0	$153.6	$52.0	$44.8

SOURCE: Data from Elizabeth T. Boris and Carol Hooper, *1986 Foundation Management Report* (Washington, D.C.: Council on Foundations, 1986).

Table 11.5 Average Salaries (in $ thousands) for Men and Women by Position and Foundation Size (1986 data)

Position	Women	Men	Ratio (men/women)
Assets of $100 Million or More:			
Chief Executive Officer	$110.0	$130.5	0.84
Associate Director	58.2	89.4	0.65
Vice President (General)	66.1	100.7	0.66
Vice President (Program)	93.2	93.4	1.00
Vice President (Administration)	71.9	90.7	0.79
Program Director	60.1	81.9	0.73
Program Officer	44.6	54.0	0.83
Program Associate	31.6	32.1	0.98
Other Executives or Professionals	39.1	53.0	0.74
Support Staff	21.5	24.7	0.87
Assets of $25–99 Million:			
Chief Executive Officer	61.2	77.3	0.79
Associate Director	42.3	48.5	0.87
Vice President (General)	56.0	70.0	0.8
Vice President (Program)	45.2	66.6	0.68
Vice President (Administration)	44.5	52.2	0.85
Program Director	39.4	47.2	0.83
Program Officer	34.2	40.4	0.85
Program Associate	28.0	33.1	0.85
Other Executive or Professionals	35.0	38.6	0.91
Support Staff	19.2	18.9	1.01
Assets of $10–24.9 Million:			
Chief Executive Officer	41.7	55.8	0.75
Associate Director	34.3	28.5	1.2
Program Officer	24.3	28.1	0.86
Other Executives or Professionals	39.5	34.9	1.1
Support Staff	16.5	12.0	1.4
Assets of Under $10 Million:			
Chief Executive Officer	36.0	41.1	0.8
Associate Director	26.4	26.3	1.01
Vice President (General)	—	35.6	—
Program Officer	22.8	28.0	0.82
Other Executive/Professionals	25.0	48.0	0.52
Support Staff	16.5	—	—

SOURCE: Data from Elizabeth T. Boris and Carol Hooper, *1986 Foundation Management Survey* (Washington, D.C.: Council on Foundations, 1986).

RECRUITMENT

High-level grant-making positions are rarely advertised but are acquired, as one CEO reported, "through a version of the old boy network." The most prestigious candidates are tapped; they do not need to apply. Over half of the staff members we interviewed were recruited for their founda-

Table 11.6   Direct Application by Gender (Interview Sample)

Application Source	Women	Men	Total
Advertisement or Agency	4	0	4
	16%	0	8%
Letter or Personal Visit	7	1	8
	27%	4%	16%
Network Contacts	5	2	7
	19%	9%	14%
Recruited	10	20	30
	39%	87%	61%
Total	26	23	49
	53%	47%	100%

NOTE: Administrative assistants are excluded.

tion positions without initiating a job search. Foundation employees who actively sought grant-making work were also assisted by personal contacts. Older employees were more likely to have been recruited; younger employees, often women, were more likely to have applied for their jobs.

Gender is a major explanatory variable that accounts for differences between those recruited for a grant-making job and those applying directly. When administrative assistants are excluded, almost 87 percent (all but three) of the men in the sample were invited to apply for their jobs by a member of the board. In contrast, 39 percent (ten) of the women were recruited for their positions (see Table 11.6). All but one of the twenty-two men in the sample were either referred to the foundation or learned of an opening through a network, whereas 43 percent (eleven) of the women used other methods. Women actively pursued foundation employment and then attempted to advance within the field, whereas men were often brought into the top positions at large foundations after service in other career areas.

EDUCATION

Despite the stated preference of CEOs for staff members who are bright, articulate generalists with the breadth of a classical liberal arts education, an advanced degree is an implicit requirement for many foundation positions. Almost all of the foundation chief executives and program officers in the sample are college educated, and over 65 percent have earned advanced degrees. Men in the interview sample are somewhat more highly educated than the women. Only two women have Ph.D.s, compared with eight men.

The highest percentage (39 percent) of participants had degrees in the social sciences, followed by the arts and humanities (25 percent). The possession of an advanced degree is significantly correlated with higher salaries among CEOs in the 1982 compensation survey data. Of the twenty-two CEOs with the highest salaries in that survey, fifteen (68 percent) had JD, MD, or Ph.D. degrees, and over three-quarters worked for the largest foundations.

CAREER PATHS

The foundation field is a relatively new occupational area that people enter from a variety of backgrounds. However, several career paths could be discerned among those interviewed. The first grows out of building opportunities, particularly for women. In small, less-structured foundations, status lines are not strong. By knowing someone, a person can enter the field and create a job or work up to a more responsible position.

Volunteers form a second career path. Men in this mold have been social activists and committed Peace Corps or Vista workers who later enter foundation employment. Women volunteers were more likely to have been grantees who came to the attention of the foundation through work they did in the community. The third career path is that of a distinguished academic or national leader who is recruited from a prestigious position to lead one of the largest foundations. These are almost exclusively men, often close to retirement. The newest career path is found primarily among women who choose philanthropy as a profession and actively seek to rise to positions of leadership.

Chief executive officers are older, thus tend to have had more employment experience than other foundation staff members. Most foundation employees in the study have had four or five prior jobs and changed their occupational areas at least once before entering philanthropy. Men were more likely than the women to have made a major career change. All of the men interviewed said they had changed career areas at least once, and over half made two or more changes. More than half of the women had made one career change but only eight women (23.5 percent) had made two or more career changes. A fifth of the women had always worked in philanthropy, none of the men reflected this pattern. This phenomenon appears to have arisen in the 1970s.

Participants worked in foundations from one to thirty years, with an average of eight years, yet half worked for their current employer for five years or less. On average each had held two jobs within the same organization. There appears to be a fairly high level of turnover in foundation work, which may reflect both institutional policy and the conventional wisdom of the field, that one should not stay in grantmaking for more than five to ten years.

When combined with the divergent interests of donors, the diversity of career paths among staff helps to explain the wide variety of operating styles and grantmaking programs of U.S. foundations. The influx of women careerists, however, is probably a major force in the field's growing sense of professionalism.

WORK AND FAMILY LIFE

The increasing numbers of women who work in foundations, 70 percent of staff members in the latest survey, indicate that major changes have occurred in foundation employment during the past decade. Foundations provide a congenial setting for a woman generalist to develop a satisfying career, but women in foundations report the same frustrations as women in other fields in trying to combine careers and family life. Child care is a major problem. Striking differences in marital status of interviewees illustrate this point. All but one of the men we interviewed were currently married. Over half of the women were not currently married. Women are clearly making trade-offs involving their personal lives in order to pursue a career. As in the larger society, women are expected to shoulder their dual roles with little institutional support. One CEO explained:

I've witnessed with so many of my other friends, the women, that dual role that they have is very schizophrenic and so tension-producing. The need to be always competent, and on top of it, and working as hard as possible, and sometimes that translates into hours. There's a certain competitiveness within every office. I think any of these (foundation) jobs could expand to fill eighty hours a week if you wanted them to. A lot of the men have more opportunity to let it expand, because they have fewer responsibilities at home. And the women, I think there were a lot of pressures, and we just bear a lot more stress and tension. I think of my saying, "Oh, God, it's six-thirty, I can't stay any longer. I've got to go home." That sort of thing. And then putting on an entirely different hat at home.

However, because of their flexibility and resources, foundations should be capable of devising innovative work arrangements that could become models within the labor force.

Foundation work, for all the strain of meeting career and home responsibility, has the potential to offer a real niche of opportunity for women struggling with these problems.

WOMEN IN THE HIERARCHY

Although foundations have proved to be a source of opportunity for women in the last decade, compared to men in the same positions, women are not as well paid and, overall, they have lower status and less authority than men. Few women lead the largest foundations; most do not serve as members of the board, or have discretionary grant funds. As CEOs they are found mainly in the small to mid-sized foundations, where they are less likely to have the title of president or the perquisites that enable them to network in clubs and professional associations. While many have advanced degrees, few have doctoral degrees. The fact that more women are entering the field without referrals and that they are coming from positions in the nonprofit sector reflects greater openness among foundations in recent years. The fact that women command lower salaries than men and that their numbers are rapidly increasing, leads some to fear that there may be an incipient trend toward feminization of the field.

THE PARADOXES OF FOUNDATION EMPLOYMENT

Foundations provide their employees with interesting work. Paid grant-makers have the opportunity to learn new things, meet new people, and think about new ideas on the job. They usually have pleasant working conditions in an atmosphere where collegiality, informality, and flexibility are prized. It is not surprising, then, that the people we spoke to expressed satisfaction with their jobs.

One problem for the professional in this field is the realization that there is probably a limited future for careerists in the philanthropic world. First, the career ladder in philanthropy is short. Once one has become the exec-utive of a small private foundation, it is hard to find an equivalent position elsewhere. The best one can hope for is a parallel move to a similar or a slightly larger foundation. Opportunities are shrinking for movement to government, universities, and other nonprofit organizations, especially in times of policy changes and economic belt-tightening, such as those expe-rienced in recent years.

At the end of the career ladder the question becomes, What next? Some people are so content with their work that this question does not arise. They find renewal in the variety of projects and in sharing their experi-ences with newcomers to the foundation field or with grantees. For many others, with fewer institutional resources or less flexibility, horizons seem limited and discontent is evident. How does one reconcile staying in an interesting, well-paid position that has little growth potential?

Networking and the variety of programmatic and professional organi-zations that have sprung up in the last ten years may provide a partial answer. Conferences and seminars that enable foundation staff members to enrich the professional aspects of their work provide growth possibili-ties. One of the most significant developments is the considerable growth of organizations that serve the administrative, programmatic, and network aspects of foundation work.

Belonging to organizations also serves the initiation needs of newcomers to the field. In this study, for example, we noted that lower-level founda-tion staff were eager to participate in networks. But it was not clear that such participation would offer them any real advantages for career mobil-ity, although they might find them personally rewarding and interesting. Some individuals, however, might enjoy career benefits from the increased visibility that networking affords.

Grantmakers have been greatly concerned with efforts to build profes-sional organization and capability into their work. This tendency may in the future undermine the concept, common in the field, that foundation staff are generalists, that one can come to this work from many different kinds of disciplinary and occupational backgrounds. Up to now, grant-making employees have benefited from this generalist orientation. Both women and men often report that they came to their work in the foundation by accident, adapted to it very well, and developed expertise and commitment through their experience.

Pressure to hire a staff in order to professionalize giving by smaller foundations leads to hiring individuals, usually women, willing to accept the modest salaries they can afford to pay. These women may then gain experience and go on to better-paying positions in other foundations. However, our study suggests that both horizontal and vertical mobility require participation in a special kind of network, in which a background from elite schools and the possession of advanced degrees are important for entry.

Here lies another paradox of foundation work: that the aim for many is social reform, help for the disadvantaged, or the provision of opportunity so that society may become more egalitarian. Yet, the people who provide services are expected to be highly educated and are often from elite schools and backgrounds. Foundations devoted to social equality and solving the problems of society reflect the dominant patterns. The irony is that the field prides itself in path breaking, but has not been breaking any paths at home. Foundations generally have not been innovators in child care, job sharing, or job enrichment programs for women and minority staff members.

Throughout our discussion of the work in philanthropy, we have stressed the paradoxes that the employees themselves have called to our attention. Perhaps the most clear paradox in their eyes is that they have some of the power of philanthropists but are not themselves rich and powerful donors. A somewhat less stressed theme is that if foundation workers, particularly CEOs, have nominal power to give or withhold grants, it is also true that boards of directors (their employers) have authority over them. In short, the paradox for the executive is that he or she may appear arrogant and intractable to grant seekers and yet have to appear humble or at least tractable to board members. The executive may assume an extremely liberal or reform-oriented posture toward candidates and yet need to appear conservative and very deliberate or cautious before a board.

This study of foundation work clarifies several issues beyond the advantages and limitations that await those who chance to find foundation jobs: the shifting, sensitive, and ambiguous relation of philanthropy to the local community as well as to society at large; the delicate balance between the perspectives of the professional staff and board; the maintenance of more or less hierarchical relations within the staff structure.

REFERENCES

American Association of University Professors. 1985. "Report on the Economic Status of the Profession 1984–85." *Academe* 71, no. 2.

———. 1983. "The Annual Report on the Economic State of the Profession 1982–83." *Academe* 69, no. 4.

Boris, Elizabeth T., and Carol Hooper. 1986. *1986 Foundation Management Report.* Washington, D.C.: Council on Foundations.

———. 1984. *1984 Foundation Management Report.* Washington, D.C.: Council on Foundations.

———. 1982. *Compensation and Benefits Report.* Washington, D.C.: Council on Foundations.

Boris, Elizabeth T., and Patricia A. Unkle. 1981. *1980 Compensation Survey.* Washington, D.C.: Council on Foundations.

Boris, Elizabeth T., Patricia A. Unkle, and Carol Hooper. 1981. *1980 Trustee Report.* Washington, D.C.: Council on Foundations.

Bucher, Rue, and Joan G. Stelling. 1977. *Becoming Professional.* Beverly Hills, Calif.: Sage, pp. 20–37.

Charlton, Joy C. 1983. *Secretaries and Bosses: The Social Organization of Office Work.* Doctoral dissertation, Northwestern University.

Collins, Randall. 1975. *Conflict Sociology: Toward an Explanatory Science.* New York: Academic Press.

Coser, Rose Laub. 1982. "Stay Home, Little Sheba: On Placement, Displacement and Social Change." In R. Kahn-Hut, A. K. Daniels, and R. Colvard, eds., In *Women and Work Problems and Perspectives.* New York: Oxford University Press.

Dykstra, Gretchen. 1979. *Survey of Community Foundations.* New York: Women and Foundations/Corporate Philanthropy.

Garson, Barbara. 1975. *All the Livelong Day.* Kingsport, Tenn.: Kingsport Press.

Hertz, Rosanna. 1983. *Dual Career Couples in the Corporate World.* Doctoral dissertation, Northwestern University.

Hughes, Everett C. 1958. *Men and Their Work.* Glencoe, Ill.: Free Press.

Kanter, Rosabeth Moss. 1977a. *Men and Women of the Corporation.* New York: Basic Books.

———. 1977b. *Work and Family in the United States: A Critical Review and Agenda for Policy and Research.* New York: Russell Sage Foundation.

Lofland, J. 1971. *Analyzing Social Settings: A Guide to Qualitative Observation and Analysis.* Belmont, Calif.: Wadsworth.

Margolis, Diane Rothbard. 1979. *The Managers Corporate Life in America.* New York: Morrow.

Marting, Leeda. 1976. "What We Now Know . . . What We Need to Do." New York: Women and Foundations/Corporate Philanthropy. (unpublished report)

McPherson, J. Miller, and Lynn Smith-Lovin. 1982. "Women and Weak Ties: Differences by Sex in the Size of Voluntary Organizations." *American Journal of Sociology* 87.

Merton, Robert K., Marjorie Fiske, and Patricia Kendall. 1956. *The Focused Interview.* Glencoe, Ill.: Free Press.

Nason, John W. 1977. *Trustees and the Future of Foundations.* New York: Council on Foundations.

Nie, Norman H., C. Hadlai Hull, Jean G. Jenkins, Karin Steinbrenner, and Dale H. Bent. 1975. *Statistical Package For the Social Sciences.* New York: McGraw-Hill.

Odendahl, Teresa J., and Elizabeth Boris. 1983. "A Delicate Balance: Foundation Board-Staff Relations." *Foundation News* 24. Washington, D.C.: Council on Foundations.

———. "The Grantmaking Process." *Foundation News,* 24. Washington, D.C.: Council on Foundations.

Odendahl, Teresa J., Elizabeth T. Boris, and Arlene K. Daniels. 1985. *Working in Foundations: Career Patterns of Women and Men.* New York: The Foundation Center.

Odendahl, Teresa, Phyllis Palmer, and Ronnie Ratner. 1980. "Comparable Worth: Research Issues and Methods." In Joy Ann Grune, ed. *Manual on Pay Equity.* Washington, D.C.: Committee on Pay Equity.

Ostrander, Susan A. 1984. *Women of the Upper Class.* Philadelphia: Temple University Press.

Renz, Loren, and Olson, Stanley, eds. 1987. *Foundation Directory*, 11th ed. New York: The Foundation Center.

Rosenberg, Rosalind. 1982. *Beyond Separate Spheres*. New Haven: Yale University Press.

Rossiter, Margaret W. *Women Scientists in America: Struggles and Strategies to 1940*. Baltimore: Johns Hopkins University Press.

Selden, Catherine, Ellen Mutare, Mary Rubin, and Karen Sacks. 1982. *Equal Pay for Work of Comparable Worth: An Annotated Bibliography*. Chicago: American Library Association.

Simmel, Georg. 1955. *Conflict and the Web of Group-Affiliations*. New York: Free Press.

Stouffer, S.A., et al. 1949. *The American Soldier: Adjustment During Army Life*. Vol. 1. Princeton, N.J.: Princeton University Press.

Treimann, Donald J. and Heidi I. Hartmann, eds. 1981. *Women, Work, and Wages: Equal Pay for Jobs of Equal Value*. Washington, D.C.: National Academy Press.

U.S. Department of Commerce, Bureau of the Census, 1983. *Money Income of Households, Families, and Persons in the U.S.* Series P-60. Washington, D.C.: Government Printing Office.

U.S. Department of Labor, Bureau of Labor Statistics. 1983. *Employment and Earnings*. Vol. 30. Washington, D.C.: Government Printing Office.

Wilensky, H.C. 1964. "The Professionalization of Everyone?" *American Journal of Sociology* 70; no. 2: 137–58.

Yankelovich, Skelly and White, Inc. 1982. *Corporate Giving: The Views of Chief Executive Officers of Major American Corporations*. Washington, D.C.: Council on Foundations.

Zurcher, Arnold J., and Jane Dustan. 1972. *The Foundation Administrator: A Study of Those Who Manage America's Foundations*. New York: Russell Sage Foundation.

V

CORPORATE GIVING: MOTIVATION AND MANAGEMENT

12

Business Giving and Social Investment in the United States

PETER DOBKIN HALL

This essay examines the eleemosynary role of business in the United States, delineating its unique contributions to philanthropy and outlining some of the complex ties between for-profit and nonprofit activity. It concludes by identifying some of the core values that underlie both charity and commerce in the American setting.

BUSINESS AND CHARITY IN EARLY AMERICA, 1780–1860

Although corporations of various kinds—townships, congregations, schools, and colleges—were familiar to colonial Americans, virtually all of them were publicly supported entities. Not until after the Revolution did the corporation come to be commonly used as a device for organizing private commercial and charitable activity.[1] Simultaneously, the older corporations were privatized. In Massachusetts, for example, lay control of Harvard's governing boards (1780), the establishment of the Massachusetts Medical Society (1780), the Boston Atheneum (1870), Massachusetts General Hospital (1810), and a host of other eleemosynary enterprises coincided with the chartering of the state's first business corporations: the Massachusetts Bank (1784), the Charles River and Essex Bridge companies (1784, 1787), and a flotilla of other insurance, transportation, and banking enterprises.

In these early corporate ventures, for-profit and nonprofit activity overlapped significantly. The same individuals appeared in both types of organizations as officers, directors, donors, and stockholders with a frequency

The research on which this paper is based was funded by grants from the American Council of Learned Societies, the AT&T Foundation, the Ellis Phillips Foundation, the Exxon Education Foundation, the General Electric Foundation, the Equitable Life Assurance Society of the United States, the Teagle Foundation, and the Program on Non-Profit Organizations, Yale University.

that suggests an important degree of interdependence. Because capital markets were primitive in the early republic, charitable endowments served as
important devices for pooling financial resources for investment in commercial and industrial ventures. By the 1840s, the Massachusetts Hospital
Life Insurance Company, a nonprofit corporation set up to generate income for the Massachusetts General Hospital through the sale of life insurance and the management of endowment and testamentary trusts, was
the major source of capital for New England's growing textile mills and
railroads.[2]

The close ties between business and charity involved more than direct
financial interdependence. Service on charitable boards helped merchants
and manufacturers modulate the centrifugal forces generated by competition in the marketplace. This enabled them to deal with matters of common concern, such as education and public health, on which government
was unwilling to act. Further, private control of nonprofit organizations
counterpoised the rising tides of popular democracy, permitting the "wealthy,
learned, and respectable" to influence the masses over whom they no longer
had political power.[3]

The westward movement of population was spearheaded by New Englanders.[4] Wherever they settled, merchants and manufacturers took the
lead in establishing and supporting academies, colleges, churches, libraries,
hospitals, lyceums, temperance organizations, and Bible societies.[5] These
enterprises brought order and opportunity to the frontier and maintained
common values in a population that was increasingly dispersed and het-
erogeneous.

Charity was only one dimension of business philanthropy in the Antebellum period. Businessmen also used their companies to provide medical,
social, and cultural services.[6] To some extent this was an expansion of the
traditional relationship between masters and their journeymen and apprentices. But, even major industrial innovators like Eli Whitney, who funda-
mentally changed the labor process itself, recognized that his obligations
to his workers—and to society—extended far beyond the cash nexus. In
addition to imparting skills to his workers, Whitney taught them to read
and write and provided them with housing, health-care, and moral oversight. Other pioneer industrialists followed his lead in creating factory villages, complete with schools, churches, and housing.[7]

Although these new industrial relationships were paternalistic, they were
no more so than those prevailing in schools, colleges, or other institutions.
Moreover, unlike the relationship of apprentices and journeymen to their
masters, employees were not legally bound to their employers. Free to come
and go as they pleased, taking with them valuable skills learned from their
employers, the early industrial workers were economically and politically
empowered by the factory setting.[8]

Investing in health, welfare, and education for workers was expensive.
But, in the labor-short economy of the early nineteenth century, it was
regarded as part of the cost of doing business. By the 1850s, however,

many employers began to curtail these social investments. With the spread of mechanical literacy, it was no longer necessary to educate the work force. Increasing European immigration lowered the cost of labor. And intensifying competition encouraged industrialists to reduce costs by simplifying production, so that most tasks could be performed by cheap unskilled labor. In addition, as workers became more politically active, they resisted industrial paternalism, preferring to obtain social services through mutual benefit societies, ranging from savings banks and schools through burial societies, and through government.

BUSINESS AND CHARITY IN THE AGE OF ENTERPRISE, 1865–1900

The turbulence of the decades following the Civil War presented businessmen immediate and compelling challenges. Economic instability—deflation, panics, securities manipulation, and cutthroat competition—determined the basic conditions of daily decision making. The growth of cities raised issues of political corruption, regulation, and access to transportation and municipal services affected profits. Immigration affected the character of markets and cost of labor. Poverty and disorder tended to translate themselves into trade union activity.

If business leaders turned toward public life out of self-interest, their activism was welcomed by the public. Fueled by such events as the attempted impeachment of President Andrew Johnson, the corruption of the Grant administration, the stolen presidential election of 1876, and revelations of pervasive municipal corruption and mismanagement, Americans grew increasingly distrustful of government and politicians. They did not, however, turn to nonprofit organizations for solutions—for neither the universities nor the churches had, up to this point, begun to address themselves to public policy issues. Instead, they turned to leading businessmen—and business methods—to restore honesty and order to American institutions.[9]

In spite of the public's expectations, business had no master plan to solve the nation's problems. It took up the challenge, but did so in ways as diverse as the scope and scale of the thousands of private enterprises operating across the country and as varied as the imagination, ambition, and resources of those who ran them. Some businessmen believed that the solution to the problems of community and nation lay within their firms. These, the pioneers of "scientific management," concentrated their energies on technologies and accounting systems that would make production more efficient and wage rates more equitable.[10] Others, the founders of "welfare capitalism," concentrated on delivering education, welfare, and health-care services to their employees, using their firms as instruments of social reform.[11] Many, whom we could call the founders of corporate liberalism, donated generously to promoting social services through both traditional nonprofit organizations like the churches, and innovative ones such as settlement houses and federated charities.[12] A handful understood that the nation's problems required national solutions. They became active in national reform politics, in organizing such policy-oriented enterprises as the American Social Science Association and the Civic Federation and in funding the birth of social science disciplines in the universities.[13] Not until the end of the century would this "methodless enthusiasm" begin to coalesce into unified reform strategy.

If there was a single common element underlying the various forms of business social activism in this period, it was an awareness of the importance of education. Though American educators had been preaching about the importance of the higher learning for decades, businessmen did not begin to pay attention to them until after the Civil War. Until then, most colleges had featured classical curricula that seemed largely irrelevant to the needs of business. Without curricula including the physical and social sciences, modern languages, and history; without capacities for research and postgraduate training; and without an increased business presence on governing boards, higher education was unlikely to attract significant support from business.

By the late 1860s, it was clear that education was too important a matter to leave in the hands of the educators. Businessmen in New York and Boston began pushing for change at institutions like Harvard, Yale, and Columbia. Their goals were summed up in a series of articles entitled "The New Education," which appeared in the *Atlantic Monthly* during the spring and summer of 1869. Though their author, Charles W. Eliot, came from a prominent Boston business family, he had taught at Harvard and had spent three years in Europe studying the ties between education and economic development.

"The New Education" was a devastating critique of American colleges and universities. "The American people are fighting the wilderness, physical and moral on the one hand," Eliot wrote, "and on the other are trying to work out the awful problem of self-government. For this fight they must be trained and armed."[14] The struggle required a new kind of fighter: not the ingenious Yankee who could run his hand to anything, but the trained expert, who combined specialized knowledge with administrative skills. "What the country needs," Eliot continued,

is a steady supply of men well trained in recognized principles of science and art, and well informed about established practice. We need engineers who thoroughly understand what is already known at home and abroad about mining, road and bridge building, railways, canals, water power, and steam machinery; architects who have fully studied their art; builders who can at least construct buildings that will not fall down; chemists and metallurgists who know what the world has done and is doing in the chemical arts, and in the extraction and working of metals; manufacturers who appreciate what science and technical skill can do for the works they superintend.[15]

Top-heavy with ministers, the colleges had generally resisted such vocationalism. But even where they had accepted endowments from industrialists to establish "scientific schools," these schools were at a decided disadvantage. "The foundling," Eliot wrote,

has suffered by comparison with the children of the house. Even where there had been no jealousies about money or influence, and no jarrings about theological tendencies or religious temper, the faculty and students of the scientific schools have necessarily felt themselves in an inferior position to the college proper as regards property, numbers, and the confidence of the community. They have been in a defensive attitude. It is the story of the ugly duckling.[16]

But this inferior position was not unwarranted. The schools had neither credible standards for admission nor coherent curricula. Harvard's Lawrence Scientific School was, he charged, no more than "a group of independent professorships, each with its own treasury and methods of operation." There was no connection between fields of study. Because of this, there were no accepted degree requirements. A student could study a single subject, chemistry or engineering, be "densely ignorant of everything else," and yet qualify for a degree. The result was a profoundly deficient educational product: "it is quite possible for a young man to become a Bachelor of Science without a sound knowledge of any language, not even his own, and without any knowledge at all of philosophy, history, political science, or of any natural or physical science, except the single one to which he had devoted two or three years at the most."[17]

It was not enough, in Eliot's view, to create science curricula that paralleled the colleges' traditional classical offerings. The colleges needed to entirely reformulate their goals in terms of national needs. Once a classical education or "the seven league boots of genius" were sufficient to prepare men for the bench, the bar, and the counting house. But the increasing scope, scale, and complexity of enterprise and government now required something more. "Only after years of the bitterest experience," Eliot asserted,

did we come to believe the professional training of a soldier to be of value in war. This lack of faith in the prophecy of a natural bent, and in the value of discipline concentrated upon a single subject amounts to a national danger.[18]

The principle of divided and subordinate responsibilities, which rules in government bureaus, in manufactures, and in all great companies, which makes the modern army a possibility, must be applied in the university.[19]

Unlike earlier reformers, Eliot did not propose to turn the colleges into vocational schools, for the understood that bureaucratized business and government would need more than men with specialized skills.[20] Under the conditions of rapid growth and change that he foresaw, specialists would have to be both flexible and personally committed. They would need both broad knowledge and values that tied personal fulfillment to specialized professional achievement.

On the undergraduate level, Eliot called for a broad course offerings in which the social and physical sciences, mathematics, modern languages, and history would standing on equal footing with the classics. Students would be free to "elect" whatever course they wished, a process enabling them to determine where their real talents and interests lay. This experience would provide the basis for pursuing postgraduate studies and subsequent careers with a genuine sense of calling. He further proposed major reforms in the graduate and professional schools, shifting their focus from mere vocationalism to a research orientation. This would not only tie these programs to the real needs of government and business but would enable students to participate in the pursuit of genuinely useful knowledge.

Eliot's proposals were not only more radical than any previous efforts to reform higher education, but more important, they represented more than an educational viewpoint. Evolved from years of discussion with some of the most influential leaders of Boston's business community, neither the time nor the place of his proposal's appearance was accidental. Harvard's governing boards, of which Eliot's patrons were key members, were meeting to select a new president. The *Atlantic Monthly* was the city's most important periodical. Within months of the publication of "The New Education," young Eliot, who had six years earlier been denied tenure at Harvard, became its president. Massive financial and political backing from business permitted him to implement the most thoroughgoing curricular and pedagogical reform ever undertaken by an American university.[21]

Business-spurred education reform was not welcomed everywhere. When Yale's business alumni attempted to take control of the college from the Connecticut ministry in the late 1870s, they were stoutly resisted.[22] The Yale Corporation elected Reverend Noah Porter, an outspoken opponent of Eliot's reforms, as president. The only concession "Young Yale" could wring from its enemies was an agreement to permit laymen to sit on the sixteen-member governing board. Yale's indifference to business was answered by business indifference to Yale. The college's efforts to raise funds failed without business support.[23] The nation's wealthiest and most influential families began to send their sons elsewhere. Only in 1890, with the election of Arthur Twining Hadley, a layman noted for his work on railroad economics and his sympathy to education reform, did Yale begin to change.

In the end, resistance to business and the reforms it favored was futile. Institutions could not remain immured in the classics, in recitations, and in prescribed studies and also remain in the front ranks of American education. The nationalization of American life in the late nineteenth century subjected education to the same market forces that transformed business corporations and political parties. While business shaped the priorities of education reform and underwrote its implementation, the universities began to supply the educated managers, the technical experts, and the social, economic, and technological intelligence that enabled corporate business to become the dominant actor in American life after 1900.

Although educational investment was only one of many varieties of business philanthropy, it was the catalyst that brought the others together into a comprehensive reformation of political, cultural, and economic life. As the graduates of the new universities fanned out into the professions, business, and public administration, they brought with them not only their expertise, but a new conception of their place in the American polity. These were not "specialists without spirit or sensualists without heart": these were men with a calling who viewed their specialized expertise as a way of serving the nation and humanity.[24] Just as the Protestant Reformation led people to embrace the world, pursuing their religious duty in crafts and commerce rather than the church, so the products of the New Education first found their callings in the private sector rather than in government. This would lay the foundation for the eventual reform of public life.

By concentrating on training graduates for business and the professions, the reformed universities were not withdrawing from public life. Rather, they were proceeding on an understanding of the interdependence of public and private sectors. Thus, when reformers like Louis Brandeis urged graduates to enter business, calling it a field "rich in opportunity for the exercise of man's finest and most varied mental faculties and moral qualities," they were only affirming their belief that political reform was rooted in the reform of private action.[25]

In the last decade of the nineteenth century, the educational investments of a few businessmen began to unify the diverse strands of earlier business reform efforts. The growing scope and scale of American enterprise made advanced technology and managerial efficiency increasingly important determinants of corporate success. Because these required more skilled and better motivated workers, they were inseparable from changes in the work place and in industrial communities sought by other business reformers.[26] With the new century it began to be clear that education was the key not only to making American business more productive and competitive but also to reconciling the unavoidable inequalities of advanced capitalism with traditional democratic values. In this setting, universalizing educational opportunity was not a disinterested act; it was a practical requirement of both corporate success and political stability.

THE EMERGENCE OF CORPORATE PHILANTHROPY, 1900–1935

Until the turn of the century, the individual businessman, not his firm, was the major corporate charitable actor. This was in line with the proprietary management style that characterized most American companies. But as the titans of post-Civil War industry began to retire from business, their places were taken by professional managers. This separation of ownership and control led to major changes in business philanthropy. When owner-managers like Carnegie, Vanderbilt, or Pullman diverted corporate assets for charitable purposes, they were accountable to no one because they were, in effect, giving away their own money. Usually modest gifts from their firms were accompanied by generous ones from their private fortunes. The professional managers enjoyed no such discretion. They had no private fortunes. And, because they were dealing with other people's money, they were not free to make corporate gifts unless these could be justified on grounds of corporate well-being.

This constraint posed particular difficulties for professional managers at a time when political and economic pressures required a strong corporate presence. On the one hand, public clamor against the avarice of the corporations grew through the 1890s. In 1896, the Democratic party was captured by a populist faction that called for government control of banking, transportation, and communications. The attack on corporations spread from radical newspapers to mainstream periodicals, as muckraking journalists revealed the immense power and frequently corrupting influence of big business. Theodore Roosevelt, a president closely tied to big business, initiated active antitrust and other regulatory policies.

On the other hand, business faced the challenge of helping the communities in which they operated to deal with inefficient and corrupt government, inadequate health care, welfare, and educational facilities, and growing illiterate and unskilled immigrant populations. As managers tried to deal with these problems, they were greeted by a rising tide of litigation challenging the extent to which philanthropy lay within the powers of business corporations. The earliest cases, involving the efforts of corporations to provide schools, churches, hospitals, and libraries for employees living in company towns, upheld the power of management to engage in charity. With the case of *Dodge v. Ford* in 1919, litigation took a more serious turn.[27] Rather than considering whether specific charitable acts lay within the powers of corporations, this case considered the general question of the corporation as a charitable actor.

The Ford Motor Company had been noted not only for its liberal wage and personnel policies, but also for its proprietor's radical ideas about political economy. Ford was among the first to understand the possibilities of a mass consumer market as both a source of profits and as a source of social reform.[28] For him, the low-priced automobile was more than a means of transportation: it was a means of socially, economically, and culturally empowering the Americans, particularly rural Americans. In pursuit of this agenda, Ford followed a policy of reinvesting a substantial proportion of his firm's earnings in research, creating new product lines, in marketing, and in his work force.

Had the company been privately held, Ford could have proceeded with a free hand. Unfortunately, this was not the case. Among his shareholders were John and Horace Dodge, who were, in 1919, launching an automobile company of their own. Though their motive in taking Ford to court was self-interest—they wanted to prevent him from further lowering the price of his cars—the grounds on which the case was argued had to do with the use of corporate funds for philanthropic purposes. Citing the 1883

British case, *Hutton v. West Cork Railway Company*, a Michigan court affirmed the doctrine that "charity has no business to sit at boards of directors *qua* charity":

The difference between an incidental humanitarian expenditure of corporate funds for the benefit of employees, like the building of a hospital for their use, and the employment of agencies for the betterment of their condition, and a general purpose to benefit mankind at the expense of others is obvious. . . . A business corporation is organized and carried on primarily for the profit of the stockholders. The powers of the directors are to be used to that end. The discretion of directors is to be exercised in the choice of means to attain that end, and does not extend to a change in that end itself, to the reduction of profits, or to the nondistribution of profits among the stockholders, in order to donate them to other purposes.[29]

Although *Dodge v. Ford* did not directly affect the actions of managers in other states, the issues it raised undoubtedly influenced federal tax authorities to challenge the efforts of corporations to write-off charitable contributions as business expenses. Notwithstanding, the first generation of professional managers, imaginative individuals, were well equipped to make the corporation an effective social and philanthropic actor.

The new type of business manager was epitomized by men like Walter S. Gifford and Gerard Swope, who took the reins at AT&T and General Electric after World War I.[30] Both products of the New Education and deeply committed to Progressive social reform, they were able to institutionalize the ties between mass production and social reform that Ford had merely intuited. Using techniques of statistical analysis and market research drawn from the social and sanitary survey activities of the Progressive reform movement, they were able to demonstrate that high volume sales of low-priced items to the mass of consumers would yield greater profits than low-volume sales of high-priced products to other corporations. Mass production, mass distribution, mass education through advertising, and the creation of credit machinery that made possible a middle-class life-style for the mass of Americans would not only yield greater profits but would, at the same time, make it possible to realize many of the concrete goals of the Progressives. Improved sanitation and nutrition, for example, could be achieved as effectively by enabling Americans to purchase modern home appliances at reasonable prices as by implementing cumbersome and expensive government programs. At the same time, the strategy would help eradicate the more obvious class distinctions that had come to characterize industrial society.

Progressive managers like Gifford and Swope understood the relation between for-profit and nonprofit enterprise with remarkable clarity. Although they created major research laboratories within their own firms, they also recognized the dependence of their companies both for technical and administrative expertise on research universities and both took leading roles in developing ongoing and mutually beneficial relations between their alma maters and major industries. These included regular corporation gifts in support of higher education, service by top executives on the governing boards of universities and foundations, preferential hiring of university graduates by their companies, and the formation of early corporate foundations (such as GE's C. A. Coffin Foundation, which was established in 1922 to underwrite postgraduate study in fields of interest to the company).

The impact of the new managers was not restricted to large oligopolistic enterprises like AT&T and General Electric. Their greatest impact appears to have been on the community level, particularly in medium-sized cities. In such places as Cleveland, Indianapolis, Minneapolis, and Detroit, the Community Chest and community foundation movements first took root. These efforts were not only led by business people, business corporations were major contributors to them and, in the case of community trusts, managed their funds. The Civic Federation, Americanization, Better Homes, Beautiful Cities, and Regional Planning movements, all spearheaded by business people and their firms, played key roles in changing and improving American living standards. All proceeded from the recognition that economic growth depended on a variety of noneconomic "quality of life" factors: the availability and cost of housing, a skilled and contented work force, up-to-date public utilities, recreation, and freedom from political corruption. These required not only public expenditures for such things as education, recreation, and planning, but also the nurturing of privately supported nonprofit institutions: libraries, hospitals, YMCAs, family welfare societies, and service clubs. Business people, serving on municipal boards and mustering public support through voluntary organizations, played leading roles in the transformation of community life in the first decades of the twentieth century.

The thrust of these activities was articulately summarized by Herbert Hoover's *American Individualism*. Noting that the years since the end of the Great War had witnessed "the spread of revolution over one third of the world," Hoover urged Americans to embrace "progressive individualism:"

while we build our society upon the attainment of the individual, we shall safeguard to every individual an equality of opportunity to take that position in the community to which his intelligence, character, ability, and ambition entitle him; that we keep the social solution free from frozen strata of classes; that we shall stimulate effort of each individual to achievement; that through an enlarging sense of responsibility we shall assist him to this achievement; while he in turn must stand up to the emery wheel of competition.[31]

Institutionally, Hoover worked to foster these goals through government, which gathered and disseminated information about public issues, intermediary business groups like the National Industrial Conference Board, the National Bureau of Economic Research, and other trade associations and through business itself. This work went on not only on the national level but also in towns and cities across the country.

Hoover's ideas were widely accepted not because they were new or original but because they eloquently and concisely summarized the previous half-century's ideas of socially concerned business people. Hoover was familiar with these ideas because he had been a successful businessman and, as such, very much a member of the managerial generation epitomized by Gifford and Swope. Like many of his contemporaries, he had entered public service during the war and had seen firsthand both the horrors of war and revolution and the exciting possibilities of corporate social action, such as those initiated by AEG, the German counterpart to America's General Electric.[32] These seemed to offer alternatives to state-funded welfare programs.

With Hoover's election to the presidency in 1928, his vision of an "associative state" seemed well on its way to implementation. Many firms were offering their employees—and the communities in which they operated—a surprising range of social and cultural services. Corporations were making record profits while enabling Americans to live healthier and richer lives than they ever had before. Public education systems, especially in industrial centers, expanded dramatically, encouraged by Chambers of Commerce and other business-led civic organizations. Business initiatives were not always welcomed. When General Electric tried to persuade its employees to contribute to an unemployment fund in 1926, its efforts were rejected by labor leaders, who viewed it as a paternalistic scheme; four years later, they had reason to regret their stance. In spite of such resistance, the conception of business as being on the front lines of social action remained strong. Even after the onset of the Depression, private sector action formed the conceptual core of the NRA, the program of government-fostered business cooperation that Franklin Roosevelt hoped would return prosperity to the country.

We have no clear idea of the scale and scope of either business giving or corporate contributions before 1936, when charitable donations became tax deductible. However, the fragmentary data that exist are highly suggestive. The first systematic study of charitable giving undertaken by the Cleveland Chamber of Commerce's Committee on Benevolent Associations in 1909 showed that of 3537 large donors to the city's charities, 800 (23 percent) were business firms. A 1930 study conducted for the National Bureau of Economic Research, *Corporation Contributions to Organized Community Welfare Services,* shows that corporations gave over $300 million to community chests between 1920 and 1929. Analysis of business contributions to Harvard in the mid-1920s shows that, through a combination of gifts by individual businessmen, their firms, the trade associations to which their firms belonged, and the charitable foundations on whose boards they sat, over 35 percent of gifts came from the business community.[33] The value of corporate contributions, however, hardly encompasses the full range of business social action. As noted, businesses also initiated education, recreation, health care, and retirement plans. In-kind giving was common, especially making available company facilities for community events. And service clubs—the Rotary, Kiwanis, and the Lions—pooled the energies and resources of smaller businessmen into the civic task.

THE ERA OF CORPORATE LIBERALISM, 1936–1980

Corporations like GE, Eastman Kodak, Ford, US Rubber, US Steel, and others that had demonstrated serious social concerns responded to the Depression by attempting to sustain their social commitments. Many set up loan and unemployment benefit funds, while pledging to maintain wage levels and, where not possible, to share tasks among their workers. Corporate contributions to community chests reached record levels in 1932. But it became increasingly clear, as Charles Schwab of Bethlehem Steel admitted in the spring of 1932, "None of us can escape the inexorable law of the balance sheet."[34] In the deepening economic crisis, the resources of corporations could no longer be devoted to purposes not essential to their survival.

Recognizing this, Roosevelt's first efforts to promote recovery attempted to salvage Hoover's associational system. The creation of the NRA in 1933 was in many ways less a departure from the business thinking of the 1920s than a fulfillment of it. To be sure, many of the pillars of the New Era, including Hoover himself, disavowed Roosevelt's system of government-fostered cooperation, just as they decried direct federal aid to the unemployed. But many, most notably Swope, were enthusiastic backers of the plan. The most important aspect of the NRA from the standpoint of business social investment was its emphasis on the preeminence of the role of the private sector in the provision of welfare services: the rationale of the NRA codes was the desire to avoid the dole by keeping people at work. This could only be done by ensuring the maintenance of wages and working conditions through voluntary agreements between firms and organized labor. The NRA was not concerned with charity per se, though it was assumed that the code provisions would enable socially responsible firms to carry on the programs they had implemented before the Depression.

Until the Supreme Court declared the NRA unconstitutional in 1935, it was generally believed that the corporate role in the realm of public welfare would remain as central it had been through the preceding decade. The court's decision, combined with the absence of economic recovery, prompted the Roosevelt administration to turn sharply to the left. It became more directly involved in alleviating the distress of the unemployed. The radicalization and growing strength of the labor movement encouraged Roosevelt to mount a rhetorical offensive against business and the rich. This found expression not only in the president's attacks on the "malefactors of great wealth," but in policies directed at redistributing the national wealth. The two most important of these were a sharply progressive income tax and a national system of social insurance.

Throughout the 1920s, charitable fund raisers, particularly from the Community Chest movement, had been lobbying for the inclusion of a regulation in the Internal Revenue Code to permit corporations to take charitable deductions,[35] but the effort had not gotten far. Legal strictures on corporation giving discouraged such a provision. Congressional opposition to big business, while politically in the minority, was sufficiently vocal through the 1920s, to block the corporate charitable deduction. Even had such a provision been passed, the Revenue Act of 1926, which sharply reduced tax rates, would have reduced tax-driven incentives for large-scale giving. But FDR's shift toward redistributional policies after 1935 gave the corporate charitable deduction a new lease on life. The Community Chest leaders, aided by the efforts of some of the nation's most influential business people put unremitting pressure on the White House and on Congress to include a 5 percent charitable deduction provision in the 1935 tax act. After much debate, the provision was passed "in conjunction with some of the most sweeping and fundamental New Deal reform legislation."[36]

The deductibility of corporate charitable contributions led neither to dramatically increased levels of business giving nor to revival of welfare capitalist doctrines and practices, primarily because of the historical context in which the corporate charitable deduction came into being. Sharply increased corporate and personal taxation, the growth of federal social spending, the rise of a powerful labor movement, and pronounced public and legislative hostility to private enterprise—all these added up to a set of disincentives for corporate social activism. Why should the private sector voluntarily implement expensive social initiatives when the government and the unions were coercing firms into underwriting expensive benefits programs? And why, given the 5 percent limit on the deduction, should corporations give any more than that? Further, in the adverse political climate, why shouldn't business shift its commitments away from spending that provided general benefits to a variety of entities to programs that promised more direct and calculable returns?

There is reason to believe that the corporate charitable deduction may have limited rather than increased the scope and scale of corporate social investment. This appears particularly likely in view of the failure of legal doctrines affecting corporation giving to change significantly during that period. The old rule that charity had no place in the boardroom and that giving, when it took place at all, should be related directly to the corporate interest, still remained very much in place: as late as 1934, the Supreme Court had ruled in the case of *Old Mission Portland Cement* v. *Helverling* that a firm had acted improperly in making charitable contributions where no direct corporate benefit had been demonstrated.[37] Though the Progressive managers of the 1920s had been able to avoid litigation during times of prosperity and business optimism—often through hiding contributions in their firms' operating budgets—increased government oversight and the scrutiny of dividend-hungry stockholders during the 1930s further increased the restrictiveness of the law of corporate charity.

No major efforts to change the legal framework of corporation giving occurred between 1935 and the late 1940s, although, thanks to the unusually heavy taxation of the war years, both levels of corporation giving and the nature of giving mechanisms changed dramatically: corporate charitable contributions increased from .38 percent of net corporate income in 1936 to 1.45 percent in 1945, while the number of corporate foundations increased from 19 in 1938 to 208 by 1945.

If industrial managers found their philanthropic energies increasingly absorbed in the task of corporate survival during the 1930s, another factor came into play to transform the character of business giving. The 1930s was the decade in which the founders of many of the early twentieth century's great industrial fortunes began to die off. In increasing numbers, these "founding fathers" turned to charitable foundations, which enabled them both to perpetuate family control of their firms while avoiding increasingly onerous estate taxes. Unlike corporations, foundations were subject to a minimum of regulation: they held their assets free of taxation, donations and bequests to them provided significant tax benefits, they were not required to make public reports, their charitable disbursements were restricted only by the collective consciences of their boards. Typical, although not the earliest of these efforts, was the Ford Foundation, established in 1936. Although its charter purpose was purely philanthropic, its immediate goals had much more to do with the desire of the Ford family to perpetuate its control over the Ford Motor Company in the event of its founder's death. Accordingly, when Henry Ford died in 1947, the foundation acquired 90 percent of the shares of the firm, making it one of the nation's major holding companies. Other notable instances combining genuine philanthropy with dynastic purposes were the Altman Foundation (1913), which controlled the assets of the Altman department stores, and the W. K. Kellogg Foundation (1930), which had been established by one of the pioneers of corporate philanthropy and which to this day controls the assets of the nation's largest maker of breakfast foods.

The consolidation of vast financial resources under the virtually unregulated control of charitable foundations was controversial. The initial major criticisms of these organizations was based not on their potential economic power but, ironically, on their funding of unpopular liberal causes. By the early 1950s, as a part of the McCarthyite hysteria, congressional committees were beginning to question the power of foundations to influence public policy.[38] By the late 1960s, under the leadership of Congressman Wright Patman, inquiries began into the economic issues raised by the share of the national wealth controlled by foundations.[39] Ironically, the close relationship between firms and foundations that came to draw so much criticism in the 1960s was initiated half a century earlier not by venial and hypocritical businessmen, but by leading business idealists who had pioneered the corporate responsibility and welfare capitalist activities of the Progressive era. But it was a relationship that, without oversight, invited abuse.

If some corporate leaders were concerned with tax avoidance and dynastic control, others had more serious preoccupations. These were the aging spokesmen for the New Era concepts of corporate responsibility. The failure of welfare capitalism early in the Depression and the defeat of the NRA in 1935 had left these men in a quandary. They were seriously committed to the survival of the private sector not only because of their personal stake in it but also for ideological reasons. Like Carnegie, they recognized the remarkable dynamism of capitalism. While they acknowledged the tendency of large-scale enterprise to generate social and economic inequities, they also understood its capacity to maximize opportunities for individuals of talent. Their understanding of these issues was not merely abstract: most of the top executives of the late 1940s were men like Swope of GE, Gifford of AT&T; Sloan, Wilson, and Kettering of General Motors; and Abrams and Teagle of Standard Oil, who had come from modest backgrounds and had worked their way up through the system. For them, the free enterprise system was integral to the survival of the American way of life. And, because they owed their successes in large measure to privately financed institutions of higher education, they also recognized the relation between the private sector in culture and the private sector in commerce.

These men worried about the future. The power of the federal government, which had been vastly increased over every aspect of American life during the New Deal and, even more so, during the war, showed no signs of diminishing. Some feared that a welfare state and central economic and social planning were inevitable. Even if this could be avoided, it was clear that the price of American leadership of the postwar world would be an enlarged federal role in areas like higher education, where until the war, federal presence had been largely absent.

Under Hoover, business had come to accept the importance of partnership between government and business. After the war, business had to accept that government had become the dominant partner. Under the circumstances, the best it could do was to work strategically through private philanthropy to ensure that its own interests were served. Led by such men as Arthur W. Page, a former progressive journalist and Harvard classmate and AT&T colleague of Walter Gifford's, America's corporate leaders began to design a comprehensive postwar strategy for business philanthropy.[40] One component was the aggressive promotion of efforts to educate Americans to the virtues of free enterprise and the distinctiveness of the American way of life.[41] Another involved coordinating corporate support for the private universities, which would include funding for new programs in fields like American studies that would underline the uniqueness of American values and institutions.[42] By the late 1940s, ideas about the philanthropic role of business had become institutionalized through groups like the Council for Financial Aid to Education and the National Planning Association, which educated the business community about the legal and tax advantages of contributions programs.

But there were major obstacles to such large-scale corporate philanthropy. The most significant of these were legal doctrines requiring that corporate charity demonstrate direct benefit to the profit-making purposes of donor firms. Led by Frank Abrams of Standard Oil of New Jersey, the corporate philanthropists initiated a test case to broaden the definition of "direct benefit." Prompted by Abrams and other leading business people, the New Jersey legislature amended its corporation law to permit corporations chartered in the state to make contributions to educational institutions. In 1951, the board of directors of the A. P. Smith Company, a New Jersey manufacturer, voted to donate $1500 to Princeton University. A group of the company's stockholders, "in what may have been a maneuver planned by the New Jersey corporations themselves to provoke a test case," challenged the legality of the donation.[43] The New Jersey courts ruled in favor of corporate givers, acknowledging the importance to business of a " 'friendly reservoir' of trained men and women from which industry might draw" and declaring that there was no "greater benefit to corporations in this country than to build and continue to build, respect for and adherence to a system of free enterprise and democratic government, the serious impairment of either of which may well spell the destruction of all corporate enterprise."[44]

Smith v. *Barlow* had a dramatic impact on corporation giving. IRS figures show that corporate contributions increased from .95 percent of pretax net income ($252 million) to 1.2 percent ($495 million) by 1953. During the same period, the rate of establishment of corporate foundations tripled, from 172 in 1946–1949 to 620 in 1950–1953.[45] By the late 1950s, businesses were devising new forms of corporate philanthropy, most notably the matching of employee contributions, pioneered by General Foods, in 1959.

Although corporate philanthropy increased after 1950, its growth was not uniform.[46] Giving tended to be concentrated in a handful of large firms; except in cities like Minneapolis and Cleveland, which fostered a charitable ethos in the business community, thousands of small and medium-sized firms gave nothing at all. This suggested that corporate philanthropy was not widely appreciated by the generation of managers who succeeded the Giffords and Swopes. One indicator was the unwillingness of the Business Roundtable, a group of the nation's largest 200 companies, to commit itself to even a 2 percent level of giving.[47] Surveys of CEOs suggested that most viewed their contributions programs as forms of public relations and executive patronage. Few believed that their firms had any broad role in society. A 1981 study of the nation's 200 largest firms concluded that corporation giving was "a relatively underdeveloped, poorly understood function in most companies."[48]

This is not to say that corporations were unimportant social actors. By the late 1960s, business corporations were giving more than a billion dollars a year to charity. Responding to rising public concern about the environment, poverty, and racial discrimination, some firms went beyond philanthropy, supplementing their contributions programs with affirmative action hiring programs and permitting social considerations to influence plant location and marketing decisions. In most cases, these gestures were responses to public demand or expressions of well-intentioned executives to "do good." They were, in essence, exercises in "community relations" that lacked any clearly defined connection to productivity, profitability, or the other core purposes of business firms.

CORPORATE GIVING IN THE POSTLIBERAL ERA

The essentially reactive character of corporate giving began to change in the late 1970s, impelled both by shifts in the economic and political environment and by forces within business itself. Responding to the tide of protest and disorder, some conservative business leaders had begun questioning the undiscriminating quality of corporate philanthropy, especially support for organizations that appeared to be outspokenly hostile to business.[49] Corporate philanthropy, they argued, should be a means for advancing the interests of business not the interests of government. The election of Ronald Reagan in 1980 broadened this effort to recast the corporate social role. In 1981, the president convened a Task Force on Private Sector Initiatives in the hope of tapping the resources of the private sector to offset the effects of the revenue shortfall that would result from federal spending cuts.[50] Although it became clear, early in the task force's work, that the private sector could not hope to take the place of government, the group did succeed in focusing attention on the important and relatively undeveloped potential of corporate philanthropy.

Calls for greater and better-focused corporate giving coincided with growing concerns about declining American leadership in the world economy. Commentators from outside the corporate community saw us "managing our way to economic decline" due to failures in executive vision. They criticized the unwillingness of managers to adopt long-term strategies involving the replacement of "labor and other scarce capital equipment" and the development of "new products and processes that open new markets and restructure old ones."[51] From within the business community, other critics blamed government. Tax and regulatory policies, they argued, discouraged economic growth. More important, public education, in spite of huge tax subsidies, had failed to create a work force with skills and motivation comparable to those of Japan and other international competitors.[52]

Through the early 1980s, a variety of groups sought to influence the corporate giving agenda. Some philanthropic trade associations, most notably the Council on Foundations and Independent Sector, sought to strengthen the commitment of corporate donors to traditional liberal goals.[53] Other groups of a more conservative bent urged corporations to shape their contributions policies in a more self-interested direction. The Task

Force on Private Sectors Initiatives, while not delineating funding priorities, emphasized the importance of community-based giving and volunteering.

Although the business community's response to this onslaught of attention has only begun to emerge, there appear to be a number of important changes in the character of corporate philanthropy. There appears to be an active effort to rethink the rationale for corporate contributions and to tie them more closely to issues of profits and productivity. In some firms, the mere desire to "do good" is being replaced by an investment orientation that not only seeks to fund activities beneficial both to company and community but also seeks to assess the productivity of such charitable investments.[54] Others have gone beyond mere money giving. Corporations have again begun operating social, cultural, and educational programs. Encouraging executives to serve as board members and volunteers in community-based nonprofit organizations has become a matter of corporate policy for many firms.

A few companies have explored the ways in which their pursuit of profits and markets could also serve social goals. In 1984, Ronald Grzywinski of Chicago's South Shore Bank pointed out the failure of public efforts to deal with major urban problems. Public programs, he wrote, tended to be insufficiently funded, too bureaucratic and centralized, and not self-sustaining. Most important, they failed "to release the energy of local residents in a way that rebuilt community self-confidence, induced other individuals, institutions, and local government to invest their capital and savings in the development process, or encourage a wholistic, bottom-up approach to renewal."[55] Banks, he suggested, could serve as "pivotal points in the fight against spiraling neighborhood deterioration," not only by bringing capital and managerial expertise to inner-city neighborhoods but by mobilizing the entrepreneurial and economic resources of these communities. Though his own institution was successful, Grzywinski was keenly aware that most bankers preferred the more conventional task of serving the affluent and that a private sector alternative to federal redevelopment efforts would require regulatory encouragement:

The challenge now for public policymakers is to creatively blend financial and nonfinancial incentives for specific performance and adopt a new ordering of the privileges of all of America's financial institutions, depositories, and others. Such a realignment could assure a close relationship between the privileges that are conferred through public policy and the public entrepreneurship that is needed to rebuild our neighborhoods, cities, and economic capacity. That is the reason the national banking system was created in the first place: to finance a pluralistic, market-disciplined approach to economic development.[56]

Although business has pursued a variety of approaches in expanding its social role, the single most important initiative in corporate philanthropy has been in the field of education. In 1980, a group of Boston businesses, concerned about the declining quality of the local work force, decided to work with the city's public school system in an attempt to reduce truancy and dropout rates and improve educational quality.[57] As the Boston Compact evolved, it moved from conventional philanthropy on the policy level toward direct involvement of executives in the schools. A keystone of the program was the willingness of cooperating firms to guarantee employment to all graduates of the city's schools, as well as a commitment to "last-dollar" scholarship funding of graduates who wished to obtain higher education. The Boston Compact set the standard for other cities.[58] By 1985, corporate communities in Los Angeles, Seattle, Minneapolis–St. Paul, and New York had mounted similar programs. In other cities, individual businesses and executives had begun "adopting" schools and, on a lesser scale, classrooms, in an effort to improve educational quality.

Corporate involvement in public education has not been without controversy. Some executives blamed the deficiencies of public education on the "entrenched educational bureaucracy" and the teachers' unions—hardly an auspicious starting point for an effort that, by its very nature, had to be a cooperative one.[59] In some school systems corporate involvement was resisted by politicians and unions, who feared that business would disrupt their mutually beneficial relationships. In a few communities, citizens' groups questioned the appropriateness of corporate involvement in the public enterprise of education. Over time, however, these conflicts have disappeared with the rising public recognition of the ties between educational quality and national economic growth and the corporate acknowledgment that its educational agenda could only be implemented through direct, hands-on participation.

Although these educational initiatives suggest a new and exciting future corporate social role, other changes in the economy raise disturbing questions about whether these promises will be fulfilled. In the 1980s, the financial markets have been increasingly dominated by speculators who specialize in buying and dismantling established companies. To accumulate the cash reserves needed to fight off such raiders, many firms have had to reduce or eliminate their philanthropic commitments.[60] Other companies, spun off from take-overs and reshufflings, have found themselves too burdened with debt to engage in philanthropy. Another feature of the dramatic economic changes of the 1980s has been the growing role of foreign corporations in the United States. Surprisingly, some of them have become major philanthropic actors. In 1972, the Japan-based Sony Corporation established a charitable foundation. It was followed in 1980 by the Japanese shipbuilding industry, which set up the $38 million United States–Japan Foundation, based in New York. In 1984, the Honda Motor Company, established a $5 million charitable foundation, based in California.

CONCLUSION

In a crudely quantitative sense, business philanthropy has a short history, dating only from 1936, when corporations first began reporting charitable contributions to the federal government under the provisions of the 1935 Tax Act. In fact, the scale and scope of corporate social action has been far greater.

Beginning with the Industrial Revolution of the late eighteenth century, business has had to reach beyond the realm of production and distribution to create the political, social, and cultural conditions essential to its prosperity. After the Civil War the combined challenges of large-scale economic development and governing an increasingly diverse and geographically dispersed population pushed business to the forefront of efforts to create a new national order. By the beginning of the twentieth century, this "corporate reconstruction of American life" had created the basis for a Progressive-liberal political consensus.

Because they understood that capitalism must be self-renewing in order to survive, business leaders through the 1920s strove to reconcile the intrinsic inequalities of America's advanced industrial economy with traditional democratic ideals. By equalizing opportunities to learn and consume, they hoped not only to forestall the class divisions that had led to revolutions abroad but to help Americans understand the identity of interests that bound together capital and labor.

The growth of the federal role in all areas of American life after 1933 displaced business from its leading role. The Depression forced firms to abandon long-term social commitments, which were taken up by the public sector. The revival of social activism did not occur until the late 1940s, when business leaders became convinced that the preservation of free enterprise would require major philanthropic investments in education and community life. Although the years after 1950 brought major increases in the scale of corporate charitable giving, these changes were selectively concentrated in certain large firms and in particular metropolitan regions. On the whole, the corporate social role remained poorly developed and was seldom viewed as an integral component of corporate operations.

The underdeveloped and poorly understood character of corporate philanthropy only changed in the 1980s, when American business began to be faced with major competitive challenges from abroad. Foreign competition highlighted failures of managerial vision and worker productivity that appeared to have their roots in the educational system. Responses to this have led not only to changes in the methods of corporate philanthropy but also in its rationale as managers rediscovered the extent to which their firms' profits depended on the social and cultural infrastructure. Though recent developments are encouraging, it is impossible to predict whether the development of proactive corporate philanthropy will mature into a full understanding of the social role of the corporation.

The relation between doing well and doing good is far from simple. Market theorists like Milton Friedman have argued that the social impact of business has had less to do with its desire to do good than its desire to do well. Thus, business should be free to engage its productive task as efficiently and profitably as possible—and let social consequences take care of themselves; but history suggests otherwise. The successes of Whitney, Carnegie, Ford, and Swope appear to have stemmed not from their willingness to ignore their social setting but from their awareness of it. Eli Whitney understood from the beginning that his system of manufacturing was more than just a better way of making muskets, just as Steven Jobs in our own time understood that the personal computer was more than an improved typewriter, filing system, and calculator. Each would not only change the way work was done, but would transform the nature of work itself—and in so doing, change society. Their ability to anticipate and calculate these consequences justified risk taking and innovation on a scale that would have been precluded by more conventional and circumscribed conceptions of the productive task.

Foresight of this kind is not altruistic; it is intensely self-interested. As long as the calculation of social costs and consequences remains in the margins of corporate policy making and planning, social investment will be no more than "window dressing." Only when more managers rediscover that the survival of capitalism depends on its capacity for self-renewal can we expect to see major changes in the scope and scale of corporate philanthropy.

NOTES

1. J. S. Davis, *Essays on the History of Earlier American Corporations* (Cambridge, Mass.: Harvard University Press, 1917); E. M. Dodd, *American Business Corporations until 1860, with Special Reference to Massachusetts* (Cambridge, Mass.: Harvard University Press, 1960); P. D. Hall, *The Organization of American Culture, 1700–1900: Institutions, Elites, and the Origins of American Nationality* (New York: New York University Press, 1982).

2. On the use of charitable endowments as capital pools, see M. S. Foster, *"Out of Smalle Beginnings": An Economic History of Harvard College in the Puritan Period* (Cambridge, Mass.: Harvard University Press, 1962); G. T. White, *The Massachusetts Hospital Life Insurance Company* (Cambridge, Mass.: Harvard University Press, 1955); P. D. Hall, "The Model of Boston Charity," *Science and Society* (Winter 1974).

3. Tocqueville provides the best overview of the use of voluntary organizations as a counterpoise to electoral power in *Democracy in America,* vol. 2 (New York: Random House, 1981), pp. 403–408, 412–13. For the best detailed study of how voluntary organizations were used by the Federalists in the early nineteenth century, see Charles I. Foster, *An Errand of Mercy: The Evangelical United Front, 1790–1837* (Chapel Hill: University of North Carolina Press, 1960).

4. Tocqueville, ibid., 1: 304, comments on the fact that in 1830, thirty-one Connecticut natives sat in Congress, only five of whom were representing the state of their birth; Lee Soltow, *Men and Wealth in the United States, 1850–1870* (New Haven, Conn.: Yale University Press, 1975), pp. 148, 185, notes the extraordinary economic success of New Englanders who had moved to the South and West.

5. For the philanthropic role of business in an urban setting, see Ronald Story, *The Forging of an Aristocracy: Harvard and Boston's Upper Class.* (Middletown, Conn.: Wesleyan University Press, 1981); in the provincial setting, see Mary P. Ryan, *The Cradle of the Middle Class: The Family in Oneida County New York, 1790–1865* (New York: Cambridge University Press, 1982); and Paul E. Johnson, *Shopkeeper's Millenium: Society and Revivals in Rochester, New York, 1815–1837* (New York: Hill and Wang, 1978).

6. Early factory owners retained the workers using standard apprenticeship indentures, which specified the provision of such services, for example, "Agreement between Amos Bradley and Eli Whitney" (May 14, 1813); "Agreement between Eli Whitney and Daniel Tallmadge" (December 22, 1814) (Eli Whitney Papers, Yale University Archives). A Middletown, Connecticut, physician's account book details payments for worker medical care from manufacturer Johosophat Starr over a ten-year period, 1799–1809 in W. B. Hall, "Account Book, 1807–1809," (author's collection). The Boston-based textile manufacturers adapted these traditional relationships to a larger-scale setting, Robert F. Dalzell, Jr., *Enterprising Elite: The Boston Associates and the World They Made* (Cambridge, Mass.: Harvard University Press, 1987), pp. 31 ff.

7. K. L. Hall and Carolyn Cooper, *Windows on the Works: Industry on the Eli Whitney Site, 1798–1979,* (Hamden, Conn.: Eli Whitney Museum, 1984), pp. 35 ff; John P. Coolidge, *Mill and Mansion: A Study of Architecture and Society in Lowell, Massachusetts, 1820– 1865* (New York: Columbia University Press, 1942).

8. Much of the Eli Whitney correspondence is devoted to efforts to retain skilled mechanics who threatened to leave his Armory, and to luring away those working for other arms makers: James Carrington to Roswell Lee (October 31, 1822); Eli Whitney to Nathan Starr (December 13, 1817; December 16, 1817; December 26, 1817); Nathan Starr to Eli Whitney (December 18, 1817), Eli Whitney Papers, Yale University Archives.

9. On the role of businessmen in the post-Civil War reform movement, see Robert Weibe, *Businessmen and Reform: A Study of the Progressive Movement* (Cambridge, Mass.: Harvard University Press, 1962); John G. Sproat, *"The Best Men:" Liberal Reformers in the Gilded Age* (New York: Oxford University Press, 1968); Arthur Mann, *Yankee Reformers in the Urban Age: Social Reform in Boston, 1880–1900* (Cambridge, Mass.: Harvard University Press, 1954). The most important—and overlooked—study of business thinking about reform is Morrell Heald, *The Social Responsibilities of Business: Company and Community, 1900–1960* (Cleveland: Press of Case Western Reserve University, 1970).

10. Although most historical studies emphasize the exploitative components of scientific management, the papers of the early management reformers make clear the links they saw between reorganizing the work place and reconciling conflicts between labor and management. See the *Transactions* of the 1886 Annual Meeting of the American Society of Mechanical Engineers, which featured papers on the subject by Henry Towne and Henry Metcalfe, with comments by W. E. Partridge, Frederick W. Taylor, and other manufacturing notables. In 1889, Towne presented a paper on "gain sharing" to his colleagues. Though known as the "father of scientific management," Frederick W. Taylor did not present any original work on the subject until the annual meeting of 1895. His famous book, *Scientific Management,* did not appear until 1908.

11. The best summary of this dimension of reform is Stewart Brandes, *Welfare Capitalism* (Chicago: University of Chicago Press, 1976).

12. Among the studies that discuss the importance of business philanthropy in metropolitan settings in the late nineteenth century are Kathleen McCarthy's *Noblesse Oblige: Charity and Cultural Philanthropy in Chicago, 1849–1929* (Chicago: University of Chicago Press, 1982); and Thomas Bender, *New York Intellect* (New York: Alfred A. Knopf, 1987).

13. On the role of businessmen in the ASSA, see Thomas Haskell, *The Emergence of Professional Social Science: The American Social Science Association and the Nineteenth Century Crisis of Authority* (Urbana: University of Illinois Press, 1977); Gordon M. Jensen, "The National Civic Federation: American Business in an Age of Social Change and Social Reform" (Ph.D. dissertation, Princeton University, 1956). On the role of Boston financiers in the teaching of political economy at Harvard, see Paul Buck, ed., *The Social Sciences at Harvard* (Cambridge, Mass.: Harvard University Press, 1965).

14. Charles W. Eliot, "The New Education," *Atlantic Monthly* (February 1869), p. 203.

15. Ibid., p. 225.

16. Ibid., p. 207.

17. Ibid.

18. Ibid., p. 208.

19. Charles W. Eliot, "Inaugural Address," in Richard Hofstadter and Wilson Smith, eds., *American Higher Education: A Documentary History,* vol. 2 (Chicago: University of Chicago Press, 1961), pp. 205, 210.

20. Eliot was critical of vocationalism in undergraduate education because he believed that it was impossible to foresee what kinds of knowledge would ultimately be most useful in a rapidly changing world. On Eliot's conception of educational utility, see Hugh Hawkins, *Between Harvard and America: The Educational Leadership of Charles W. Eliot* (New York: Oxford University Press, 1972), p. 201.

21. The year before Eliot's election to Harvard's presidency, the university received $26,000 in gifts and bequests, none of which came from donors who can be identified as businessmen. By the end of his first year, giving had increased to $100,000, almost all of which came from business donors. By the end of his first decade at Harvard, he had attracted more than $1.5 million, an amount equal to the university's entire endowment in 1869. Harvard University, *Report of the Treasurer* 1867/68–1880.

22. To date, no detailed account has been written of the Young Yale Movement. Though Brook M. Kelley's *Yale—A History* (New Haven, Conn.: Yale University Press, 1968) sheds some light on the subject, the best narrative is still contained in Harris E. Starr's *William Graham Sumner* (New York: Henry Holt and Co., 1925), pp. 81 ff. Conservatives on the Yale Corporation regarded the pioneer social scientist Sumner as a "mole" for the New York businessmen who sought to reform the university. This led to an effort in 1874 to purge him from the faculty for teaching from the works of Herbert Spencer, which were very much in favor with intellectually inclined businessmen. (Andrew Carnegie regarded himself as Spencer's chief American spokesman.)

23. Yale's 1871 Woolsey Fund Drive, which sought to raise $500,000 brought in only $172,452. Because of its failure, the university prudently avoided further fund-raising efforts until 1901. The latter succeeded, thanks in large part to a lay president, economist A. T. Hadley, and the substantial representation of businessmen on the Corporation. S. R. Betts, *Alumni Gifts to Yale* (New Haven, Conn.: Yale University, 1922).

24. Max Weber, *The Protestant Ethic and the Spirit of Capitalism* (New York: Charles Scribner & Sons, 1948), p. 184.

25. Louis Brandeis, "Business—A Profession: An Address Delivered at Brown University, Commencement Day, 1912," in *Business—A Profession* (Boston: Hale, Cushman and Flint, 1914). Similar sentiments were expressed by Herbert Croly's definitive statement of Progressive ideology, *The Promise of American Life* (New York: Macmillan Co., 1909).

26. Although disaggregating the process of production resulted in "deskilling," Taylorism did not reduce workers to the status of "trained gorillas." As Taylor's descriptions of the labor process make clear, literacy, numeracy, and the ability to read and interpret engineering drawings actually constituted additions to the fund of skills of many workers. Frederick W. Taylor and Sanford E. Thompson, *Concrete Costs: Tables and Recommendations for Estimating the Time and Cost of Labor Operations in Concrete Construction and for Introducing Economical Methods of Management* (New York: John Wiley & Sons, 1912).

27. 204 Mich. 459, 170 N.W. 668 (1919).

28. For the background of this case, see Allan Nevins and Frank E. Hill, *Ford: Expansion and Challenge, 1915–1933* (New York: Charles Scribner's Sons, 1957), pp. 86 ff.

29. 23 *The Law Reports, Chancery Division* 1883, 654.

30. Gifford, a 1905 Harvard graduate, took his first job at Western Electric's Chicago plant. While there, he was a resident of Jane Addams' pioneer social settlement, Hull House. When his statistical talents caught the eye of the AT&T headquarters staff, he was transferred to New York to organize a statistical department for the company, which soon became the core of its planning capacity. Made an AT&T vice-president in 1919, he became president in 1925, serving in that position until 1949. Though no biography has been written, Gifford's work and ideas can be found in Harvard College, *Class of 1905, Sixth Report* (1930) and in Walter S. Gifford, *Addresses, Papers, and Interviews* (New York: American Telephone and Telegraph Company, 1928–49).

Swope graduated from MIT in 1895 and, like Gifford, went to work at Western Electric.

Also a Hull House resident, he married a colleague and, when transferred by his company to St. Louis, set up the city's first settlement house. Swope left Western Electric for General Electric when his proposals for high-volume, low-priced product lines were resisted by its management. He took charge of General Electric's foreign operations in 1916 and became president of the company in 1921, serving until 1941. Swope's career is summarized in David Loth, *Swope of GE* (New York: Arno Press, 1976).

31. Herbert Hoover, *American Individualism* (Garden City, N.Y.: Doubleday and Co., 1922). The best analysis of Hoover's ideas and their influence is Ellis Hawley's article, "Herbert Hoover, the Commerce Secretariat, and the Vision of an 'Associative State'," in Edwin G. Perkins, ed., *Men and Organizations* (New York: G. P. Putnam's Sons, 1977), pp. 131–48. See also Hawley, ed., *Herbert Hoover as Secretary of Commerce: Studies in New Era Thought and Practice* (Iowa City: University of Iowa Press, 1974).

32. On AEG's commitment to social engineering, see Tilmann Buddensieg, *Industriekultur: Peter Behrens and the AEG, 1907–1914* (Cambridge: MIT Press, 1984). It was no coincidence that Walter Rathenau, the son of AEG's founder, was a leader of the German Social Democratic Party during the 1920s. Swope seems to have attempted to play a similar role in the United States.

33. Pierce Williams and Frederic E. Croxon, *Corporate Contributions to Organized Community Welfare Services* (New York: National Bureau of Economic Research, 1930).

34. Raw donor data is contained in Harvard University, *Report of the Treasurer, 1924–25* (1925). The identification of donors and their affiliations was done by the author with the assistance of Andi Donovan. The 35 percent figure is probably an underestimate of the business contribution, since only gifts of more than $1000 were analyzed.

35. The best account of lobbying for the deduction is in Scott M. Cutlip, *Fund Raising in the United States: Its Role in Philanthropy* (New Brunswick, N.J.: Rutgers University Press, 1965), pp. 318 ff. Swope was one of the leaders of the lobbying effort.

36. Ibid., p. 328.

37. 293 U.S. 289 (1934).

38. F. Emerson Andrews, *Patman and the Foundations* (New York: Foundation Center, 1968).

39. Ibid.; and Ferdinand Lundberg, *The Rich and the Super Rich* (New York: Lyle Stuart, 1969).

40. For an overview of this, see Merle Curti and Roderick Nash, *Philanthropy in the Shaping of American Higher Education* (New Brunswick, N.J.: Rutgers University Press, 1965), pp. 238 ff. The Arthur W. Page Papers at the Wisconsin State Historical Society contain important correspondence and memoranda detailing the thinking of top managers on the subject of philanthropy in the postwar period. Page, a Harvard classmate of Walter Gifford's, went from a career as a Progressive journalist to being vice president in charge of public relations at AT&T.

41. The Page Papers include correspondence and memoranda regarding corporation-funded campaigns to promote the American free enterprise system in the postwar years. In one particularly important letter to the Carnegie Corporation's Charles Dollard, Page expresses his fear that government economic regulations put in place during the war will become a permanent feature of American life unless the corporations mount a concerted campaign against them.

42. The Page Papers not only outline the overall strategy of focusing attention on American values through foundation funding of American studies programs but also contain early prospectuses for *American Heritage* magazine—an important, popularized dimension of the American studies movement.

43. Curti and Nash, *Philanthropy in the Shaping of American Higher Education*, p. 242. The key case in legitimating corporate philanthropy was *A. P. Smith Manufacturing Company v. Barlow, et. al,* 26 N.J. Superior Court Reports (1953), 114; and 13 N.J. Supreme Court Reports (1954) 147–49.

44. Ibid., 114.

45. Thomas Vasquez, "Corporate Giving Measures," in Commission on Private Philan-

thropy and Public Needs, *Research Papers* (Washington, D.C.: U.S. Department of the Treasury, 1977), pp. 1838 ff.

46. The most lucid analysis of corporate giving trends since 1936 is Hayden W. Smith's *A Profile of Corporate Contributions* (New York: Council for Financial Aid to Education, 1983).

47. E. B. Knauft interview (1983). On this, see Knauft, *Effective Corporate Giving Programs* (Washington, D.C.: Foundation Center, 1985), p. 4.

48. Arthur White and John Bartolomeo, *Corporate Giving: The Views of Chief Executive Officers of Major American Corporations* (Washington, D.C.: Council on Foundations, 1982).

49. William E. Simon, "Reaping the Whirlwind," *Philanthropic Monthly* 13 (1980), pp. 5–8; and Institute for Educational Affairs, *Independent Philanthropy* (New York: Authors, 1983).

50. On the task force, see Marvin N. Olasky, "Reagan's Second Thoughts on Corporate Giving," *Fortune* (September 20, 1982), pp. 130 ff.

51. Robert H. Hayes and William J. Abernathy, "Managing Our Way to Economic Decline," *Harvard Business Review* 58, no. 4: 67–77.

52. The failure of the educational system was pointedly summarized by Gilbert Sewell's paper, "Issues of Conflict and Cooperation between the Business and Education Communities," presented at the U.S. Department of Education–American Enterprise Institute Conference on Business Involvement in School Reform, Washington, D.C., November 20, 1986.

53. On the Council on Foundation's liberal social agenda and conservative response to it, see "James Joseph Memo Provokes Council Controversy," *Independent Philanthropy* (Fall 1984).

54. Adam Stern and Mark Vermilion, "Corporate Social Investment," *Foundation News* (November–December 1986): 38 ff. For an historically based discussion of an investment rationale for corporate giving, see P. D. Hall, "Philanthropy as Investment," *History of Education Quarterly* 22, no. 2 (1982): 185–91.

55. Ronald Grzywinski, "The Role of Banks in Neighborhood-Based Development Strategies," *The Entrepreneurial Economy* (March 1984): 7. See also "Social Change and Finance Interest Groups," in *Investing in Social Change* (New Haven, Conn.: Yale School of Organization and Management, 1986) for an excellent anthology of essays on social investments by for-profit firms.

56. Ibid., p. 9.

57. On the Boston Compact, see Eleanor Farrar, "Case Study," presented at the U.S. Department of Education–American Enterprise Institute Conference on Business Involvement in School Reform, November 20, 1986; and the sessions on the "Business Community's Commitment" and "Beyond Business-Education Partnerships" presented at the Conference Board's meeting, The New Agenda for Business, Washington, D.C., February 12, 1987.

58. For an excellent overview of business-education partnerships, see Roberta Trachtman, "Education Reform and the States: A Case in Point," presented at the U.S. Department of Education conference, November 20, 1986.

59. Sewell's was one of the more pointed critiques of the education establishment presented at the U.S. Department of Education–American Enterprise Institute conference.

60. The impact of economic uncertainty on corporate giving is discussed in Thomas R. Buckman and Loren Renz, "Introduction," *Foundation Directory* (Washington, D.C.: Foundation Center, 1987), pp. xxiii–xxv. See also Joseph Foote, "The Name of the Game Is Change," *Foundation News* 8, no. 2 (March–April 1987): 51–53; and Anne L. Bailey, "Corporate Giving to Charities Slated for Cuts as Result of Merger Mania," *Chronicle of Higher Education* (April 1, 1987): 23–24.

13

Corporate Contributions to Charity: Nothing More than a Marketing Strategy?

JOSEPH GALASKIEWICZ

Of the many different kinds of corporate public service activities, corporate contributions have received the most attention. Critics on the left such as Perrow (1972) and Nader and Green (1971) have argued that contributions are nothing more than strategies to coopt those opposed to the ways of corporate America. Critics on the right such as Friedman (1982) and MacAvoy and Millstein (1982) have argued that the business of business should be business, and if business insists on meddling in the supply of public or merit goods it will soon lose its license to do business. There recently has been considerable research on corporate contributions that allows us to begin to sort out the facts from rhetoric. This chapter is an attempt to synthesize this literature (see also Useem 1987), identify the problems that currently plague research in this area, and select findings from the literature that may be particularly enlightening to policy makers and practitioners.

CORPORATE GIVING: A REVIEW OF THE LITERATURE

Contributions as Marketing

There are variations on the contributions-as-marketing theme. At one extreme is "cause-related marketing": the American Express campaign to donate a certain amount of money to a worthy cause for each time a

Funding for this research came from the National Science Foundation (800-8570) and the Program on Non-Profit Organizations, Yale University. The paper is an extensive revision of "The Environment and Corporate Giving Behavior" presented at the Independent Sector's Research Forum, March 13–14, 1986, New York.

customer uses its services or opens a new account has received the most attention (Maher 1984, p. 86). In contrast to the long-term strategy of building up good will towards the firm, cause-related marketing is said to result in immediate sales. Customers use American Express not because they "feel good" about the firm but because they can thereby control the company's contributions. They not only receive the service that goes with a credit card but also help fund a nonprofit organization at no cost to themselves.

Another example is the donation of computer hardware to colleges and universities by manufacturers such as IBM and Apple. Maher (1984, p. 86) points out that the benefit to the manufacturers is the establishment of a user base. Students, faculty, and administrators become trained on the donor's machines and thus are more likely to buy them.

More broadly conceived, marketing strategists view contributions as a way to coopt or win the good will of prospective customers. Murray and Montanari (1986, p. 819) argue that all forms of socially responsible behavior help to hold and attract customers. Not only do companies deliver goods and services, they also provide society with philanthropic contributions, ecological-environmental by-products, new technologies, tax revenue, and so on. To the extent that firms are living up to the expectations of various social interest groups, they ensure their legitimacy that in turn should have some bottom-line payoff. Granted, enhancing the firm's reputation is only a small part of a company's overall marketing strategy. However, it could give them an edge in the marketplace by making the firm seem less self-serving and thus more trustworthy.

Testing this theory Burt (1983, pp. 197–221) hypothesized and found that the market position of the firm determines the level of contributions to charitable organizations. Using data for 1967, he found that the amount of industrywide contributions measured in absolute dollars, per capita dollars, or as a proportion of profits was directly associated with the percentage of sales to households. This association held when controlling for both income and the price of a contribution. Burt argued that firms that were dependent upon people (as opposed to companies) for sales sought to impress them favorably by their generosity and public spiritedness.

Several authors, finding an empirical association between advertising and contributions expenditures, have argued that the two are really indistinguishable. Schwartz (1968) found this association looking at data for 1948, 1959, and 1960, as did Levy and Shatto (1978, 1980), who looked at the 1971 corporate tax returns of fifty-six Standard Industrial Classification (SIC) industries and the philanthropic activities of America's fifty-five largest investor-owned electric utilities; Fry, Keim, and Meiners (1982), who looked at IRS data for the years 1946–1973 for thirty-six industry groups. They found that (1) marginal changes in advertising expenditures and marginal changes in contribution expenditures were significantly related; (2) firms with more public contact spent more at all income levels on advertising and contributions than did firms with little public contact; and (3) changes in contributions and changes in other business expenses usually considered to be profit motivated (such as officer compensation, dividends, and employee benefits) were highly correlated.

Contributions as Public Relations

The strategy of contributions as public relations similarly attempts to enhance the image of the firm. However, it differs from contributions as marketing in targeting the entire public and not aiming specifically at increasing sales. Its goal is to show that the firm is a good corporate citizen. In this respect the contributions as public relations is a political as opposed to an economic strategy; ultimately, it seeks to ensure the autonomy of the firm from external formal controls (see Pfeffer and Salancik 1978).

Ermann (1978) argues that contributions to the Public Broadcasting System (PBS) between 1972 and 1976 were efforts by upwardly mobile companies to produce good will for themselves among the national elite and the general public. He cites PBS literature that suggests repeatedly that a gift is an investment that can produce important public-opinion gains. Although his research design did not allow him to assess whether elites felt more positively toward firms that had donated to PBS, he did find that many oil companies and firms that had recently increased their profits were among the biggest contributors. Miles (1982) showed how the tobacco industry, when challenged by the Sloan-Kettering Commission and the Surgeon General's Report on smoking's health hazards, immediately responded by funneling millions of dollars to universities and research institutes that did work on cancer-related topics. Indeed, this put the tobacco companies in touch with research that was of immediate interest to them, but the contributions also gave a signal to the public that the industry wanted to shed light objectively on cigarette smoking.

That companies donate a large proportion of their contributions budget in the communities where their headquarters are located is often taken as evidence supporting the contributions-as-public relations thesis (White and Bartolomeo 1982, p. 41). Here, company giving is a way to demonstrate its citizenship to local residents and local elites. It could be rooted in loyalty to locale; alternatively, it is a political strategy to win the good will of local influentials, who sooner or later will deliberate over zoning ordinances, property taxes, environmental regulations, disclosure, affirmative action, and other matters affecting business in the community.

If the contributions-as-public relations thesis is true, we should find that companies that give more are viewed as being "more responsible" or "better corporate citizens" by relevant segments of the public but that this good will does not necessarily result in superior market performance. Research on the relationship between company giving and public opinion is scanty and inconclusive. In a study by Yankelovich, Skelly, and White (White 1980), a large sample of Americans was asked which industries were most socially responsible. The authors correlated these rankings with Conference Board data on level of corporate giving by industry and found a positive, although not perfect, relationship (see also Brooks 1976).

There have been many more studies on the relationship between socially responsible behavior and profitability, but the results are inconclusive: Aldag and Bartol (1978, p. 172); Preston (1981); Arlow and Gannon (1982, p. 240); Cochran and Wood (1984, p. 55).

Contributions as Enlightened Self-Interest

The business press and the management literature suggest a much different explanation for why firms give money to charity. Donations are not intended to increase short-term profits or improve a firm's public relations, rather American business philanthropy is motivated by enlightened self-interest. In 1984, the *Wall Street Journal* declared that firms give so that their long-term interests will be served (Wall 1984, p. 1). In 1981 an article in *Fortune* stated, "few corporations engage in philanthropy because others need money, as though a corporation were a well-heeled uncle who should spread his good fortune around the family. For the most part, corporations give because it serves their own interests—or appears to" (L. Smith 1981, p. 121). The argument is familiar: "Society expects business to accomplish a variety of social goals, and it must accomplish these goals if it expects to profit in the long run. The firm which is most sensitive to its community needs will, as a result, have a better community in which to conduct its business . . . . a better society produces a better environment for business" (Davis 1973, p. 313).

Surveys of company executives confirm that enlightened self-interest is a dominant ideology in motivating corporate social responsibility. In their review of the literature, Arlow and Gannon (1982, pp. 235–36) cite a number of studies that demonstrate how enlightened self-interest dominates the thinking of corporate executives. For example, Bowman (1977) studied executives' attitudes toward socially responsible behavior with respect to the environment. Business executives in his sample felt it was necessary to do more than required by law, even if it meant a short-term reduction in profits, because otherwise government will impose regulations that will hurt the *entire* industry (see also Ostlund, 1977). Sacrifices need to be made in the short term so that the industry can prosper in the long term.

The most impressive survey of the ideologies related to company giving was conducted by White and Bartolomeo (1982) for the Council on Foundations. After intensively questioning respondents, they concluded:

About 7 in 10 claim to be motivated by a desire to help the needy in the communities in which their company has plants/locations and by a desire to do what is ethically correct. But 2 out of 3 also emphasize the goals of improving local communities in order to benefit their own employees and of protecting/improving the environment in which to work and do business. Then, too, about a third hope to improve their company's public image; and about a quarter expect that their corporate giving effects will result in increased revenues/profitability and in an enhanced ability to recruit quality employees. The more self-interested goals are especially important to CEOs of Fortune 1300 companies. (White and Bartolomeo 1982, pp. 62–63)

White and Bartolomeo argue that corporate giving surely has a double agenda—one altruistic and the other selfish. They also note that business is very candid about this double agenda, suggesting that it is now an acceptable and legitimate way to rationalize corporate giving programs.

Skeptics in the management literature correctly point out that it is impossible to measure the impact of responsible behavior on *future* sales and public opinion and that espousing the ethic of enlightened self-interest is not rational from an economic point of view. For example, McGuire struggled with the problem of measuring and evaluating the effect of enlightened self-interest on profits. He finally concluded that enlightened self-interest at best represented "a crude blend of long-run profit and altruism" (1963, p. 143). Keim (1978) concentrates on the unresolved free-rider problems. He cites Wallich and McGowan (1970), who argue that firms may rationalize enlightened self-interest on the basis that stockholders now have diverse portfolios and thus a broad interest in the benefits that a large group of firms, even an industry, might realize if companies acted in a socially responsible way. However, as Keim (1978, p. 34) points out, in actuality investors hold stock in more than one but not in every corporation. Thus, for them to advise managers to use a social or group rate of return for criteria in decision making is irrational.

Contributions as Tax Strategy

The tax codes can affect company giving in several ways. Most research has focused on the complement of the marginal tax rate or, more euphemistically, the "price" of a contribution. The two definitive studies on this effect were done by Schwartz (1968) and Nelson (1970), and their findings are similar. Using time-series data at the industry level, both used the price of giving as a predictor variable with contributions as the dependent variable in simple econometric analyses. Schwartz (1968), looking at data extending from 1936 through 1961, did analyses including all industrial groups together, and one for each of nine industry categories. Controlling for the average after-tax income and then for cash flows, the complement of the average tax rate consistently had a statistically significant negative effect on contributions. Nelson (1970) looked at industry-level data between 1936 and 1963 and analyzed aggregate after-tax corporation income, the complement of the marginal tax rate, and aggregate contributions of corporations. He, too, found a price effect, but his analysis produced a lower price elasticity coefficient.

Given these empirical findings, it is surprising that surveys have not always found that executives always give high priority to tax matters in making company contributions. In interviews with 219 chief executive officers in 1981 and 1982, White and Bartolomeo (1982, pp. 54–55) asked if tax incentives led the company to make cash contributions to nonprofit organizations. Only 26 percent of their respondents said that tax laws provided "great" or "substantial" incentives, 36 percent said "some" incentives, and 36 percent said "very slight" or "no" incentives. When asked to evaluate several objectives related to company contributions, only 12 percent of the CEOs said that to "alleviate the tax burden on the company" was an "extremely" or "very important" goal of their corporate cash contributions (White and Bartolomeo 1982, p. 72). However, Harris and Klepper (1976, p. 41) found that 53 percent of the 408 Fortune 1300 chief executives surveyed said their firms would increase the percentage of contributions to taxable income by 50 percent or more if incentives increased.

The ceiling on the amount of charitable contributions that firms can deduct from their taxable income has a very limited effect on company giving. In 1986 it was 10 percent of pretax net income; prior to the Economic Recovery Tax Act of 1981 it was 5 percent. In their survey of Fortune 1300 firms, Harris and Klepper (1976, p. 41) were told by 86 percent of 439 corporate giving officers that the 5 percent limit had no effect. Furthermore, looking at aggregate Internal Revenue Service and Department of Commerce data for 1936 through 1981, Smith (1983, p. 8) found that contributions as a percentage of pretax corporate income had never exceeded 1.34 percent in any one given year. Although 11.6 percent of the companies reporting made contributions of 5.0 percent or more of their net taxable income in 1977, these tended to be smaller companies whose average yearly contributions totaled $7,400 and whose average yearly net income was only $130,000 (H. Smith 1983, p. 16).

The tax codes regarding company contributions of property should also affect the amount that companies deduct at tax time, but research on this matter is sketchy. In this limited space only a few watershed bills can be noted. For example, prior to the Tax Reform Act of 1969, business corporations making charitable gifts in kind in the form of inventory were allowed a tax deduction on the basis of the fair market value of such property. By altering the valuation of such gifts the 1969 act reduced considerably the attractiveness of making donations of inventory. The Tax Reform Act of 1976 liberalized the law somewhat by providing a special incentive for contributions of inventory or depreciable trade or business property if the property is used solely for the care of the ill, needy, or infants. Although donors still could not deduct the fair market value of the gift, they could deduct an amount equal to the basis of the donated property plus 50 percent of the difference between its fair market value and its basis (Milani and Wittenbach 1983, p. 322). The Economic Recovery Tax Act of 1981 provided a special incentive for contributions of inventory to colleges and universities for research. It allowed a corporate taxpayer to deduct its basis in the property contributed plus 50 percent of the unrealized appreciation (Milani and Wittenbach 1983, p. 323).

Unfortunately, no one has been able to rigorously measure the impact of these changes on company giving. Hayden Smith (1986, p. 126) estimated that $350–400 million worth of inventory was donated in 1985, which he estimated to be roughly 10 percent of total corporate contributions. He contended that prior to the 1981 act inventory donations were of only minor importance (H. Smith 1986, p. 125). An article in *The Chronicle of Higher Education* (Turner 1984) said that equipment donations due to the change in the tax law in 1981 accounted for a significant part of IBM's 50 percent increase in its support of education between 1982 and 1983. However, it is difficult to document the effects of tax legislation on gifts of inventory, because companies are often reluctant to disclose the value of the inventory, donees are typically only told the fair market value of the gift, and at tax time the amount deducted is added into cash contributions, thus making IRS data on corporate tax returns an unreliable source for this information (H. Smith 1986, p. 126).

Contributions as Social Currency

Finally, company contributions could be made to elicit the applause and approval of business peers and local philanthropic elites. This often has been derisively referred to as *country club philanthropy, old boy network philanthropy,* or *corporate back scratching.* However, it is better understood as status competition among very powerful actors within an economic elite.

Unfortunately, very little research has been done on the topic. Only in Sheehan (1966, p. 148) and Useem (1984, pp. 123–24) do we get a glimpse at the role that status rewards (and punishments) play in motivating company contributions. Through the institution of peer pressure, executives learn the expectations of their peers, are solicited, and are awarded certain status benefits. Giving is the norm in many business elite subcultures, and those who want to remain in the inner circles had best conform and make the appropriate contributions. In one of the few studies, Useem (1984, p. 127) looked at the impact of these networks on company giving. Studying 212 American business firms and 196 British firms, he found that companies with more "inner circle" directors on their boards were more likely to be (1) recognized as generous contributors to the arts, (2) members of arts or educational organizations, or (3) larger contributors in general.

STUDYING AN URBAN GRANTS ECONOMY: THE MINNEAPOLIS–ST. PAUL CASE

In 1980 and 1981 Galaskiewicz (1985a) conducted an exhaustive case study of Minneapolis–St. Paul business firms and nonprofit organizations in order to better understand why some firms gave more money to charity than others. Interviews were conducted with executives of 150 publicly held companies. He also interviewed a sample of 229 public charities and approximately 140 community leaders. The results described here are for sixty-nine publicly held firms that had 200 or more employees. He first tested the propositions that company giving is prompted by the desire to win favorable public relations, respect, or legitimacy from three constituencies: customers and employees, the public, and business leaders. Next, he tested the proposition that giving can be spurred by a firm's commitment to an ideology of enlightened self-interest.

Controlling for pretax earnings, he found that the ratio of employees to assets, the percentage of employees living in the Twin Cities area, and the percentage of sales to consumers were all unrelated to the level of company contributions. The contributions-as-marketing and the contributions-as-public-relations theses would suggest that firms more dependent upon various stakeholders would give more money to charity, but this was not the case. However, companies that gave more money to charity *or* supported charitable organizations valued by the local community elite were recognized by more members of the elite as being very generous to non-profit organizations. Companies received good public relations for their charitable contributions whether they wanted to or not.

These results were surprising because firms that could have benefited most from good local public relations were not taking full advantage of the situation. It may have been that firms were acting foolishly, or public relations simply may not be worth that much to these companies and the emphasis on it overplayed. Certainly, the literature has found little or no correlation between business performance and a reputation as socially responsible. Although they know they could derive PR benefits, companies may also have realized that impressing customers, employees, and the public with their largesse has no ultimate payoff.

He next tested the contributions-as-social-currency hypothesis. After reviewing the local business press, it was clear that a set of business leaders had taken it upon themselves to organize Twin Cities businesses and increase their levels of corporate contributions. Using reputational measures Galaskiewicz identified twenty-nine members of this elite and interviewed twenty-seven. They tended to be from Minnesota or the Upper Midwest and current or former top executives of Fortune-cited firms.

Interestingly, companies gave more money to charity if their officers and directors were in the networks of locally prominent business persons active in philanthropic affairs. This effect was independent of pretax earnings, percent of sales to consumers, and the birthplace of the CEO. Furthermore, companies that were better integrated into the social circles of the corporate philanthropic elite tended to give more money to charities that the elite either supported or used themselves. Executives' social positions influenced the specific allocations that their companies made, as well as the overall amount.

He also found that companies that contributed more money to charity or supported nonprofit organizations that the philanthropic elite itself patronized were recognized by more members of the corporate philanthropic elite as being very generous. Also, companies that gave more money to charity were recognized by more members of the corporate philanthropic elite as being very successful businesses, even controlling for pretax earnings and performance ratios. A reputation as a successful business appears to have been a function of either how much a company earned or how much it gave away.

Essentially, there was a trilateral exchange among local businesses, nonprofit organizations, and business leaders. At the risk of oversimplification, business executives who were in the social circles of the elite were pressured into giving more money to the nonprofit organizations that were providing services to members of the elite. In turn, business leaders bestowed recognition and status upon firms that supported these nonprofit organizations. Everyone received something: the nonprofits received funding, the elite received services, and the firm achieved status in the business community.

Finally, he tested the proposition that firms could be motivated by the belief that giving was in their enlightened self-interest. By including a dummy variable in his equation indicating if a firm rationalized giving this way, he found that it had a significant independent effect on the level of contributions, controlling for the firm's average annual pretax earnings and the proximity of the company's executives to the philanthropic elite. In other words, rationalizing contributions on the grounds that they help to protect the long-term interests of the company or the free enterprise system appears to have led some companies to give more money to charities.

Curiously only this rationalization was positively correlated with the size of contributions. Neither rationalizations based on favorably impressing customers, employees, or the public-at-large nor rationalizations based on religious principles or morality had any effect on the level of company giving. Collectivist sentiments apparently can move companies to contribute money to charity. However, the collectivity to be benefited is not primarily the larger community; it is the business community.

Pursuing this analysis further, he discovered that the correlates of enlightened self-interest were (1) a firm's broad-gauge dependency upon different sectors of the economy and (2) company participation in local educational programs aimed at socializing executives into an ethic of corporate responsibility. Firms that were larger and involved in industries highly dependent upon other industries for sales and purchasing tended to see contributions as serving their long-term self-interest. He also found that firms whose executives participated in local seminars and workshops sponsored by the Minnesota Project on Corporate Responsibility rationalized contributions as enlightened self-interest. Thus, organized educational exposure to an ethic of corporate responsibility could broaden a firm's perspective on contributions and could lead to greater contributions. This is an important finding because it suggests that corporate cultural variables can be as important as selective incentives like recognition and prestige in motivating company contributions to charity.

PROSPECTS FOR FUTURE RESEARCH

Future research needs to refine and retest the propositions currently in the literature. It also needs to develop and test new hypotheses related to company giving. In retesting the contributions-as-marketing and contributions-as-public relations theses, it is important that analysts demonstrate clearly that customers, employees, or the public at large change their opinions of a firm because of greater contributions to charity or more strategic giving. This requires that a sample be polled prior to a change in corporate giving policy as well as after the change takes place. Polling customers twice is difficult and costly; today's customers are not necessarily tomorrow's. Polling employees is somewhat easier, and polling the general public is not a problem. However, without a pre-post design it is difficult to isolate the effect of giving on stakeholders' perceptions.

Several other independent variables should predict the level of contributions if the contributions as-public-relations thesis is correct. Analysts might do a content analysis of the business press to see which firms are being chided for unethical or unlawful business practices. Firms in such circumstances would be motivated to pay "dues" to the court of public opinion in the form of greater corporate contributions. Firms involved in take-over activities or that earn extraordinarily high profits might also be giving more to charitable causes in order to make up for real or alleged transgressions.

As giving becomes more strategic, it becomes more important to study the contributions professional or giving officer. Troy (1982) provides an excellent overview of the giving officer's role, and contributions professionals in Minneapolis-St. Paul are discussed in detail in Galaskiewicz (1985a). A central thesis of Galaskiewicz's work is that the giving officer has become a key functionary in the Twin Cities grants economy, playing a gatekeeper role for corporate dollars. He describes the networks among these professionals and their influence on the perceptions and evaluations of giving officers (1985b). More attention must be given to the way that contributions professionals intercede in the grants economy, their career paths, backgrounds, networks, interests, and ties to others in the firm.

Research is also needed on the role of ideology and corporate culture in perpetuating contributions. Although couched in terms of what benefits the firm, an ideology of enlightened self-interest focuses on *long-term* benefits—benefits that neither the CEO nor current board of directors will ever see or get credit for. Our hope is that research on corporate culture will examine how ideologies related to corporate contributions come to be institutionalized in the firm thus ensuring their survival beyond one generation. Alternatively, Galaskiewicz (1985a) argues that organizational structures outside the firm may be important in "socializing" the more barbaric elements within the firm. One such is the Minnesota Project on Corporate Responsibility, which has counterparts in other cities and at the national level. Analysts need to examine the mechanisms by which these organizations transmit values of corporate responsibility to executives.

Analysts must also address Useem's (1984) argument that corporate contributions cannot be understood completely without reference to social class. There are several ways to test his thesis. The simplest is to get extensive background information on the officers and the directors in the firm and to see if this is correlated to contributions. Class analysis views contributions as motivated by an ethic of noblesse oblige found in traditional upper-class social circles and institutions that now is carried forward in the modern corporation. Even though firms might be publicly owned, as long as control remains in the hands of upper-class offspring, the practice continues.

Stock ownership may have an equally important effect on contributions. Another premise of class-control theory is that a network of "old family" interests still controls the major firms and financial institutions. Companies whose major shareholders have upper-class origins should be giving more money to charity than firms whose major shareholders are first generation wealth or large institutional investors. The rationale is simply that the former are less "greedy," will not press the firm for the greatest possible short-term return on investments, and will recognize (and have an interest in) the long-term social investment necessary to perpetuate the legitimacy of the free enterprise system.

Apart from class theory, the research on contributions as social currency needs to discover whether and how local and national business communities institutionalize social controls in order to check the opportunism of actors who are overly concerned with bottom-line indicators and quarterly earnings. The work of Ouchi (1984) is highly suggestive. Citing efforts by Japanese firms and the Japanese government to institutionalize social controls, he challenges U.S. firms to coordinate and plan their strategies in conjunction with government agencies. It is important to understand how a system of actors can curb the overly ambitious within their ranks for the benefit of the collective interest.

Moving to a macro-structural level of analysis, we suggest that analysts examine the various agency roles that influence the allocation of resources among nonprofit organizations in a grants economy. These roles include brokers such as the United Way and the United Arts Fund, fund raisers and contributions professionals whose networking can affect the distribution of grants, and business elites who often mediate between the nonprofit and corporate worlds. We believe that these agency roles are critical in matching supply to demand in unstructured organizational fields such as a grants economy, which lack prices to discipline the behavior of donors and donees.

Galaskiewicz (1985a) showed that agency roles were critical in allocating dollars in the Twin Cities grants economy. For example, studying a sample of 229 nonprofit organizations, he found those that were more widely known and respected among corporate contributions professionals tended to receive more corporate donations independent of their size, activities, and degree of professionalization. He also found that corporations tended to give larger contributions to nonprofits if the company had a professional corporate giving officer *and* the nonprofit organization had at least a part-time fund raiser. Furthermore, gifts tended to be larger if more members of the Twin Cities philanthropic elite had social ties to the firm's officers *and* themselves supported or used the services of the nonprofit organization.

Although these findings are suggestive, there is still no sound theoretical rationale that can explain when brokers, fund raisers, and contributions professionals become necessary and why one type of agent is selected over another. The number of donors and donees in the field and the degree to which philanthropic dollars are concentrated in the hands of a few donors could explain why brokerages like the United Way emerge. The size of the firm's contributions budget and the heterogeneity of prospective donees in the field could explain why contributions professionals are more attractive to some donors than to others. This, though, is all speculative, and work needs to be done to derive credible and testable theories of agency.

Future research should also take into account both the vertical and horizontal dimensions of corporate activity. Most research has looked at a sample of firms defined at the national level or at the industry level and at total contributions and other variables measured at the level of industry or firm. This, though, ignores the fact that, since companies are headquartered in particular locales, the firm and its executives often become integrated into local institutions and status systems. We believe that integration into these local control systems have an effect on the volume of company giving and the type of gifts that are made. Thus, we hope that analysts will design future studies to ensure that firms are selected in enough different locales to make feasible comparative community analyses.

Once a comparative community design has been introduced, many new hypotheses are possible. For example, average company contributions should be greater in communities where business organizations such as the Chamber of Commerce are stronger; where corporate executives are members of local upper-class institutions; where there is greater criticism of the free enterprise system and the legitimacy of capitalism is questioned; where voluntary associations such as the United Way involve business people in setting priorities, fund-raising, and evaluating nonprofit organizations; and where there are enough corporations to provide the critical mass for the creation of a business subculture. A comparative approach would also show whether these effects are independent of community needs (for example, percent of families below the poverty line, morbidity, and so forth), community structure (for example, population size, occupational distribution, and so forth), and institutional response (for example, government expenditures on health, education, and welfare). We expect that a considerable amount of the variation in corporate contributions across communities should be explained in terms of the social organization of the local business community and upper class.

POLICY AND PRACTICE IMPLICATIONS

In addition to setting an agenda for future research, we believe that several lessons for practitioners and policy makers are in the research we reviewed. That our list of suggestions for practitioners is so short reaffirms our earlier observation that research on corporate giving is only in its infancy.

Nonprofit Practitioners

- Be aware that companies give to charity for a number of reasons and that it is often difficult to tell which is most important for any given firm.
- An effective appeal rationalizes the solicitation in terms of the long-term benefits to business; for example, a more attractive community to do business in, a better educated labor force, a higher standard of living for executives.
- An effective appeal assures the firm that it will get publicity and credit for its contribution to your organization.
- Be aware that decisions on contributions are influenced by the CEO, the contributions officer, and the marketing department; however, at first, it is often difficult to tell which one is most important in any given firm.
- Retain the services of a professional fund raiser to gain access to professional contributions staff.
- Recruit the support of prestigious executive officers to gain access to other prestigious executive officers.

Corporate Practitioners

- A professional contributions staff exerts considerable influence over the allocation of company funds, even when contributions decisions are made by committees.
- As marketing criteria become more important in evaluating contributions, the contributions function will likely move from the public relations or community affairs department to the marketing department.
- Cause-related marketing is not offensive to the general public and increases sales.
- Giving money to charity enhances the firm's reputation as a generous and successful firm in the eyes of elite populations.
- Gifts made in the spirit of serving the long-term interest of the firm and free enterprise system cannot be subject to rigorous evaluation and thus should not be held accountable to bottom-line indicators.
- Firms should specify beforehand the percentage of the contributions budget that will be used to market the firm to customers, for public relations in general, to placate the peers of the CEO and other top officers, and to serve the long-term interests of the firm and the free enterprise system.

Policy Makers

- Adjusting the marginal tax rates of firms will affect the level of contributions significantly. Putting more money in the hands of corporations will likely reduce the overall amount given to charity.
- It is unclear whether changes in the tax laws with respect to donated property alone result in greater contributions to charity.
- Adjusting the maximum amount deductible appears to have had no effect on the amount donated to charity except among the very smallest donors, who appear to give more as the ceiling rises.

REFERENCES

Aldag, Ramon J., and Kathryn M. Bartol. 1978. "Empirical Studies of Corporate Social Performance and Policy: A Survey of Problems and Results." In Lee E. Preston, ed. *Corporate Social Performance and Policy* Vol 1. Greenwich, Conn.: JAI Press.

Arlow, Peter, and Martin J. Gannon. 1982. "Social Responsiveness, Corporate Structure, and Economic Performance." *Academy of Management Review* 7: 235–41.

Bowman, J. S. 1977. "Business and the Environment: Corporate Attitudes, Actions in Energy-Rich States." *MSU Business Topics* 25: 37–49.

Brooks, John. 1976. "Fueling the Arts, or, Exxon as Medici." *New York Times* (January 25), Sect. pp. D1 ff.

Burt, Ronald S. 1983. *Corporate Profits and Co-optation: Networks of Market Constraints and Directorate Ties in the American Economy.* New York: Academic Press.

Cochran, Philip L., and Robert A. Wood. 1984. "Corporate Social Responsibility and Financial Performance." *Academy of Management Journal* 27: 42–56.

Davis, Keith. 1973. "The Case for and against Business Assumption of Social Responsibilities." *Academy of Management Journal* 16: 312–22.

Erman, M. David. 1978. "The Operative Goals of Corporate Philanthropy: Contributions to the Public Broadcasting Service." *Social Problems* 25: 504–14.

Friedman, Milton. 1982. *Capitalism and Freedom.* Chicago: University of Chicago Press, [1967].

Fry, Louis W., Gerald Keim, and Roger E. Meiners. 1982. "Corporate Contributions: Altruistic or For Profit." *Academy of Management Journal* 25: 94–106.

Galaskiewicz, Joseph. 1985a. *Social Organization of an Urban Grants Economy: A Study of Business Philanthropy and Nonprofit Organizations.* Orlando, Fla.: Academic Press.

———. 1985b. "Professional Networks and the Institutionalization of the Single Mind Set." *American Sociological Review* 50: 639–58.

Harris, James, and Anne Klepper. 1976. *Corporate Philanthropic Public Service Activities.* New York: Conference Board.

Keim, Gerald D. 1978. "Corporate Social Responsibility: An Assessment of the Enlightened Self-Interest Model." *Academy of Management Review* 3: 32–39.

Levy, Ferdinand K., and Gloria M. Shatto. 1980. "Social Responsibility in Large Electric Utility Firms: The Case for Philanthropy." In Lee E. Preston, ed. *Research in Corporate Social Performance and Policy* Vol. 2. Greenwich, CT: JAI Press, pp. 237–49.

———. 1978. "The Evaluation of Corporate Contributions." *Public Choice* 33: 19–28.

MacAvoy, Paul W. and Ira M. Millstein. 1982. "Corporate Philanthropy vs. Corporate Purpose." In *Corporate Philanthropy.* Washington, D.C.: Council on Foundations, pp. 25–27.

Maher, Philip. 1984. "What Corporations Get by Giving: Finding Success with Donations to Symphonies, the Statute of Liberty, and the Puffin." *Business Marketing* 69: 80–89.

McGuire, Joseph W. 1963. *Business and Society.* New York: McGraw-Hill Book Co.

Milani, Ken, and James L. Wittenbach. 1983. "A Charitable Contribution Deduction Flow-chart for Corporations—An Update." *Taxes* 61: 319–24.

Miles, Robert H. 1982. *Coffin Nails and Corporate Strategies*. Englewood Cliffs, N.J.: Prentice-Hall.

Murray, Keith B., and John R. Montanari. 1986. "Strategic Management of the Socially Responsible Firm: Integrating Management and Marketing Theory." *Academy of Management Review* 11: 815–27.

Nader, Ralph, and Mark J. Green. 1971. *Corporate Power in America*. New York: Grossman Publications.

Nelson, Ralph. 1970. *Economic Factors in the Growth of Corporation Giving*. New York: National Bureau of Economic Research and Russell Sage Foundation.

Ostlund, L. E. 1977. "Attitudes of Managers toward Corporate Social Responsibility," *California Management Review* 19: 36–49.

Ouchi, William G. 1984. *The M-Form Society: How American Teamwork Can Recapture the Competitive Edge*. Reading, Mass.: Addison-Wesley.

Perrow, Charles. 1972. *Radical Attack on Business: A Critical Analysis*. New York: Harcourt, Brace and Jovanovich.

Pfeffer, Jeffrey, and Gerald Salancik. 1978. *The External Control of Organizations: A Resource Dependence Perspective*. New York: Harper and Row.

Preston, Lee E. 1981. "Corporate Power and Social Performance: Approaches to Positive Analysis." In *Corporate Social Performance and Policy*, Vol. 3. Greenwich, Conn.: JAI Press, pp. 1–16.

Schwartz, Robert A. 1968. "Corporate Philanthropic Contributions." *Journal of Finance* 23: 479–97.

Sheehan, Robert. 1966. "Those Fund Raising Businessmen. . . ." *Fortune* (January), pp. 148 ff.

Smith, Hayden W. 1986. "Problems in the Measurement of Corporate Contributions." In *Philanthropy, Voluntary Action, and the Public Good*. Spring Research Forum Working Papers, March 13–14. New York: Independent Sector, pp. 121–40.

———. 1983. *A Profile of Corporate Contributions*. New York: Council For Financial Aid to Education.

Smith, Lee. 1981. "The Unsentimental Corporate Giver." *Fortune* (September), pp. 121 ff.

Troy, Kathryn. 1982. *The Corporate Contributions Function*. New York: Conference Board.

Turner, Judith. 1984. "IBM Increased its Support of Education More than 50 Percent in 1983." *Chronicle of Higher Education* (January 25): 9.

Useem, Michael. 1987. "Corporate Philanthropy." In Walter W. Powell, ed. *The Handbook of Non-Profit Organizations*. New Haven, Conn.: Yale University Press, pp. 341–59.

———. 1984. *The Inner Circle: Large Corporations and the Rise of Business Political Activity in the U.S. and U.K.* New York: Oxford University Press.

Wall, Wendy L. 1984. "Companies Change the Ways They Make Charitable Donations." *Wall Street Journal*, (June 21): 1 ff.

Wallich, H., and J. J. McGowan. 1970. "Stockholder Interest and the Corporation's Role in Social Policy." In *A New Rationale for Corporate Social Policy*. New York: Committee for Economic Development.

White, Arthur H. 1980. "Corporate Philanthropy: Impact on Public Attitudes." In *Corporate Philanthropy in the Eighties*. Washington, D.C.: National Chamber Foundation, pp. 17–19.

White, Arthur, and John Bartolomeo. 1982. *Corporate Giving: The Views of Chief Executive Officers of Major American Corporations*. Washington, D.C.: Council on Foundations.

14

The Management of Corporate Giving Programs

E. B. KNAUFT

Although there has been an increase in research on corporate philanthropy in recent years, the literature in the field is still relatively limited. Previous studies (see the References) have generally explored such areas as the collection of statistical information on the giving levels of companies, the staffing of the contributions function, and descriptions of the rationale for corporate giving. Only the studies of Levy and Shatto (1977) and Siegfried and Maddox (1981) investigated in some depth the giving programs of a number of companies.

This chapter describes research that examined a larger number of variables, studied the decision-making process in detail, examined the extent to which contributions programs reflect the culture and environment of a company, and analyzed information about contributions managers' opinions of the factors characterizing excellence in a contributions program.

DESCRIPTION OF SAMPLE OF COMPANIES STUDIED

A sample of 48 American corporations provided the basis for the study. The primary selection criterion was evidence that a company possessed a well-developed contributions program and employed at least one professional staff person devoting full time to this activity. The sample was weighted in favor of large corporations. To reduce travel expenses, the companies were selected in seven cities in the East, Midwest, and West. Table 14.1 lists the major data on the companies.

This paper describes research supported by Yale University's Program on Non-Profit Organizations and by grants to that Program by Aetna Life and Casualty Foundation, Atlantic Richfield Foundation, Exxon Corporation, General Electric Foundation, General Mills Foundation, Levi Strauss Foundation, and Metropolitan Life Foundation.

Table 14.1   Description of the Companies Studied ($N = 48$)

Types: 30 manufacturing, 18 nonmanufacturing; 15 different industry classifications.
Geographic location: East, 20; Midwest, 12; Far West, 16.
1983 earnings: From $14 million to $6.2 billion; mean = $779 million; median = $187 million.
Five-year annual earnings growth: From −54% to +43%; mean = 8%; median = 10.5%.
Number of U.S. employees: From 1500 to 450,000; mean = 54,000; median = 27,500.
1983 contributions: Range: $98,000–53 million; mean = $9.2 million; median = $4.3 million; first quartile = $2.3 million.
1983 contributions as a percent of pretax income: Range: 0.6–5.0%; mean = 1.9%; median = 1.3%.
Percent of company's contributions allocated to United Way: Range: 3–44%; mean = 16.5%; median = 14%.
Percent of company's contributions allocated to employee matching gift program: Range: 2–33%; mean = 9.8%; median = 7%.
Companies using a company foundation: 77%.
Public disclosure: 54% publish an annual list of contributions.
Size of professional contributions staff: Range: 1–20; mean = 3.8 persons; median = 3.
Size of administrative contributions staff: Range: 1–26; mean = 3.6 persons; median = 2.
Sex of contributions manager: 52% men; 48% women.

ANALYSIS OF QUANTITATIVE DATA

Quantitative data were collected from each company to describe its size, financial performance, and contributions program. In addition, subjective numerical ratings were made by the author on ten aspects of each company's program. Various analyses helped identify relationships among variables that might have statistical significance.

Several factors were found to affect the relationships among contributions and corporate earnings, contributions as a percentage of pretax net income and as a percentage of stockholder dividend, and five-year earnings growth.

- There is a high positive relationship between a company's earnings and its level of contributions.
- There is a negative, but not statistically significant, relationship between a company's earnings and its contributions expressed as a percentage of pretax income. That is, there is a slight trend for companies with higher earnings to give a lower *proportion* of their income than companies with lower earnings.
- There is no relationship between a company's total contributions and the ratio of contributions to pretax income.
- Contributions as a ratio of dividends is highly related to contributions as a ratio of pretax earnings, but contributions as a ratio of dividends is not related either to total dollar contributions or to earnings.
- The five-year earnings growth rate of a company is not related to either the amount of contributions or to contributions as a percentage of pretax income. The fastest growing companies in the sample do not tend to have the larger or the "more generous" programs.

The existence or absence of a foundation was found to have no statistically significant impact on any of these variables:

- Total contributions;
- Contributions as a percentage of pretax income;
- Size of professional contributions staff;
- Subjective rating of overall quality of contributions program (as will be explained later).

In addition, no significant relationship was found between the existence of a foundation and publication or nonpublication of an annual list of grants.

Relationship between proportion of company contributions to the United Way and other variables was not correlated significantly with any of the following variables:

- Total contributions;
- Contributions as a percentage of pretax income;
- Percent of total contributions made to a matching gift program;
- Size of professional staff;
- Subjective rating of the uniqueness or specificity of the company's giving priorities or the company CEO's involvement in individual grant decisions.

However, a significant negative relationship was found between percent of United Way giving and a subjective rating of overall quality of the company's program. The larger the proportion of United Way giving, the lower the program quality.

In none of the following variables was there a statistically significant relationship between contributions to a matching gift program and the proportion of all giving:

- Total dollar contributions;
- Contributions as a percentage of pretax income;
- Size of professional staff;
- Subjective rating of uniqueness of priorities, CEO's involvement in grants, or overall quality of contributions programs.

The following variables were significantly related to the gender of the contributions manager:

- The larger contributions programs tended to be managed by men.
- The "more generous" programs relative to a company's earnings were managed by women.
- Women managers tended to be subjectively rated in interviews as more "astute."

The following variables were not significantly related to the gender of the manager:

- Five-year earnings growth rate of the company;
- Percent of total contributions in "arts and culture";
- Size of professional staff;
- Subjective rating of extent to which "liberal" organizations are funded;
- Existence of a foundation;
- Publication of a grant list.

Drawing from interviews and data, the author ratings were made on eleven attributes of each company's program. Ratings were made on a 1–

Table 14.2 Descriptive Statistics on Subjective Rating Items
(N = 48 Companies)

Item	Range of Scores	Mean Score	Standard Deviation
Specificity of Priorities	2–5	3.12	0.93
Uniqueness of Priorities	1–5	3.15	1.04
Adherence to Priorities	2–5	3.37	0.70
Autonomy of Staff Decisions	1–5	3.10	1.06
Involvement of CEO in Grants	1–5	3.57	1.27
Senior Management Interest and Support of Contributions Program	2–5	3.75	0.99
Breadth of Input in Contributions Decisions	1–5	3.29	1.37
Specific Target for Giving Exists	1–5	3.49	1.47
Liberal Organizations Funded	1–5	3.32	1.01
Astuteness of Interviewee	1–5	3.54	1.01
Overall Program Quality	2–5	3.73	0.86

5 scale (see Table 14.2). An analysis of intercorrelations among all variables indicated that there were higher relationships between "overall program quality" and each of the other variables than in other cross comparisons between pairs of these variables. For example, overall program quality correlated highly with certain variables: "uniqueness of priorities," "adherence to priorities," "senior management interest and support," "funding liberal organizations," and "astuteness of interviewee."

The use of subjective ratings, included in the project for experimental reasons, appears to have serious limitations in that there are such high intercorrelations, suggesting a "halo" effect. In making ratings of overall program quality, the investigator may have been unduly influenced by (1) the astuteness and articulateness of the interviewee, and (2) the uniqueness of the company's priorities.

Several multiple regressions were conducted between selected groups of quantitative variables. In both of the following, the dependent variable is 1982 contributions as a percentage of pretax income. With the following independent variables

- 1982 and 1983 earnings;
- Five-year earnings growth;
- United Way as percent of total giving;
- Matching program as percent of total giving;
- Size of professional staff;
- Extent of senior management interest and support (subjective rating);
- Existence of giving target (by amount or percent)

the multiple correlation was .742, corrected to .633, producing a significance level of 2 percent, indicating a reasonable probability that contributions as a percentage of pretax income can be predicted with these variables. Interestingly, the independent variable with the largest negative weight was 1982 earnings, again corroborating the thesis that companies with higher earnings tend to give at a lower level when contributions are expressed as a percentage of pretax income.

With the following independent variables

- 1982 total dollar contributions;
- Size of professional staff;
- Percent of total contributions to United Way;
- Percent of total contributions to matching gift program;
- Percent of total contributions to education;
- Percent of total contributions to health and human services;
- Percent of total contributions to community causes or organizations;
- Percent of total contributions to the arts and culture.

the multiple correlation was .712, corrected to .585, with a significance level of 4 percent, indicating a significant prediction of percent giving level is possible with these variables.

ANALYSIS OF INTERVIEW DATA—THE DECISION-MAKING PROCESS

The research plan was predicated on the assumption that the decision-making process could be analyzed on the basis of three factors: (1) the total annual dollar level of contributions to be made, both long range and for the particular year; (2) the establishment of policies and priorities of giving; and (3) whether to fund individual grants and the dollar amount for each grant. But, the majority of companies did not follow such a process. Although decisions as to level of giving were usually made at the start of each year, the character of the contributions programs of many companies was being determined by the sum total of case-by-case decisions rather than through a systematic adherence to well-developed priorities. Similarly, there often was no explicit statement of rationale or objectives for the contributions program. Grant making in many of the companies studied was "reactive" rather than "proactive," in that many decisions were made in response to a wide range of unsolicited grant requests. In contrast, companies that had well-defined grant priorities used them to attract and select grants that met company objectives.

A second major finding about decision-making was that a number of management levels tend to be involved and at least one committee plays a key role in grant decisions. Such committees may be composed entirely of company employees or of members of a committee of the company's board of directors or both.

The Decision-Making Hierarchy

Hierarchy and committee involvement vary from company to company and with the kind of decisions being made. Figure 14.1 portrays the two

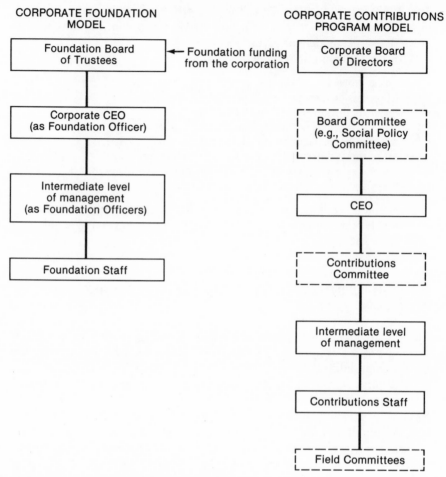

Figure 14.1 Hierarchy of decision making.

hierarchies typically involved in contributions decisions. The solid boxes represent levels found in all companies studied, but boxes drawn with dashed lines were found only in some companies. Although more than three-fourths of the companies had foundations, generally they operated a corporate program in tandem with the foundations, with the same company personnel serving on both staffs. Similarly, where both foundation boards and company contributions committees exist, the same individuals typically serve on both. With very few exceptions, foundation boards are composed entirely of company officers and management, and for practical purposes the two models are quite similar. The major exception is that contributions to the foundation's corpus are by action of the company's board of directors, whereas the use of those funds in grant making is the legal responsibility of the foundation board. The following analysis deals with the corporate contributions program model.

Types of Decisions

Total level of giving

Typically, the annual contributions budget is approved by the Chief Executive Officer (CEO) of the company. The contributions manager first develops a recommendation, which is reviewed by the next level of management and often by a contributions committee. This process gives greatest weight to the amount budgeted the previous year and to current and projected corporate profits. If profits are improving, there is a tendency to increase the contributions budget modestly.

Relatively few of the companies established a long-range target. The most sophisticated approach involved a target derived from a percentage of pretax corporate profits, based on the earnings of the past year, or an average of the preceding three years. Where such a target was used, it was set between 1.5 percent and 2 percent of pretax earnings. Very few companies publicize the target externally. Some establish an internal target amount but no target period. Contributions rarely rise dramatically because companies are generally reluctant to increase the size of the contributions staff and do not want to make contributions beyond a level that can be well administered by the existing staff.

Priorities

Priorities tend to be stated in broad, conventional categories, such as education, arts and culture, health and social services, or civic causes. Largely reflecting the company's past practices and the grant requests received, they are infrequently reassessed.

In contrast, some companies have augmented the conventional categories with more specific priorities and have set aside special funds for these areas. The following are examples:

- Urban public education—support of programs that develop and sustain school initiatives, including school-business relationships and responsible parent involvement;
- Reform of the civil justice system—support for development and evaluation of dispute resolution systems as an alternative to the courts;
- Neighborhood revitalization—support of national and regional organizations that provide technical assistance to neighborhood-based groups engaged in revitalization and support of innovative grass-roots efforts to reduce arson;
- Nonprofit management—support of projects to improve the management and fund-raising capabilities of nonprofit voluntary organizations;
- Minority science and engineering education—support of programs to increase the interest and knowledge of minority high school students in the sciences and scholarships and remedial programs in science and engineering at the college level.

Other guidelines

Corporations frequently establish limits by identifying organizations or activities they will not fund, such as religious organizations or those located outside cities where the company operates. More interesting are positive examples of guidelines that cut across the subject areas of giving. For example:

We overlay our priorities with three other criteria. The first is "leverage," where we want other grantmakers to join us in an innovative opportunity or where our grant is structured to encourage the donee to get additional funds from other sources. The second is "partnership," where we seek a joint venture with the public sector, such as a school system, as the best way of addressing some community problems. And third, we want to make sure that a fair share of our grant money goes away from our headquarters city to other places where we have . . . employees.

The Role of Boards and Committees

Corporate boards or committees of the board give final approval to the annual contributions budget and, where a foundation exists, approve the company's lump sum contributions to the foundation. Typically, contributions objectives and policy are developed by the staff, reviewed by an internal committee, and then given perfunctory approval at the board level. Grants over a fixed amount are approved by the corporate or foundation board. Board decisions are heavily influenced by the CEO.

Internal committees are key decision-making bodies. They vary greatly in composition, status, and mode of operation. In some companies, committees provide active and creative direction to contributions practices, and members may even make on-site investigations of potential grantees. Committee membership is usually restricted to upper- and middle-level line and staff managers. Several companies include rank-and-file employees—usually smaller firms that support employee participation in other types of decisions. Internal committees generally operate as follows:

- They are vested with considerable authority, take their work seriously, and genuinely assist the contributions manager.
- They spread the work of thinly staffed contributions units and capitalize on the technical expertise and interests of the company staff.
- They are sometimes used to expose "fast track" operating managers to external experiences and issues in preparation for senior management positions.
- Some companies rotate their committee members regularly. It takes time to orient new members but rotation enlarges the number of employees who are knowledgeable about contributions, and many of them come to enjoy this activity.

The Role of the Company's Environment

Every company has a unique culture, reflecting its history, size, line of business, and the impact of past and present CEOs. Since contributions activity is subjective, discretionary, and more visible to the local community than many business functions, the company's culture can have a significant effect on grant making. For example, it may influence a newly promoted CEO who was initially unenthusiastic about contributions but later acknowledges its value. Other manifestations of company culture include the following:

- Smaller and newer companies often have a more open, participatory management style, which results in a contributions committee that assumes more initiative and is characterized by more give-and-take in its meetings. They are more likely to include lower-level management and rank-and-file employees on contributions committees, especially in plant locations.
- Companies with highly decentralized decision making will delegate more contributions authority to divisions and field units. In such a company, the corporate contributions unit operates as a counselor and monitor, and restricts its grant making to the headquarters community or to national projects.
- Companies with greater concern for employee relations respond more generously in times of adversity. Take two companies that had about the same level of corporate profits. The first continued to make grants in a small community after it closed its plant because the company assumed local needs were greater at such a time. By contrast, the second company, which had laid off a large proportion of employees in another small town, turned down a grant to an important local agency because they felt the grant would give employees a conflicting message.

CONTRIBUTIONS MANAGERS' APPRAISAL OF FACTORS DESCRIPTIVE OF CONTRIBUTIONS PROGRAMS

The author obtained the opinions of fifty corporate contributions managers on (1) the performance of their own contributions program on each of eighteen factors, and (2) the extent to which such factors were important in describing the "model" contributions program. All were experienced managers in large companies that had well-established contributions programs.

Appraisal of the Relative Importance of Factors of Their Programs

The form used to collect the data is shown in Figure 14.2. Eighteen factors, grouped under four major headings and representing items describing various aspects of a contributions program, were appraised. The author hypothesized that the clear presence of these factors in combination might constitute the "model" or "excellent" program. Table 14.3 shows the distribution of "scores" of the fifty companies on the questionnaire. An analysis of the average scores assigned to each of the eighteen items is included in the next section.

Table 14.3 Distribution of Scores on Questionnaire

Scores	Number of Companies Receiving Score	Percent of Companies	Cumulative Percent
2.0–2.4	1	2	2
2.5–2.9	5	10	12
3.0–3.4	7	14	26
3.5–3.9	18	36	62
4.0–4.4	11	22	84
4.5–4.9	8	16	100
Total	50	100	

Below are a number of factors that may be present in a company contributions program. Please indicate in the column to the right the extent to which each of these factors is descriptive of *your* program on a scale of 1 to 5 where:

1 = the factor is not at all descriptive of your program
5 = the factor accurately describes your program

(The numbers 2, 3, and 4 may be used to indicate items that fall between the two extremes.)

Factors	Rating (1,2,3,4, or 5)
1. Total Level of Giving:	
a. Budget constructed at start of each year	_____
b. Numerical, long-range goal established for internal purposes	_____
c. Numerical, long-range goal announced publicly	_____
2. Pattern or Character of Giving:	
a. Written objectives and rationale of program are in existence	_____
b. Areas of giving have been identified and clearly defined	_____
c. Special high-priority areas of giving have been identified and clearly defined beyond or within the broader categories	_____
d. Areas of giving are formally reassessed at least once every three years	_____
e. Exceptions to defined areas of giving do not represent more than 5 percent of the annual dollar pay out of the program	_____
f. The proportion or dollar value of grants in nonheadquarters locations to total grants fairly reflects the representation of those locations in terms of number of employees, revenue produced, etc.	_____
3. Decision Making and Review Process:	
a. The contributions budget contains a contingency, discretionary, or unallocated category representing at least 20 percent of the total budget	_____
b. Defined dollar approval levels exist for unbudgeted grants at each organizational level	_____
c. The chief executive officer (CEO) is an active supporter of the contributions program	_____
d. The CEO desires to have a contributions program that is comparable to or better than the best programs of peer companies	_____
e. A formal mechanism (such as membership on a contributions committee) exists for line managers to have an impact on the contributions program	_____
4. Grant Seekers' Accessibility to Program	
a. Published statement of contributions guidelines, priorities, and application procedures is available	_____
b. Published grant list, including amounts, is available	_____
c. Grant seekers submitting formal proposals consistently receive prompt response as to status of their request, including, as appropriate, the time frame for further steps	_____
d. Grant seekers are automatically notified of status of any grants still pending within three months after original response	_____

Figure 14.2 Describing your company's contributions program.

Appraisal of the Relative Importance of Factors Descriptive of a "Model" Program

The second step was collection of ratings from the same fifty managers as to the extent each of the eighteen factors described a high-quality or "model" contributions program. The same five-point scale was used. Table 14.4 shows the factors listed in rank order, with those evaluated as most descriptive of a "model" program at the top of the list. The two factors receiving the highest rating were "areas of giving have been identified and clearly defined," and "the CEO is an active supporter of the contributions program." The numerical goal-setting factor came out a little above the middle (with an average score of 3.8), whereas "numerical, long-range goal announced publicly" received the lowest rating (1.7). This finding confirms the general impression that many larger companies do not establish a long-range giving goal and are even more reluctant to "go public" with any goal that may exist.

A final step involved a comparison between the average scores on each item for one's own program and the scores the managers assigned to the same items in a "model" program. This comparison is shown in Table 14.5 with the right-hand column showing the differences between average ratings of "own" and "ideal" for each item. A positive value in the "difference" column indicates that "own" program received a higher value on the particular item or factor; a negative value, vice versa.

The factors on which the managers felt most clearly that their own program came out ahead of a "model" program were the following:

- Areas of giving have been identified and clearly defined;
- Exceptions to defined areas of giving do not represent more than 5 percent of the annual dollar pay out of the program;
- The proportion of dollar value of grants in nonheadquarters locations fairly reflects the representation of these locations in terms of number of employees, revenue produced, and so forth;
- Defined dollar approval levels exist for unbudgeted grants at each organizational level.

The factors managers felt most clearly indicated their program was not achieving the standards of a "model" program were as follows:

- Numerical long-range goal established for internal purposes;
- Areas of giving are formally reassessed at least once every three years;
- The contributions budget contains a contingency, discretionary, or unallocated category representing at least 20 percent of the total budget;
- Published statement of contributions guidelines, priorities, and application procedures is available.

In the following section, I attempt to integrate all data from the study and consider more fully the factors that might be used in a composite definition of excellence in corporate contributions.

Table 14.4   Importance Ranking of Factors Describing a Company
Contributions Program

Each factor was rated by each contributions manager on a scale of 1 to 5 where 1 = no
importance to the quality of the program, and 5 = of great importance to the quality of the
program.

The factors are listed in order of relative importance, the highest first, based on their
average ranking by the fifty managers:

Rank	Factor	Average Score
1	Areas of giving have been identified and clearly defined	4.7
2	The chief executive officer (CEO) is an active supporter of the contributions program	4.6
3	Budget constructed at start of each year	4.53
4	Written objectives and rationale of program are in existence	4.48
5	Areas of giving are formally reassessed at least once every three years	4.4
6	Grant seekers submitting formal proposals consistently receive prompt response as to status of their request, including, as appropriate, the time frame for further steps	4.3
7	Special, high-priority areas of giving have been identified and clearly defined beyond or within the broader categories	4.28
8	The CEO desires to have a contributions program that is comparable to or better than the best programs of peer companies	4.2
9	Published statement of contributions guidelines, priorities, and application procedures is available	4.1
10	Numerical, long-range goal established for internal purposes	3.8
11	A formal mechanism (such as membership on a contributions committee) exists for line managers to have an impact on the contributions program	3.8
12	Grant seekers are automatically notified of status of any grants still pending within three months after original response	3.7
13	Defined dollar approval levels exist for unbudgeted grants at each organizational level	3.6
14	The contributions budget contains a contingency, discretionary, or unallocated category representing at least 20 percent of the total budget	3.4
15	Exceptions to defined areas of giving do not represent more than 5 percent of the annual dollar pay out of the program	3.3
16	Published grant list, including amounts, is available	3.2
17	The proportion or dollar value of grants in nonheadquarters locations to total grants fairly relects the representation of those locations in terms of number of employees, revenue produced, etc.	2.8
18	Numerical long-range goal announced publicly	1.7

Table 14.5 Comparison of "Own" Ratings with "Model" Ratings

Item Number (on Figure 14.2)	Own Average Rating	Model Average Rating	Difference (Own-Model)
1 a	4.8	4.5	.3
b	3.4	3.8	−.4
c	1.8	1.7	.1
2 a	4.4	4.5	−.2
b	4.5	4.7	−.2
c	4.2	4.3	−.1
d	4.1	4.4	−.3
e	4.0	3.3	.7
f	3.2	2.8	.4
3 a	3.1	3.4	−.3
b	4.0	3.6	.4
c	4.4	4.6	−.2
d	3.9	4.2	−.3
e	3.6	3.8	−.2
4 a	3.8	4.1	−.3
b	3.4	3.2	.2
c	4.1	4.3	−.2
d	3.6	3.7	−.1

CHARACTERISTICS OF AN EFFECTIVE CONTRIBUTIONS PROGRAM

Discussions with many contributions managers suggest that the "excellent" program has three components: the "mechanical" management, the character and philosophy as reflected by the nature of grants made, and the program size in relation to the size and resources of the business. Cutting across the three components is the program's ability to reflect the needs and nature of the business while responding appropriately to the needs of the community where the company is located and of the nation. The factors that describe the program management techniques and performance objectives that apply to the contributions function are listed in Figure 14.2, and these will not be discussed further here.

Character of the Program

The factors that make up the character of an effective contributions program include the following:

- Existence of a rationale, addressing such questions as What is the program's basic purpose? How does the company want employees, community people, and nonprofit organizations to view the program? What should be the content of the program five years from now? To what extent should the program actively seek and encourage grant requests in target areas?
- Existence of a process to involve company personnel in the program.

- Responsiveness of the program to the nature of the company's business and product line by relating company interests or expertise to grant making; for example, support of literacy training programs by a chain of book stores.
- Responsiveness of the program to the needs of community where the company is located; for example, a bank's sponsorship of a "Handi-Van" that provides supplies and expertise for senior citizens to make home repairs.
- Concentration of a portion of the program in several major grants sustained for several years, as contrasted with scattering all funds in small grants to many organizations.

Size of the Program

The program's dollar size, its most objective characteristic, often is an area of controversy among contributions professionals. Conventional management practice suggests that the contributions function should have long- and short-range numerical objectives, like most other business operations. The clearest example is the 2 percent and 5 percent "giving clubs" that started in Minneapolis and later spread to a dozen cities. The corporate members pledge to attain a giving level representing 2 percent or 5 percent of their pretax profits. Most major corporations have resisted the clubs because they are perceived as imposing an external goal on management and would often result in doubling or tripling a company's giving. Currently, the corporate world has not accepted any "ideal" level of giving; even the advocates of the concept of a target agree that no single index or ratio applies equitably to all corporations at all times.

Recognizing the difficulty, but adhering to the principle that "excellence" should include a quantitative objective, some contributions professionals have suggested the following guidelines for program size:

- A dollar contributions objective should be carefully developed and communicated to appropriate persons within the company. Final and intermediate target dates should be established. The goal should be developed after the purpose and content have been defined. It should be accompanied by staffing and committee assignments to produce a program that will achieve both qualitative and quantitative objectives.
- Companies with highly erratic profits should strive for a modicum of consistency in their giving levels. Many local agencies supported by the company will experience an increase in case loads and a decrease in support from other sources during times of economic downturn.
- The statement of a quantitative goal should specify categories. The first consideration should be achieving the desired level of cash contributions as distinct from gifts of equipment and the estimated value of time volunteered by employees and employee services donated by the company, important as they are as a means of corporate involvement.

CONCLUSION

The key findings of the study included the following. In many companies, grant making is "reactive" rather than "proactive": decisions are made in response to a wide range of unsolicited grant requests. Companies that developed well-defined grant criteria used them to attract and select grant requests that helped meet company objectives.

The annual level of giving is influenced most heavily by profits for the year just completed and projected profits for the coming year. The CEO plays a key role in determining the level of giving. The larger the company, the less the CEO is involved in decisions about individual grants. Internal committees play an important role in decisions on individual grants. These committees involve a wide range of line and staff employees at various levels. Effective use of such committees tends to broaden the "ownership" and support of the contributions program throughout the company.

There is a negative, but not statistically significant, relationship between a company's earnings and its contributions expressed as a percentage of pretax income. Contributions as a ratio of dividends are highly related to contributions as a ratio of pretax earnings but are not related either to total dollar contributions or to earnings. The five-year earnings growth rate of a company is related neither to the amount of contributions nor to contributions as a percentage of pretax income: the fastest growing companies in earnings do not tend to have the larger programs. The existence of a corporate foundation as a vehicle for giving was found to have no relationship to (1) the size of contributions budget, (2) contributions as a percentage of pretax income, (3) the size of the professional contributions staff, or (4) the annual publication of a list of grants.

Contributions managers in the study were about equally divided between men and women. The larger programs tended to be managed by men, but the "more generous" programs (measured by contributions as a percentage of pretax income) tended to be managed by women. The major characteristics of the effective contributions program are hypothesized to fall in three broad categories: (1) the "mechanical" management of the program, (2) the character and content of the program, and (3) the size of the program relative to the company's resources.

The components of excellence are hypothesized to be attained most fully when they reflect the commitment and support of the CEO in combination with a contributions manager, who structures a program that reflects the needs, culture, and resources of the company and responds creatively to relevant public needs.

REFERENCES

Galaskiewicz, Joseph. 1984. *An Empirical Study of the Community Influences that Shape Patterns of Charitable Giving by Corporations.* PONPO Draft Working Paper. Program on Non-Profit Organizations, Institution for Social and Policy Studies, Yale University.

———. 1983. *Professional Networks and the Institutionalization of the Single Mind Set.* Privately published paper. Minneapolis: University of Minnesota.

Harris, James F., and Anne Klepper. 1976. *Corporate Philanthropic Services Activities.* New York: Conference Board.

Knauft, Edwin B. 1984. "Recent Trends in Corporate Giving." *Foundation News* (August–September).

———. 1983. "Corporate Giving in Non-Headquarters Communities." *Foundation News* (September–October).

Lahn, Seth M. 1980. *Corporate Philanthropy: Issues in the Current Literature.* PONPO Working Paper 29, Institution for Social and Policy Studies, Program on Non-Profit Organizations, Yale University.

Levy, Ferdinand K., and Gloria M. Shatto. 1977. *The Evaluation of Corporate Contributions.* Working Paper E-77-12. Atlanta: Georgia Institute of Technology.

Muckler, Alice, ed. 1982. *Corporate Philanthropy.* Washington, D.C.: Council on Foundations.

Sigfried, John J., and Katherine Maddox McElroy. 1981. *Corporate Philanthropy in the U.S.: 1980.* Working Paper 81–W26. Nashville: TCS Management Group.

Smith, Hayden W. 1983. *A Profile of Corporate Contributions—1977.* New York: Council for Financial Aid to Education.

Troy, Kathryn. Annual. *Annual Survey of Corporate Contributions.* New York: Conference Board.

———. 1982. *The Corporate Contributions Function.* New York: Conference Board.

VI
CONTROL AND OUTGROWTH

15

Donor Control and Perpetual Trusts: Does Anything Last Forever?

FRANCIE OSTROWER

To one who is ambitious of exercising by means of a disposition of his property the greatest and most enduring influence on human affairs, charities offer the widest opportunity (Scott 1920).

Unlike other trusts, a trust for charitable purposes may last forever. Although the testator who leaves property to other beneficiaries is legally permitted to control the disposition of that property for only a generation or two, the testator who leaves wealth to charity can control the uses of that wealth for an unlimited period, in principle, possibly forever. From this conjunction of donor restrictions and perpetual life arises the issue of mortmain, or "dead hand control."

At various periods in the histories of both English and American charities, strong concern has been expressed over the potential control exerted by the "dead hand" over the living. Critics warned that perpetuity promoted the existence of obsolete trusts and wastefulness and protected the wishes of foolish and possibly malicious donors. Declaring that "property is not the property of the dead but of the living," Sir Arthur Hobhouse catalogued a series of obsolete and often rather bizarre gifts to demonstrate the harmful social and economic consequences of perpetual endowments in England (Hobhouse 1880). The case of a gift left by one testatrix to propagate the "sacred writings of the late Joanna Southcote," whose "one prediction," that she would give birth to the Messiah, failed to materialize when she died childless (ibid., 7), is but one of the many examples that led Hobhouse to conclude:

I would like to extend my thanks to Paul DiMaggio, Susan Kelley, Frank Romo, and John Simon for their comments on drafts of this paper. I would also like to express my appreciation to John Simon for bringing the importance of the dead hand control issue to my attention. Financial assistance for work on this paper was provided by a John D. Rockefeller 3rd Fellowship from the Program on Non-Profit Organizations at Yale University, and is gratefully acknowledged. Any opinions expressed in this paper should not be taken as representing those of anyone but the author.

[It is] the most extravagant of propositions to say that because a man has been fortunate enough to enjoy a large share of this world's goods in this life, he shall therefore and for no other cause . . . be entitled to speak from his grave *for ever* and dictate *for ever* to living men how that portion of the earth's produce shall be spent. (Hobhouse 1880, p. 9)

Warning that "it is absolutely certain that the testator will be ignorant of the needs and circumstances of the generations following him," Wayland recounted the example of a bequest left to support preaching in a once flourishing New York town that had been abandoned since the death of the donor (Wayland 1890). He further told of the case of an English bequest to provide a shilling apiece to twelve poor, elderly women every Easter—with the condition that they pick up the money with their lips from the grave of the donor.

John Stuart Mill deplored the irrationality that made "a dead man's intentions for a single day, a rule for subsequent centuries" and predicted that "There is no fact in history which posterity will find it more difficult to understand, than that the idea of perpetuity, and that of any of the contrivances of man, should have been coupled together in any sane mind" (Mill 1868 [1833], p. 6).

Finally, Frederick H. Goff is said to have been "obsessed with the thought that tens of millions of dollars of social capital was being left to molder because of the 'dead hand rule' " (Nielsen 1985, p. 244). Influenced by the work of Hobhouse and familiar with outmoded trusts such as the Mullanphy Fund in St. Louis to aid pioneers traveling west, Goff proposed the community foundation as a mechanism that would allow for a more flexible management of charitable funds in the United States (ibid.).

These early observers accurately noted the potential for conflict between the wishes of a testator and the interests and goals of living generations. Objecting in principle to the idea of perpetual endowments, these observers believed that as long as such endowments could legally be established, they would remain unchanged. This paper offers a preliminary argument, based on the existing literature, that these endowments have not remained unchanged, and suggests the ways and reasons why restricted bequests have been modified or, at times, radically altered. In numerous different ways, conflicts of values and conflicts of power have led to alterations in the influence accorded the donor to charity and to changes in endowments that were, in principle, to remain unchanged forever. The paper indicates potential types of future research that would examine conflicts surrounding dead hand control, the mechanisms by which endowments are modified, and why modifications occur at certain times.

LEGISLATIVE MODIFICATIONS OF THE "DEAD HAND"

A long-standing issue concerning the power of the dead hand has had to do with competing claims on potential contributions. Objections to dead hand control arise not only because it seems to grant too much power to the wishes of individual donors, but because it diverts and protects accumulations of wealth for certain social institutions. Over and against the diversion of wealth to charity have been set the claims of the family and the state. Concern has also been expressed about the perceived potential for donated assets to create centers of power that exert influence over the political and economic life of society. These concerns have been expressed in laws that place limits on what donors may contribute and on what charitable institutions may receive and have led, at times, to changes in the management of previously donated assets. Early English attempts to modify the power of the dead hand provide one illustration.

In England the bulk of charitable bequests during the Middle Ages went to the church for "pious uses." For the medieval donor, it has been argued, the "first thought is not the transmission of a *hereditas,* but of the future welfare of his immortal soul" (Pollock and Maitland, cited in Gray 1952, p. 33; see also Douglas and Wright 1980; Willard 1894). Not only was it customary for donors to make a bequest for "pious uses," but the church could appropriate a fixed portion of the estate for charity (Gray 1952). In this situation, concerns about the dead hand had little to do with the perpetuation of the wishes of an individual donor, but were directed at the potential concentration of land and power in the Church (Bomes 1976; Friedman 1977; Raban 1981). Since the church, unlike its members, did not die, gifts of land received would not be transferred at a later point but would remain permanently under church control. Bequests made to the church, therefore, resulted in a permanent loss of revenue and services to king and nobles. Thus, "The lands were said to come to dead hands as to the lords, for a dead hand yieldeth no service" (Bristowe and Cook, p. 371).

From the thirteenth century on, a series of mortmain laws were enacted to curb the power of the dead hand (Bomes 1976; Fisch 1953a; Jordan 1959; Owen 1964). Although these laws did not necessarily prevent ecclesiastical organizations from acquiring land, they made a license in mortmain, issued by the crown, a prerequisite. The power of the dead hand in this instance then underwent continual modification in the context of the conflict between church and crown. Notes one observer, "When read together, the body of English mortmain statutes through the twentieth century highlights periods of relative strengths of the royal and ecclesiastical authorities, as well as various related social shifts" (Bomes 1976).

Later in the history of English charities, the monasteries would be dissolved, and in one fell swoop an increasingly powerful secular government would frustrate the intentions of innumerable donors who had bequeathed property to the Catholic church. The confiscation of property left to the church represents not an adaptation of the past (as expressed in the bequest) to the present but a rejection of the past and the imposition of values other than those of the original donor on the uses of the property.

By the nineteenth century, a Select Committee concluded that existing mortmain laws served only to create confusion, increased costs, and barriers to philanthropic activities. Nonetheless, new legislation was enacted with the purposes of imposing even greater restrictions on the power of the dead hand. One historian has argued that "The controlling factor was . . . the pathological fear of popery, always latent in Britain, which passed into a violent phase during the 'papal aggression' scare at that time" (Owen 1964, p. 319). It is interesting to note that members of older, elite Catholic families testified in favor of continued regulation of donations to the church. Long involved in a conflict with the church hierarchy, these families warned of the undue influence that priests might exert over individual testators (Owen 1964).

Charitable contributions in England were, at times, subject to severe scrutiny in terms of current policy. Gifts for "pious uses" when not made to the established church were regarded as gifts for "superstitious uses." The king, possessed of the power of "prerogative cy pres" could and did modify the purposes for which gifts were intended in rather dramatic ways. One example, from the middle of the eighteenth century, involves money left in trust by a Jewish testator to support readings of and instruction in Jewish law. The donor's intention was declared to go against public policy, and the trust funds were used to support a Christian minister and to instruct children in the Christian religion (Da Costa v. De Pas, discussed in Clark 1957; Fisch 1953b).

Legislation was enacted to alter secular, as well as religious, charitable endowments. In the sixteenth century, numerous charitable bequests had been left for the purpose of establishing institutions to provide instruction in Latin and Greek. By the nineteenth century, educators in these schools wanted to expand the curriculum to include subjects such as modern languages, physical science, writing, and mathematics. The dead hand, it seemed at first, would remain in control, after a judge held, in 1805, that such a change could not be made. He stated, "The question is, not what are the qualifications most suitable to the rising generation of the place where the charitable foundation subsists, but what are the qualifications intended" (cited in Scott 1920, p. 4). Soon after, however, Parliament enacted statutes that would allow changes to be made under the supervision of public officials. In the case of religious endowments, modification was accompanied by opposition to the institution that held these endowments, the church. In the case of educational endowments, no such hostility existed, and change was in the form of adaptation rather than a radical rejection of the past.

In the United States, broad changes have also occurred in the degree of power accorded the dead hand. Early laws relating to charity reflected fears about the church and about influence that might be exerted over testators, leading them to bequeath wealth to charity at the expense of family (Friedman 1977, p. 23; on the development of American charity law, see Miller 1961). Although the situation varied by state, mortmain laws were enacted that set limits on the amounts that testators could leave and charitable institutions could receive from a given testator. Some mortmain laws also required that for a charitable bequest to be valid, the testator had to survive a minimum period of time beyond the making of the will (to avoid "deathbed bequests").[1] These laws have been explained in terms of fears that so much property would be amassed in the hands of the church that it would rival the power of the secular government; that too much wealth would be taken out of circulation, to the detriment of the economy; and that close relatives would be disinherited (see Bomes 1976; Miller 1961). In the particularly prohibitive state of Mississippi, it has been suggested, such laws were the product of fears that charitable funds would be dedicated to the advancement of emancipated black slaves (Zollman 1919).

The restrictiveness of early American charity laws, argues one historian, was not a product of hostility to charitable activity, but of hostility to associated institutions (Miller 1961). In particular there were fears that donations would increase the power of the church, and in Virginia, Jeffersonians fought against the establishment of any private institutions that would be free from public accountability (Hall 1982). In addition, ideas developed in the Enlightenment promoted the view that property should belong to the living and that "a preceding generation cannot bind a succeeding one by its laws or contracts" (Jefferson 1823, cited by Wayland 1890).

New York state law also was particularly restrictive. A gift of approximately $5 million, left by former Governor Tilden "to establish and maintain a free library and reading room in the City of New York and to promote such scientific and educational objects as my said executors and trustees may more particularly desire," was ruled invalid. In reaction, the legislature acted in 1893, and reversed the state's previously restrictive policy (Zollman 1919): "What changed the law in New York was the realization that when the Tilden trust failed, no dead hand became richer; but the city itself was the poorer" (Friedman 1973, p. 370).

During the period in which restrictive laws were most prevalent, the intentions of numerous would-be donors were not allowed to be implemented. Although the current situation with regard to charity is quite different, the degree of power accorded the dead hand remains constantly open to alteration in light of changing circumstances. Recent legislative attempts to curb the perceived power of private foundations substantiates this point.

Since the earliest foundations were established, periodic concern has been expressed over the potential power concentrated in these institutions. As one observer notes, "We witness a sharply adverse reaction when foundations stray either into the business sector or the governmental sector" (Simon 1978, p. 6). In 1969, congressional legislation incorporated provisions to deal with the potential exercise of foundation power in both these sectors. As a result, restrictions were imposed on the composition and management of foundation assets. In particular, a provision was adopted prohibiting foundations from holding a controlling interest in a business, and requiring foundations with such an interest to sell their stock. In enacting the legislation, therefore, change was brought about in existing gifts that had been made, and limitations were placed on the ability of future donors to pass control of their company to a foundation.

JUDICIAL MODIFICATIONS OF THE DEAD HAND

Court decisions may bear on the potential impact of the dead hand either before or after the terms of a bequest have been implemented. The doctrine of cy pres allows the court to direct a modified application of charitable funds, a modification that, in principle, carries out the original intentions of the donor as closely as possible. Conditions that must be met for the application of cy pres are (1) the presence of a general charitable intent on the part of the donor, and (2) evidence that the original purpose has become impossible, impractical, or illegal. The cy pres doctrine, then, appears to provide a mechanism for adaptation, responding to changing circumstances while retaining the wishes of the donor as paramount.

With the testator deceased, however, the "intentions" of the donor being considered by the courts are, in fact, interpretations.[2] For instance, one testator left a bequest to Amherst College to provide scholarships for "deserving American born, Protestant, Gentile boys of good moral repute, not given to gambling, smoking, drinking or similar acts" (quoted in Scott and Scott 1966, p. 719). The college, whose charter did not allow such religious restrictions, refused to accept the bequest unless the conditions were altered. It was then up to the court to decide whether it came close to the intentions of the donor to allow the college to take the funds free of restrictions, or to design some other arrangement such as appointing a trustee to hold the funds for Amherst (subject to the donor's restrictions), or to transfer the bequest to another college. In 1961, the court applied cy pres and allowed Amherst to accept the bequest free of restrictions.

In a New Jersey case, a trust was established by a testator to support the publication of his "Random Scientific Notes." The donor left a list ranking universities to which the bequest was to be offered. Finding the work to be of little scientific merit, each of the universities designated as a potential trustee declined the gift. Princeton University, however, offered to accept the bequest with the condition that the fund be used, not to publish the testator's manuscript, but to support research and study in the Department of Philosophy, claiming that this application of funds would be in keeping with the donor's intentions. In 1949, the courts, finding evidence of a general charitable intent on the part of the donor, agreed to this arrangement (see Scott and Scott 1966, p. 681; Simes 1955, p. 123).[3]

In some cases, numerous parties may go to court offering conflicting interpretations of the donor's "true" intentions that support their respective views concerning what should be done with a charitable trust. A vivid illustration is provided by the recently settled case of the Buck Trust, left to the San Francisco Foundation to be used for charitable purposes in Marin County, one of the wealthiest areas in the country. Thought to be worth approximately $10 million at the time the donor died, the value of this gift increased about thirty times within the space of a few years. Although the foundation's managers initially administered the funds within Marin County, they later petitioned the court for permission to enlarge the geographical area in which the money could be used. Charities from surrounding areas were vocal in supporting modification of the restrictions, but the County of Marin opposed any change. While the one side insisted that had the donor known the actual size of the gift, she would never have intended that the money be restricted to Marin County, the other side accused the foundation trustees of being "grave robbers," of mismanagement and of violating a trust (on this case, see Nielsen 1985).[4]

Cases such as these demonstrate that controversies over "dead hand control" may be more a matter of conflict between living individuals than of conflict between the wishes of a deceased donor and "living generations." Although commentators have indicated the discrepancies that can arise between the wishes of the donor and efficiency or the public interest, more attention needs to be given to the way in which the wishes of donors interact with the interests and views of living individuals, leading the living either to support literal adherence to, or deviation from, the terms of a trust.

A historical example in this country is the case of the Mullanphy Trust, left by the donor in 1851 to provide assistance to travelers passing through St. Louis on their way west. Attempts to modify this trust were made by heirs trying to break the trust and obtain trust funds, and by the City of St. Louis, which, claiming the original purposes of the trust had become outmoded, wished to use the funds for other purposes. In 1934, after a series of court cases, the restrictions were modified so that trust income was used to assist travellers' aid associations in St. Louis (for discussion of parts of the history of this litigation see Simes 1955, p. 128).

To take a current example, debate has arisen over a bequest left in 1949 by Ernest Stillman, giving Black Rock Forest to Harvard University. Harvard has decided to sell the forest, a move that has been criticized by a son of the donor as well as by representatives of environmental groups. One of the donor's sons insists that his father's intention was to maintain the forest and perpetuate research projects there, and one leader of an environmental group has said "we believe Harvard is betraying a trust" (*New York Times*, April 15, 1985). Harvard, which wishes to use the proceeds from the sale to support research at another of the university's forests maintains, in contrast, that the intentions of the donor were to support Harvard forestry in general. Another of the donor's sons, moreover, agrees with the university's view of his father's intentions. Ultimately, the sale of the forest must be approved by the courts, and the approval will depend on which interpretation of the donor's intentions is accepted (*New York Times*, January 1, 1985).[5]

Such examples suggest the need for additional research to understand the factors influencing changes made to charitable bequests by the courts. Given the interpretive nature of establishing the donor's intentions, research needs to be undertaken to explain variation in the willingness of the courts to apply cy pres, and in the changing inclination of courts to apply literal or broad interpretations of the donor's intentions. In the United States, for instance, early hostility to the cy pres doctrine gave way to a more liberal attitude (Bradway 1931; Fisch 1953b). Moreover, because the fate of a bequest may hinge on the type of controversy that arises, research is needed to examine (1) when debates over dead hand control arise, (2) who challenges continued adherence to dead hand control, and (3) who acts as the living advocate of the "dead hand."

INFORMAL MECHANISMS FOR ADAPTING TO DEAD HAND CONTROL

Judicial modification of the terms of a charitable trust may be costly and time consuming. In addition, such modification is, in principle, meant to apply to extreme cases in which it is no longer possible or practicable to adhere to the wishes of the donor. What, then, becomes of those restricted trusts that do not become the subject of a court debate? Does the "dead hand" continue to rule the living? Or has there been change in the implementation of these trusts? Although the lack of data on the subject makes it impossible to draw a conclusion at this time, preliminary evidence suggests that many conflicts brought up by donor restrictions are not handled formally within the courts but are dealt with informally within nonprofit organizations. Preliminary and speculative in nature, the discussion in this section is meant to indicate the importance of examining informal mechanisms for adapting to dead hand control, rather than to draw conclusions about them.

In the case of universities, charters have been revised by trustees, and religious requirements concerning governing bodies and instructors changed: "And even when such changes have not been made, the requirements have often been evaded, particularly requirements as to the beliefs in certain religious dogmas; and such evasions have been winked at" (Scott 1920, p. 18). In the case of foundations, the introduction of a professional staff and bureaucratization can lead the organization in directions other than those intended by the original donor (Simon 1978).

One factor influencing organizational change may be the absence or presence of an "advocate" for literal adherence to the wishes of the donor. For instance, Yale received John Trumbull's art collection with the stipulation that the donor be buried below the collection. The university complied, but as Yale's art collection expanded, a new art gallery was built across the street. Yale constructed a bridge connecting the old and new buildings to remain in compliance with the will, an action, according to one account, that came at the insistence of Trumbull's heirs (*Yale Alumni Magazine*, October 1984). Yet heirs need not become interested or involved with the charitable donations of parents. For some organizations, contact with descendants pursuing what has been done with a gift may be relatively rare.

Organizations may devise ways in which the constraints imposed by donor restrictions are minimized, while technically adhering to the wishes of donors. For instance, a donor may want to benefit one specific aspect of an organization's operations. This aspect of operations, however, may not be the one for which the administrators seek funds. The organization may then accept the restricted bequest and cut back on other funding to that aspect of the organization, diverting it to other areas. Although the donor's intention to fund only one part of the organization has been complied with, the intention to provide additional support for that part of the organization has not met with success (example from personal communications from nonprofit administrators 1984, 1985).

When the degree of constraint posed by donor restrictions is perceived as being too severe, other measures may be taken. The art museum field is one in which restrictions are pervasive. According to one estimate, 90 percent of the art in American museums was given with restrictions (Thompson 1986). Examples of such restrictions include the stipulations that donated works not be sold; that donated works always be shown together; and that donated works be shown a certain number of days per year. Another problem for museums has been the unwillingness of donors to provide funds for operating, as opposed to capital, expenditures (Meyer 1979; Vladeck 1976). Compliance with restrictions on works of art, then, may be perceived as posing an additional strain on already scarce resources. According to one museum administrator, whereas some organizations accept gifts with restrictions that are perceived as onerous, hoping that some way can later be found to get around them, others, such as his, have begun to turn down such bequests (personal communication 1984). If enough organizations pursued this strategy, donors might be persuaded to stop imposing certain types of restrictions or to discuss possible restrictions with administrators before designing their bequests.[6] Although little is known about the ongoing ways in which nonprofit organizations manage restricted bequests, such information is crucial for assessing the impact of dead hand control on philanthropy. For this reason, additional research is needed both on when restrictions become problematic and on how nonprofit organization adapt to them.

THE SOCIAL FUNCTIONS OF DEAD HAND CONTROL

As we have seen, dead hand control has evoked concern because of the power it appears to give individual donors over the fate of future generations. "Piety" accorded to the wishes of the dead, it has been observed, may not coincide with "efficiency" or "contemporary relevance" (Douglas and Wright 1980), and "sanctity for the trust document" may not coincide with the "benefit of beneficiaries" (Bradway 1931). The principle according to which the wishes of donors determine the future uses of their contributions, however, remains strong. Although many have advocated the legitimacy of modifying restricted bequests on the grounds that it would be socially beneficial, actual changes have been justified on the grounds that they are in accordance with the intentions of the donor. As one legal scholar notes, moreover, the courts have been hesitant to apply the cy pres doctrine, and thus acknowledge that a change in charitable purposes is being made, preferring to regard changes as only concerning the administration of assets (Fisch, Freed, and Schachter 1974).[7] The questions that arise, therefore, are, What accounts for the strength of a belief that has been seen as frequently resulting in socially undesirable consequences? Should "dead hand control," therefore, be seen as a concession to the rights of individuals at the expense of society?

From the perspective adopted by this discussion, the strength of the dead hand principle results not solely from beliefs concerning the rights of individuals, but rather from the conjunction of attitudes concerning the right of individuals to exercise discretion over the disposition of their property, and attitudes concerning what is socially desirable; namely, support for charitable institutions. As Simes notes: "What the law has done is to say to the dying testator: If you will dispose of your property within that broad area known as charity, then a public benefit is presumed; and, in exchange for that public benefit, you are permitted to determine the future disposition of your property without limitation as to time" (Simes 1955). In this regard, it should be noted that ending the ability of donors to control in perpetuity, as well as maintaining this ability, has been seen as having potentially negative social consequences. In particular, concern has been expressed that if donors did not believe that they could control the uses of their wealth in perpetuity, they would stop making charitable bequests (see Douglas and Wright 1980; Scott 1920; Simes 1955).[8]

Second, it should be noted that given this view of dead hand control, deviation from the terms of a trust becomes a violation not only of the wishes of an individual donor, but of the terms of a socially sanctioned exchange. Scott has pointed out that whereas the wishes of donors expressed at the time they make a gift are treated "like the law of the Medes and the Persians," any subsequent wishes of donors (if they were alive) would have no bearing on the uses of their contributions. This observation supports the point that at stake in modifying the terms of a trust is not only the wishes of an individual donor, but the wishes that the law has previously agreed to honor. That respect for the wishes of donors is partly based on the perception that it is desirable for charitable resources to be made available is also indicated by the fact that a policy of allowing perpetuity entails a desire to permanently safeguard trust funds for charitable purposes against the influence of other forces. Thus, the increase of government regulation of private foundations in 1969 in the United States was justified on the basis that protected funds were being used for other than charitable purposes and not, for instance, on the basis that charitable activities ought to be curtailed. While current policy in the United States takes a favorable view of the permanent dedication of resources to charitable purposes, an important area for research on dead hand control would be to examine historically and comparatively under what circumstances there have been challenges to the protection of such resources from the claims of heirs and from claims that such resources would be better employed if utilized according to market criteria or according to criteria established by the state. Similarly, philosophical and ethical debates about dead hand control, concerning the rights of past generations with regard to the wishes of present generations, might be expanded to include consideration of the potential claim of future generations on charitable resources accumulated over time.

SUMMARY

As we have seen, the strength of dead hand control comes from beliefs concerning both the rights of individuals with regard to their property and the social desirability of charitable institutions. While the principles behind dead hand control endure, the actual restrictions imposed on charitable trusts have been modified in response to a variety of factors: conflicts over power, conflicts over values, and the process by which the living interpret and implement the intentions of the original donor. Thus, in answer to the question posed by the title of this paper: in terms of donor control and charitable trusts, at least, nothing lasts forever.

NOTES

1. Hall's research on the development of philanthropic institutions suggests that the diversity or homogeneity of local elites may be an important factor in explaining variations by state in attitudes toward the dead hand, when he argues that in Philadelphia, where the law was restrictive, ethnic and religious heterogeneity prevented the concentration of institutional resources that was accomplished in New England, where the law was favorable toward charity (Hall 1982).

2. Indeed, one legal scholar has suggested that the notion of a general charitable intention is "a fiction which constitutes a cloak behind which the court can dispose of the fund in the manner it deems most desirable" (Fisch, Freed, and Schachter 1974, p. 438).

3. Another way in which the courts impose constraint on the influence of donors is found in decisions about what constitutes a "charitable" bequest. A colorful example involves the will of George Bernard Shaw, who created a trust to examine the savings of time and expense that would result from substituting a new alphabet for the English one, and to then transliterate his plays into the new alphabet. In 1957, a court declared that the trust could not be accepted as charitable, because it was not beneficial to the community and did not advance educational purposes, since it involved no teaching (although a compromise was eventually worked out that allowed part of the funds to be used in the way Shaw intended; Scott and Scott 1966, p. 682). And when one testatrix left a trust to establish an art museum to house

her collection, the court, in 1941, accepted evidence that the collection was of little or no artistic value and the trust failed (Scott and Scott 1966, p. 680).

4. In the Buck Trust case, parties to the dispute have agreed to a settlement outside of court that transfers trust assets to a separate foundation. According to the agreement, the money would continue to be used for Marin County with a portion of the funds, approximately one fourth, going to causes that are relevant to parties outside the county.

5. Trustees may oppose as well as advocate changes in the terms of a trust. In 1723, Thomas Betton left a bequest to be used partly to rescue British citizens captured by pirates and held captive as slaves in Turkey or Barbary, and partly for educational and other purposes. When, in 1829, the Attorney General brought a suit against trustees, claiming that funds could no longer be used for this outmoded problem, trustees replied that although it could not all be used, the purpose was not entirely obsolete. In 1841, however, cy pres was applied, and the income was diverted to educational purposes.

6. Although some informal mechanisms may serve to temper dead hand control, others promote conformity to donor wishes. In particular, the fear of adverse publicity and the perception that donors will withhold support if they do not believe their wishes will be followed, provide powerful incentives for organizations to express willingness to adhere to the wishes of donors. The adverse reaction to the sale of paintings from the de Groot collection (whose donor had not legally bound the museum to keep the paintings but had only requested that is do so) by the Metropolitan Museum of Art provides an example of the type of publicity that may result. Community foundations, attempting to provide a formal institutional mechanism for change, reserve the right to modify the terms of a bequest with the vote of the distribution committee. There is some preliminary evidence, however, suggesting that an important part of public relations for community foundations consists of assuring donors that the foundation does not impose such modifications frequently or easily. For instance, one community foundation executive, interviewed as part of a study of advisors to the wealthy, said that the fear of losing control is "a constant concern to people who establish donor advised funds with the foundation" (see Ostrower 1987, pp. 247–66).

7. Fisch et al. point, for instance, to a case in which a college was permitted to transfer its lands and other assets to a state university and to change its purposes from the operation of a college, to the operation of a foundation supporting educational, literary, charitable, and scientific projects: "The court deemed these changes not to be 'material' changes in the charitable uses intended by the donors on the theory that their donations would continue to be used to advance the cause of education" (Fisch, Freed, and Schachter 1974, p. 412).

8. This idea, and the assumptions concerning the behavior of donors that it implies, however, need to be subjected to greater scrutiny in the light of additional research: on donor motivations for imposing restrictions; on the variety of restrictions that donors impose; on the reasons for which donors seek control; and on the degree of control that donors seek. It might be, for instance, that even without the possibility of perpetuity, charitable dispositions of wealth might still offer donors a greater degree of control than other options (such as having wealth pass to the government). Furthermore, most concern about the consequences of dead hand control has to do with the potential for charitable trusts to become obsolete, a situation that is at least partially dependent on the narrowness of restrictions imposed on the bequest. For this reason, research should not only examine the willingness of donors to give without a guarantee of perpetuity per se but also consider the potential willingness of donors to seek perpetuity for certain types of restrictions while foregoing it for others.

REFERENCES

Bomes, Stephen D. 1976. "The Dead Hand: The Last Grasp?" *University of Florida Law Review* 28:351–64.

Bradway, John S. 1931. "Tendencies in the Application of the Cy-Pres Doctrine." *Temple Law Quarterly* 5:289–529.

Bristowe, Leonard S., and Walter I. Cook. 1889. *The Law of Charities and Mortmain,* 3rd Ed. of *Tudor's Charitable Trusts.* London: Reeves and Turner.

Clark, Elias. 1957. "Charitable Trusts, the Fourteenth Amendment and the Will of Stephen Girard." *Yale Law Journal* 66: 980–1015.

Douglas, James, and Peter Wright. 1980. "English Charities: Legal Definition, Taxation and Regulation." Working Paper 15, Program on NonProfit Organizations, Yale University.

Escarra, Jean. 1907. *Les Fondations en Angleterre.* Paris: A. Rousseau.

Fisch, Edith L. 1953a. "American Acceptance of Charitable Trusts." *Notre Dame Lawyer* 26: 219–33.

———. 1953b. "The Cy Pres Doctrine and Changing Philosophies." *Michigan Law Review* 51: 375–88.

Fisch, Edith L., D. J. Freed, and E. R. Schachter. 1974. *Charities and Charitable Foundations.* New York: Lond Publications.

Friedman, Lawrence M. 1977. "The Law of Succession in Social Perspective." In Edward Halbach, Jr., ed., *Death, Taxes and Family Property.* St Paul: West Publishing Co.

———. 1973. *A History of American Law.* New York: Simon and Schuster.

Gray, Hamish. 1952. "The History and Development in England of the Cy-Pres Principle." *Boston University Law Review* 33: 30–49.

Hall, Peter Dobkin. 1982. *The Organization of American Culture, 1700–1900.* New York: New York University Press.

Hobhouse, Sir Arthur. 1880. *The Dead Hand.* London: Chatto and Windus.

Jordan, W. K. 1959. *Philanthropy in England: 1480–1660.* New York: Russell Sage Foundation.

Meyer, Karl. 1979. *The Art Museum: Power, Money, Ethics.* New York: William Morrow and Co.

Mill, John Stuart. 1868 (1833). "The Right and Wrong of State Interference with Corporation and Church Property." In *Dissertations and Discussions,* vol. 1. Boston: W. V. Spencer, pp. 1–41.

Miller, Howard S. 1961. *The Legal Foundations of American Philanthropy 1776–1844.* Madison: State Historical Society of Wisconsin.

Nielsen, Waldemar A. 1985. *The Golden Donors: A New Anatomy of the Great Foundations.* New York: Truman Talley Books.

Ostrower, Francie. "The Role of Advisors to the Wealthy." In T. Odendahl, ed. *America's Wealthy and the Future of Foundations.* New York: The Foundation Center, pp. 247–66.

Owen, David E. 1964. *English Philanthropy: 1660–1960.* Cambridge, Mass.: Belknap Press of Harvard University Press.

Raban, Sandra. 1982. *Mortmain Legislation and the English Church: 1279–1500.* Cambridge: Cambridge University Press.

Scott, Austin Wakeman. 1967. *The Law of Trusts.* Boston: Little, Brown and Co.

———. 1920. "Education and the Dead Hand." *Harvard Law Review* 34: 1–19.

Scott, Austin Wakeman and Austin Wakeman Scott, Jr. 1966. *Cases and Authorities on the Law of Trusts.* Boston: Little, Brown and Co.

Simes, Lewis M. 1955. *Public Policy and the Dead Hand.* Ann Arbor: University of Michigan Law School.

Simon, John. 1978. "Charity and Dynasty under the Federal Tax System." *Probate Lawyer* 5.

Thompson, Nancy L. 1986. "Financially Troubled Museums and the Law." In Paul DiMaggio, ed. *Nonprofit Enterprise in the Arts: Studies in Mission and Constraint.* New York: Oxford University Press.

Vladeck, Boris C. 1976. "Why Non-Profits Go Broke." *Public Interest* 42: 86–191.

Wayland, H. L. 1890. "The Dead Hand." *Journal of Social Science* 26: 78–90.

Willard, Joseph. 1894. "Illustrations of the Origin of Cy Pres." *Harvard Law Review* 8: 10–12.

Zollman, Carl. 1919. "The Development of the Law of Charities in the United States." *Columbia Law Review* 19: 91–111, 286–309.

16

Social Movement Philanthropy and American Democracy

J. Craig Jenkins

Upper-class philanthropy for social movements has always stirred political controversy. Recently private foundations have become significant funders of social movement organizations. In contrast to earlier patronage by wealthy individuals, foundation support is an institutionalized form of upper-class philanthropy, suggesting more profound and far-reaching consequences. Fueling this controversy has been the perennial debate over the tax exemption of foundations. Critics have argued that allowing foundations to fund social movements, in effect, subsidizes political expression. Echoing the Cox and Reece Committee hearings in the 1950s, conservative critics have charged that the foundations have become "financiers of revolution," stirring up baseless grievances and radical demands and abusing their tax exemption by diverting public tax dollars toward quixotic political organizing campaigns that misrepresent the interest of their purported beneficiaries (Brownfield 1969; Wolfe 1970; Hart 1973; Metzger 1979; McLlhaney 1980). At the other end of the political spectrum, leftist critics have dusted off the old "tainted money" charge leveled against the Rockefellers in the Progressive era, challenging the motives and the cooptative impact of this new social movement philanthropy:

These foundations . . . have a corrosive influence on a democratic society; they represent relatively unregulated and unaccountable concentrations of power and wealth which buy talent, promote causes, and, in effect, establish an agenda of what merits society's attention. They serve as "cooling out" agencies, delaying and preventing more radical, structural change. (Arnove 1980, p. 1)[1]

And, from a third viewpoint, political moderates have criticized the foundations as overly timid, avoiding potentially significant but risky social

This research was conducted in cooperation with the Program on Nonprofit Organizations, Yale University and the Center for Policy Research, New York. It was supported by grants from the National Endowment for the Humanities, the Russell Sage Foundation and the Program on Nonprofit Organizations.

change projects in favor of overcautious efforts advancing the more advantaged groups in American society (Carey 1977; Tully 1977; Nielsen 1972, 1985).

At the heart of these debates is the question of democratic control. Is this upper-class patronage compatible with our conceptions of democratic government and politics? Defenders of this new social movement philanthropy argue that it strengthens democratic institutions by providing voice for underrepresented interests and excluded groups. By providing resources to movement organizations that represent the underrepresented and promote citizen participation, foundations are creating a more representative and accessible political system.

This defense, however, has been challenged. Democratic theorists typically advance two models of democratic control: a *representation model* based on the accountability of leaders; and a *participation model* centered on direct involvement in decision making (Pitkin 1967; Bachrach 1967; Pateman 1970). If social movement leaders and organizations depend on foundations for their resources, what ensures they will actually represent the interests of their supposed constituencies? Or, turning to the participation model, what ensures these projects will actually increase grass-roots participation by previously apathetic groups? Are the critics right that these projects merely advance the perspectives and interests of movement leaders and, in effect, their foundation patrons? A related problem is foundation accountability. The tax privileges of private foundations have traditionally been justified on the grounds that private initiative and flexibility are essential for social innovation and the promotion of social plurality. Supporting projects to address the problems of political underrepresentation and inadequate citizen participation would appear to promote social innovation and plurality. Yet, most foundations are controlled by wealthy donors and their families. What ensures that they will select projects that serve these aims?

These questions have received virtually no systematic empirical research. Aside from a handful of attempts to estimate foundation funding for particular disadvantaged groups (Human Resources Corporation 1975; Carden 1977; Kanter, Muringhan, Stein, and Wheatley 1977; Tully 1977; National Council of La Raza 1977) and Carey's (1977) survey of cases of movement philanthropy, previous discussions have largely been normative, typically with a polemical thrust. The more useful empirical studies have dealt with particular movements, noting in passing their foundation support. To remedy this, several years ago I launched a project to systematically analyze foundation funding of movements in the post-World War II era. The basic method was to review annual listings of 131 foundations identified as active movement funders during 1953–1980, identifying the major movement constituency, the political issue, the type of recipient organization, its action style, and the amount of the grant. The study then set about linking these to the development of contemporary social movements and analyzing the characteristics of the foundations that funded such projects. This chapter draws on the highlights of this research and past work to answer the following questions:

1. What has been the scope and general trends in foundation funding of social movements?
2. What are the mainsprings of this funding?
3. What has been the impact of foundation funding on the movements?
4. What has been the net impact of this social movement philanthropy on the health of the American political system?

SCOPE AND TRENDS IN MOVEMENT PHILANTHROPY

Foundations that fund social movements are innovative and relatively unconventional actors within their own world. Foundation support goes overwhelmingly to established charities and nonprofit institutions. Social movements, by contrast, stand on the margins of institutionalized society, using unconventional and even unruly tactics to pursue the interests of marginal or unorganized groups. At its peak in 1977, foundation giving to movements accounted for only .69 percent of total foundation giving and averaged only .24 percent of total giving. Out of the more than 22,000 active grant-giving foundations, only 131 were found to have actually funded a social movement project.

Figure 16.1 charts the trend in the nominal amount of social movement grants and their real value (that is, adjusted for inflation). Figure 16.2 traces these movement grants as a percentage of total foundation giving. Social movement funding first became significant around 1965, rose rapidly to $18.6 million in 1971, peaked in 1977 at $25.2 million, and then slowly declined through the end of the decade. Although the nominal value of these grants remained relatively stable through the 1970s, inflation eroded

Figure 16.1   Trends in social movement philanthropy.

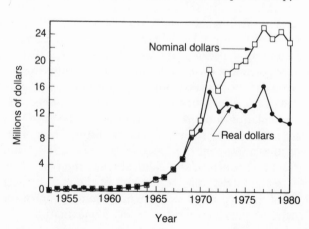

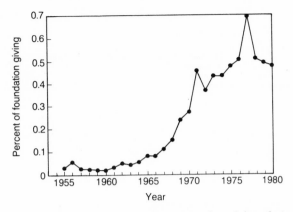

Figure 16.2   Movement grants as a percentage of total foundation giving.

over a third of the real value between 1977 and 1980. Over the same period, the number of grants rose steadily, from a mere handful through the early 1960s, to 102 grants in 1966 and a peak of 1266 grants in 1979. Confirming the Peterson Commission (1971, pp. 129–31) estimate that "controversial" undertakings constituted less than 1 percent of foundation funding, social movement funding represented a mere .24 percent of total foundation funding and stayed consistently well below 1 percent of total giving. Even among these innovative funders, movement giving is typically viewed as a small component of broader programs. Of these 131 foundations, only 39 contributed more than a third of their annual giving to movement projects and 66 kept their funding below 10 percent.[2]

What kind of movement projects were the major beneficiaries? Figure 16.3 charts the distribution of these grants across twenty-four movement constituencies and issues. The largest recipients were organizations representing black Americans, which received $56.9 million or 22.9 percent of the total; environmentalism at $25.2 million or 10.2 percent; women's rights at $16.6 million or 6.7 percent; Mexican Americans $15.8 million or 6.4 percent; and consumerism at $12.8 or 5.2 percent. Lesser causes such as peace, students, the poor, the aged, and homosexuals were on the bottom end. Table 16.1 provides greater detail on these grants. They varied enormously in size from one magnum grant of $5.5 million by the Ford Foundation to found the Mexican-American Legal Defense and Education Fund in 1971 to midget grants of $150 to grass-roots groups like feminist Take Back the Night marches and tenant union organizing. The average was $24,534, considerably larger than the model grant of $5,000.

The general ideology guiding this social movement philanthropy has been what Theodore Lowi (1979) has called *interest group liberalism*, the doctrine that the public interest is best served by allowing interest groups to directly make public policy. Guided by this ideology, the foundations set out to remedy the underrepresentation of particular groups, such as mi-

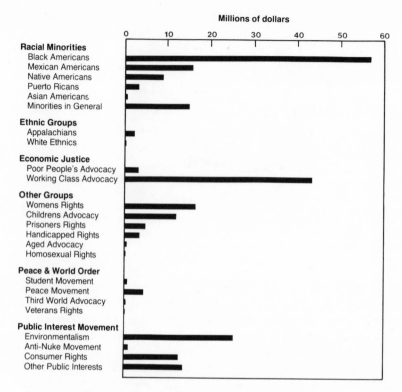

Figure 16.3   Distribution of Grants (dollars).

norities, women, or consumers, by funding advocacy organizations. The favorite model is the Black Civil Rights movement. Arguing that the victims of racial discrimination lacked the resources to be represented effectively, 40 percent of movement funds went into projects representing racial minorities, primarily black Americans and Mexican Americans. Drawing another leaf from this "interest group liberal" ideology, the public interest movement garnered 20 percent of the total funding for projects ranging from environmental research and litigation to attacks on redlining in home mortgage lending and promoting "no fault" auto insurance. The problem here was seen as the inadequate representation of large dispersed collectivities such as taxpayers and consumers. These two movements also shaped the prospects for other movements, as organizations representing women, children, prisoners, the handicapped, the aged, and homosexuals used similar claims to gain foundation support.

The recipients ranged from local community groups engaged in grassroots organizing, such as the Tenant Action Project in Philadelphia and the Philipino Bayana in Oakland, California, to associations with national membership, such as the NAACP and the National Organization for Women, to nationally prominent legal advocacy organizations, such as the Natural Resources Defense Council and the Women's Rights Project of the Amer-

Table 16.1   The Distribution of Social Movement Grants

Social Movement Projects	Number of Grants	Percent	Grant Amount	Percent	Mean Grant
Racial Minorities:					
Black Americans	1,819	18.02	$56,895,243	22.97	$25,277
Mexican Americans	271	2.68	15,833,677	6.39	56,879
Native Americans	310	3.07	9,095,165	3.67	13,721
Puerto Ricans	101	1.00	3,358,537	1.36	13,849
Asian Americans	91	.90	593,626	.24	2,248
Minorities in General	458	4.54	15,021,604	6.06	25,407
Total	3,050	30.21		40.69	
Ethnic Groups:					
Appalachians	147	1.46	2,309,105	.93	9,439
White Ethnics	4	.04	14,250	.01	446
Total	151	1.50		.94	
Economic Justice:					
Poor People's Advocacy	192	1.90	3,361,750	1.36	10,363
Working Class Advocacy	1,461	14.47	43,863,775	17.71	22,644
Total	1,653	16.37		19.07	
Other Groups:					
Women's Rights	949	9.40	16,602,912	6.70	7,890
Children's Advocacy	250	2.48	12,198,503	4.92	20,295
Prisoners' Rights	172	1.70	5,009,091	2.02	19,970
Handicapped Rights	103	1.02	3,643,163	1.47	13,873
Aged Advocacy	61	.60	668,819	.27	5,005
Homosexual Rights	58	.57	181,005	.07	1,264
Total	1,593	15.77		15.45	
Peace and World Order:					
Student Movement	53	.52	865,423	.35	5,265
Peace Movement	502	4.97	4,785,003	1.93	3,780
Third World Advocacy	101	1.00	250,610	.10	793
Veterans' Rights	9	.09	18,200	.01	306
Total	665	6.58		2.39	
Public Interest Movement:					
Environmentalism	920	9.11	25,206,367	10.18	23,035
Anti-Nuke Movement	192	1.90	1,097,078	.44	2,615
Consumer Rights	675	6.69	12,779,896	5.16	5,180
Other Public Interest	1,147	11.36	13,836,904	5.59	17,286
Total	2,934	29.06		21.37	
Unclassified:	51	.51	231,400	.09	2,344
All Grants	10,097	100.00	$245,655,812	100.00	$11,629

ican Civil Liberties Union, to technical support centers, such as the Youth Project and the Center for Community Change, to established institutions, such as the National Council of Churches and the American Friends Service Committee. The foundations strongly favored professionalized advocacy projects, such as law firms and policy research centers, investing 43 percent of their movement grant dollars in them. Technical support centers, whose primary function is providing technical advice to grass-roots groups on legal issues, leadership training, fund raising, and organizing strategies, received another third of the funding. Grass-roots groups received only 16.6 percent of the funding, with the remaining 8 percent going to established institutions, such as universities and church organizations that either worked directly on movement projects or served as pass-throughs for movement groups.

This preference for professionalism was so pronounced that it spurred two social scientists, John McCarthy and Mayer Zald, to coin a new term to define the new type of political advocacy created by these grants, *professional social movements*. In contrast to classical grass-roots movements, professional movements had outside leaders; a full-time, paid staff; non-existent or "paper" membership; mobilized their resources from outside or "conscience" constituencies; and attempted to "speak for" rather than mobilize their direct beneficiaries. In McCarthy and Zald's (1973, 1975, 1977) analysis, these professional movement organizations had become so dominant that they threatened to engulf the entire social movement arena.

This priority stems from several factors. Professional advocacy is relatively expensive and frequently necessary for certain issues. Children's rights and the interests of auto insurance customers, for example, are unlikely to mobilize sufficient grass-roots support to be represented effectively. Disorganization, poor resources, and free-rider problems make professional advocacy the only realistic alternative. By contrast, grass-roots organizations can mobilize their members and are therefore less deserving. In addition, many grass-roots groups lack a 501-c3 nonprofit tax exemption and are legally ineligible for foundation grants. They have to work with another exempt organization or consult with a technical support center.

This also reflects the foundations' political caution. Grass-roots organizations often lack a clear track record and are more likely to become involved in protests or other activities that might stir criticism. They are also more informal and decentralized, lacking the fiscal and management devices that foundations expect from their recipients. Professionalized organizations are centrally managed by a single executive or professional staff. Their hierarchical structure is more intelligible to foundation boards, who typically come from business and academia, and affords greater assurance that the money will be used prudently as specified in the grant proposal. Because grass-roots activists are often not directly accountable to a central board or a director, they represent considerable political risk to the foundations. As one foundation officer explained to me, referring to the preference for funding legal advocacy centers: "We fund *responsible militancy!*"

Funding technical support centers stands between these two alternatives. On the one hand, these centers provide assistance to grass-roots groups. Yet, the centers are professionally managed, providing a political buffer and eliminating negotiations with a large number of small groups.

Supporting established institutions is the most cautious approach. Although some institutional grants eventually go to grass-roots groups lacking 501-c3 exemption, the more typical case is an attempt by the professional staff of a church or university to move into a new issue area stirred up by a popular movement. During the late 1960s, protest by students and blacks spurred the American Friends Service Committee and the National Council of Churches to launch professional staff projects to promote pacifist criticisms of the military-industrial complex and minority group campaigns against racism and poverty (Jenkins 1977). Although these projects may be quite radical, foundation boards are more comfortable dealing with a recognized institution. In this vein, movement funders have veered away from groups with a reputation for scrappiness. Only 17 percent of the grants went to organizations with a record of organizing protests and demonstrations, the rest going to those relying exclusively on institutionalized political expression.

These foundations have also favored issues drawing on the general "civic culture" of American politics that citizens have equal rights to political and religious expression and, in return, are obligated to participate in public affairs. Civil rights concerns underlay $121.8 million or 49.2 percent of these grants, most of which was targeted at victims of racial discrimination. The second major issue was poverty, which received $58.1 million or 23.4 percent of the funding. The foundations typically took a long-range view of poverty, preferring developmental solutions, such as those of the Watts Labor Community Action Council in Watts, California, or the Woodlawn Organization in Chicago, over traditional charity. These projects combined "self-help" with efforts to alleviate poverty by small business development, consumer coops, work training, and cooperative housing. The only other issues that received significant funding were environmentalism ($26.6 million or 10.7 percent), consumer rights ($13.6 million or 5.5 percent) and work-place reform ($11 million or 4.4 percent). Such issues as peace ($4.3 million) and health care ($2.9 million) were on the margins of legitimate funding.

What then makes a social movement project fundable? In general, projects that focus on the interests of a politically underrepresented group and are operated by a professional organization that avoids unruliness and emphasizes civil rights has the greatest likelihood of support. Foundations also operate on a risk-capital principle of funding promising individuals. Personal contacts and a past track record are often essential.

THE MAINSPRINGS OF MOVEMENT PHILANTHROPY

Movement philanthropists have been variously depicted as public-spirited statesmen, misguided idealists, political meddlers, guilt-ridden limousine liberals, and cunning Machiavellians. These interpretations have typically centered on questions of motive, making them overly subjective. In their place, this study analyzes the "mainsprings" of movement philanthropy, the broader range of social and political factors that enter into arguments about *conscience, social control,* and *political opportunities.*

The study deals with the issue of motives indirectly, partially because of its inherent elusiveness. Motives, especially for something like donating money to a social cause, are subjective and complex, taking on diverse and multiple meanings. Nor are there satisfactory common sense categories for discussing philanthropy. Are self-interest and altruism actually mutually exclusive or are they not usually mixed? What about enlightened self-interest? This complexity becomes even greater if, as is contended, altruistic behavior must be seen as rooted in underlying conceptions of self-interest. For a project to be seen as public spirited, it must, after all, *interest* the beholder. Those of different social class, ethnic, gender, and religious backgrounds have different outlooks on what constitutes public spiritedness. Finally, how should these motives be tapped? Are they present in post hoc self-reports, which are likely to be colored by a desire to present a favorable image, or should the study demand private and confidential accounts, which are essentially inaccessible. Because of these problems, the study is focused on the social and political sources of movement patronage or, as they will be called, their *mainsprings*.

Conscience interpretations draw heavily on ideas of motives, arguing that patronage stems from other-regarding sentiments made possible by "slack" resources and the professionalization of social reform. Drawing on Maslow's (1962) classic theory of the hierarchy of needs, the basic idea is that the satiation of lower level self-centered material needs opens up the possibility of pursuing higher level, less egocentric concerns. Or, in a more mundane formulation, "slack" resources open up the possibility of pursuing discretionary concerns (McCarthy and Zald 1973, 1977). Professionalization of social problem solving, especially in the hands of social science–trained practitioners, provides these sentiments with a broader base of support, creating a growing cadre of public-spirited activists (Moynihan 1969; McCarthy and Zald 1975). Appalled by the plight of the disadvantaged, wealthy liberals and foundation staff members were swayed to fund social causes.

The classic case of conscience-based patronage is the early foundation support of the Civil Rights movement. Drawn by the heroism of civil rights workers and outrage at white racist violence, people such as Stephen Currier of the Taconic Foundation, Andrew Norman of the Norman Fund, and John Heyman of the New York Foundation became personally involved in the Civil Rights movement. Significantly, Kennedy administration officials Burke Marshall and Harris Wofford were the initiators of the major project, the Voter Education Project (VEP). Following a series of extensive strategy discussions, in April 1962 these foundations funded the Voter Education Project, to be administered by the Southern Regional Council, to coordinate the southern voter registration campaign and furnish legal protection and bail money for the organizers. In the summer of 1963, these patrons talked many of their friends into sponsoring the project and, pressing the civil rights organizations to form a Council for United Civil Rights Leadership (CUCRL) to reduce competition for funds, boosted their funding. When Martin Luther King, Jr. and A. Philip Randolph began planning the March on Washington, several foundations provided funds and logistic support. Meanwhile, the Congress of Racial Equality set up a direct-mail fund-raising arm, the Scholarship, Education and Defense Fund for Racial Equality (SEDFRE), which also held art auctions and solicited foundation grants. When the voter registration campaign picked up again in 1964, these patrons dug deeper in their pockets, Norman donating $60,000 in out-of-pocket support for bail money and joining Currier in an emergency loan of $100,000 (which was never repaid) to keep CORE financially afloat through the winter of 1964–1965 (Meier and Rudwick 1973, pp. 172–73, 223–24, 334–35; Wofford 1980, pp. 158–64; Garrow 1986, pp. 161–64, 272–78).[3] Equally significant were the early grants in 1961 from the Marshall Field Foundation to the Southern Christian Leadership Conference (SCLC) to transfer a citizenship training program from the Highlander Institute to the SCLC, paving the way for SCLC's move into voter registration (Morris 1984, pp. 237–39). That conscience lay behind this support was evident to all parties, later leading to ill feelings among the civil rights activists about the paternalism frequently entailed by the white patronage (see especially Wofford 1980, p. 165).

This interpretation about conscience is also supported by the study's data on foundation funding. Relative to the average foundation, the movement funders are significantly larger, typically holding $5 million or more in assets and employing roughly double the number of professional staff (cf. Boris 1987). With their "slack" resources, these foundations are more willing to take political risks and have the staff to evaluate and monitor these projects. In many cases, the staff took the initiative in identifying possible funding targets for board review.

Yet, conscience operates within a political context. With the offer of major foundation support for the VEP, black journalist Louis Lomax immediately exclaimed: "[They] are out to defang the civil rights movement" (1962, p. 284). Several analysts have advanced a social control thesis, pointing to the reactiveness of such support, its concentration on moderates, and its tendency to channel volatile protest into less disruptive actions (Piven and Cloward 1977; McAdam 1982; Haines 1988). Several events in the Civil Rights movement fit this interpretation. The major influx of patronage came after, not before, the emergence of the protest movement that spurred the donations (Jenkins and Eckert 1986). At several points, such as the Kennedy administration's support for the VEP and the last-minute rewriting of John Lewis's speech at the March on Washington, these liberal patrons and their political allies attempted to dampen the radicalism and militancy of movement activists (Garrow 1986, pp. 281–83, 345–50). Financial support has also followed a "radical flank effect" (Haines 1988), responding to rising militancy and radicalism by favoring the more moderate, less militant movement organizations.

This control argument, however, has limitations. In the absence of direct, systematic evidence about intentions, it is ultimately impossible to demonstrate an intent to "control." Timing alone is not sufficient. Reactive funding, for example, would be equally likely for conscience donors and Machiavellian controllers. More important, the underlying motives are probably complex, entailing elements of both conscience and control. Alerted by the protestors to the problem of underrepresentation and guided by interest group liberalism, the foundations set out to promote the representation of excluded groups. Because differing interests were seen as ultimately harmonious, militancy and radical demands lay outside the range of funding.

The political opportunities interpretation draws its name from the argument that under exceptional circumstances institutional leaders will find it beneficial to support the inclusion of previously excluded groups, thereby creating *opportunities* for movements (Tilly 1978; McAdam 1982; Jenkins 1985). While this argument is most relevant for political leaders, it also entered into the policies of foundation boards and staff. The argument begins with the premise that institutional elites are normally adverse to bringing new groups into the political system because this might upset the rules of the political game. But if political coalitions are volatile or normal allies are suddenly weakened politically, institutional leaders may find that the risks of bringing in new groups outweighs the status quo. The Kennedy administration, for example, saw major electoral gains from supporting civil rights that would offset southern Democratic losses. Likewise, the Johnson administration treated the consumer movement as a "motherhood" issue that would appeal to political moderates, entail few costs, and create improved relations with Congress (Nadel 1971).

The major calculation for foundations is the likelihood that the project will generate significant change, thereby rebounding to the credit of the foundation as an effective social change innovator. The clearer the likelihood of victory, the greater the probability of funding. The major outpouring for environmentalism, for example, came after the passage of the Environmental Protection Act in 1971, ensuring legal advocates a significant impact on public policies.

Foundations are also guided by particularistic interests. Of the movement funders studied, 27.7 percent of the original donors came from a Jewish background, roughly double their representation among foundations in general (Boris 1987, p. 121). These foundations were also more likely to fund civil rights and grass-roots projects, reflecting the generally progressive political views of the Jewish community and its direct stake in the battle against racial discrimination.

The opportunities for movement funding are also regulated by the tax code and IRS enforcement. Foundations are legally restricted to funding 501-c3 organizations that engage in "religious, educational, charitable, scientific, and literary purposes" and are barred from "substantial political activity." Contributions to political organizations (501-c4) are prohibited. The key problem has been what constitutes "substantial political activity." In 1955, southern Congressmen, piqued at the *Brown* v. *Board of Education* decision, convinced the IRS to revoke the tax exemption of the NAACP, forcing it to accept a 501-c4 status and formally split from the NAACP–Legal Defense and Education Fund. The Legal Defense Fund did, however, receive 501-c3 status, thereby affirming the tax deductibility of legal advocacy. In 1966, the Sierra Club lost its 501-c3 status, because of newspaper adds urging letter-writing to Department of Interior officials to block the construction of reservoirs that would flood the lower Grand Canyon, but later maneuvered around the ruling by creating the Sierra Club Legal Defense and Education Fund.

The 1969 Tax Reform Act clarified the issue by defending *political activity* as attempts to directly influence the selection of candidates for elective office, specific pieces of legislation, or administrative decisions. After a controversy over Ford Foundation funding of a CORE voter registration campaign in Cleveland, Ohio, that helped elect the nation's first black big-city mayor, Congress required that voter registration projects be "nonpartisan" and cover at least a five state area. Appearances before Congress or administrative bodies were legal so long as they were "educational"—that is, factual, and solicited by the government body in question—or constituted an insubstantial portion of the organization's activities, which most observers assumed to be 10–20 percent. Public education efforts (or grassroots lobbying) like the Sierra Club's were legal so long as they did not focus on pending congressional legislation. Still, the Nixon White House pressured the IRS to deny a tax exemption to the Natural Resources Defense Council, on the grounds that public interest law was not "charitable" because it lacked a specific constituency and hence criteria for preventing barratry. Although the IRS finally granted an exemption, the public interest law firms were barred from accepting fees from their clients and had to demonstrate annually that their practice did benefit the "public" (Goetz and Brady 1975; Harrison and Jaffe 1971; Adams 1974). Finally, in the 1976 Tax Reform Act, the meaning of *substantial* was specifically defined in terms of a sliding scale that allowed up to 20 percent of the first $500,000 of budget. Still, the uncertainty about tax deductibility and rumors of politically motivated IRS audits have generally discouraged foundation funding of movement projects.

Another mainspring of foundation funding has been the peer networks of foundation boards and staff. Foundations are often faddish, jumping on the bandwagon to share in the credit and assuming that projects have been thoroughly evaluated by their peers. Joint or coordinated grants also diffuse responsibility for risky grants (including the assumption of fiscal responsibility if the foundation provides over a third of organization's budget). Because larger foundations have more staff to evaluate projects, smaller foundations frequently wait until the larger funders have identified the "responsible" projects. Foundation board members often become personally involved, consulting their own social peers. For example, the controversial Ford Foundation grant to the Cleveland CORE chapter in 1967 was preceded by a site visit by foundation vice-president David Bell who solicited the advice of the editor of the *Cleveland Plain Dealer* and several promi-

nent leaders of the Cleveland business community before authorizing the
grant. Had any of these argued strongly against the grant, it presumably
would not have been made. Foundations also rely on consultants and technical support organizations, such as the Center for Community Change
and the Youth Project, to screen potential applicants. Well-regarded movement leaders may even gain a consulting status, steering foundation grants
toward their allies and away from potential rivals. Cesar Chavez, for example, has blocked foundation grants to the rival Texas Farm Workers
Union and other competitors, claiming that these were unreliable representatives of farm workers' interests.

The major spur to this upsurge in movement philanthropy was the ex-
plosion of grass-roots social movement activity. Figure 16.4 traces this grass-
roots upsurge of protests and demonstrations, riots, and the formation of
new movement organizations. Beginning with the Civil Rights and the Stu-
dent-Antiwar movements that began in the mid-1950s, the United States
witnessed an escalating wave of movement activity that defined new issues,
trained new movement leaders, and created new movement organizations.
It also spurred institutional leaders, including foundation trustees, to identify new social problems and, by creating political disorders, created a sense
of crisis in the higher circles that made professional social movement organizations (SMOs) appear as a "responsible" alternative to grass-roots
militancy.

This also helps explain the funding decline of the late 1970s. The average assets of the movement funders shrank by 3.2 percent between 1965
and 1972 and another 6.6 percent by 1980, limiting their discretionary
resources. Nationally, inflation and declining stock prices eroded foundation assets by almost a third. There was also a reshuffling of the movement

Figure 16.4   The grass-roots upsurge: protests, riots, and the formation of new
social movement organizations. (*Sources:* Data from Taylor and Jodice 1983 for
protests and riots; Gale Research 1982 for new social movement organizations.)

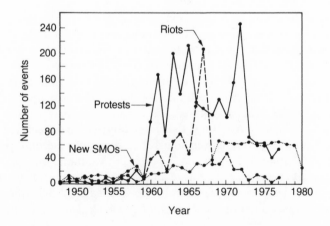

funders, several of the larger foundations trimmed their commitments while smaller funders entered the scene, delaying any serious shrinkage in overall funding until 1978. By that point, as the interpretations of social control and political opportunities would suggest, declining protest and conservative political trends dampened the pressure for movement funding.

FUELING OR COOPTING THE MOVEMENTS?

The sharpest debates have been over the political impact of movement philanthropy: Does patronage fuel the movements, providing greater resources and political access? Or does it wither their grass roots, channeling efforts into less militant, conservative actions? What is the impact on movement participation and internal democracy? Finally, what are the implications for the representativeness and accountability of the American political system?

The evidence in the study clearly refutes the most outlandish version of the conservative critique. In contrast to the notion that foundation patronage actually generated the movements, funding consistently followed the upsurge of movement activity, both for the aggregates (see earlier figures) and for the five major challenges of the post-World War II period: the Black Civil Rights movement, the Peace-Student movement, the new feminism, environmentalism, and consumerism. Drawing on movement action profiles derived from the *New York Times*,[4] the take-off of these five movements occurred before, not after, the take-off in foundation funding, typically leading by three to five years. The Black Civil Rights movement, for example, emerged as a significant force in 1955–1956 with the Montgomery bus boycott; foundation funding was insignificant until 1962. The lag in funds for the Peace movement was even greater, peace activity bobbling up and down throughout the 1950s and early 1960s (especially ban-the-bomb and antitesting campaigns in 1958 and 1962) and then mushrooming into a mass antiwar movement in 1965. Foundation funding, however, did not arrive until 1972, by which time the movement had begun to decline. As for the Women's movement, it emerged as a major public force in 1970 with the National Organization for Women's August 26 national strike and a flood of grass-roots protest. Yet, funding for women's causes did not actually take off until 1973.

Movement and funding take-off did, however, coincide for the two public interest causes, environmentalism and consumer rights. Although there were flurries of local environmental activism such as an antinuke campaign in Detroit in 1956 and the Los Angeles clean air campaign during the early 1960s, the major boost in action and funding came together in 1969 with the formation of the major environmental advocacy organizations: the National Resources Defense Council and the Environmental Defense Fund. In both cases, local groups that had been working on community problems turned to foundations to create national advocacy organizations. Grass-roots activism remained relatively low, the major upsurge coming on April 23, 1970, with Earth Day and a wave of local action projects. The Consumer movement followed a similar scenario, beginning with Ralph Nader's lone efforts in the 1966 congressional auto safety hearings, eventually leading to the formation of Public Citizen in 1970 and sufficient foundation funding for a direct-mail campaign. These public interest movements followed a different scenario from the grass-roots challenges, depending on entrepreneurial leaders to mobilize institutional patronage for new professional advocacy organizations.

The more complex question is the impact on grass-roots activism. The foundations have assumed that their support facilitates movement development by furnishing badly needed organizational resources. If welfare mothers are too deprived to mount their own challenge, then patronage will ease the way by paying the bills for professional organizers, meeting rooms, and mimeo machines (Bailis 1974). Or, without the lawyers and bail money provided by the liberal foundations, the sit-in protestors and voter registration workers of the early Civil Rights movement would have been severely hampered if not crushed (Cleghorn 1963; Watters and Cleghorn 1967). Yet the challenge comes from those who argue that organization building dampens grass-roots insurgency by diverting leaders from organizing protests and by creating the appearance of a responsive elite (Piven and Cloward 1977; McAdam 1982). These critics have also argued that patronage inevitably strengthens the professional staff, giving them greater control over decision making. By requiring the fiscal and management procedures of business corporations and universities, patrons restrict the spontaneity and participatory enthusiasm essential to mass mobilization. Constituency ties wither, leaving the organization dependent on patrons. In a complaint frequently voiced by movement leaders, the foundation preference for funding specific projects (as opposed to unrestricted grants) steers the movement organization into short-term projects that are deemed "fundable" (such as public education programs, internships, and networking conferences) but deters development of long-term programs. Stronger leadership also carries with it the specter of oligarchy, Robert Michels' classic fear that leaders will temper their tactics and goals in order to ensure their own interests in organizational survival.

A case that approximates the control argument is the Ford Foundation sponsorship of Mexican-American advocacy by the Southwest (later National) Council of La Raza. In the spring of 1966, Ford Foundation staff members approached prominent Mexican-American social workers and scholars Herman Gallegos, Ernesto Galarza, and Julian Samora to serve as consultants to devise a program for Mexican-American advocacy. The Ford Foundation's aim was to create a Mexican-American civil rights organization comparable to the NAACP and the Urban League that would overcome the "divisiveness and argumentativeness, low trust in their leaders, and an absence of a sense of priorities and programs" perceived to exist within the newly restive Mexican-American community (cited in Sierra 1983, pp. 159–60). Yet, the Ford staff favored bypassing the established Mexican-American organizations, such as the GI Forum and the League of Latin American Citizens, for the creation of a new forum. Ironically, this temporarily dovetailed with the demands of Chicano militants. After extensive discussion, the consultants recommended a grass-roots program of "community barrio efforts . . . for educational opportunities, research and social action projects" (cited in Sierra 1983, p. 152). At a tumultuous rump conference organized by Chicano militants protesting exclusion from the Johnson administration's White House Conference on Mexican-American Affairs, the militants organized a Southwest Unity Council for Social Action to serve as the conduit for the funding of local community organizing projects. Launched with a Ford grant of $630,000 in June 1968, the renamed Southwest Council set out to become a technical support center and professional advocate for four regional councils of barrio organizations.

Two developments cut short this initial focus on community organization and political action: the congressional hearings on the 1969 Tax Act, which publicized brown beret protests by a council affiliate (the Mexican-American Unity Council in San Antonio) against Congressman Henry B. Gonzalez (D.-Texas); and the decision by the new Ford Foundation director of national affairs Mitchell Sviridoff to shift to a program of community economic development. With the new legal restrictions on voter registration activities, Ford demanded a halt to political education activities and the mini-grant program that had linked the council to the barrio groups and a refocus of the program on economic development projects. Accompanying this shift from "process to hard program" was an organizational shift from a centralized umbrella organization serving neighborhood groups to a centralized professional advocacy organization. Finally, after four years of floundering, the council staff pushed through another reorganization, relocating the National Council of La Raza in Washington, D.C., as the political broker for Chicano groups attempting to influence Congress and the federal government. Despite efforts to rebuild grass-roots ties, the council currently remains a professional organization with relatively weak ties to a constituency and whose influence depends on Washington political ties. Although the Ford Foundation's overt intention had never been cooptation of the Chicano movement (in fact, funding initially strengthened the militants), the political controversies set loose by funding grass-roots activism led to a restructuring of La Raza programs to blunt militancy and temper radical demands (Sierra 1983).

A more complex picture emerges from the experiences of the Open Housing, Welfare Rights, and Nuclear Freeze movements. The first two were spin-offs of the Civil Rights movement. Formed in 1950 with one volunteer, part-time director and a donated $10 \times 15'$ office to fight a racial exclusion convenant in the Stuyvesant Town development in New York City, the National Committee against Discrimination in Housing (NCDH) functioned for over a decade as a professional advocate supported by a coalition of fifteen civil rights, religious, and liberal reform organizations. In the early 1960s, the NCDH shifted its mission to technical support for local open housing campaigns. Fueled by foundation grants, it mounted an extremely successful campaign for open housing legislation on both national and local levels. At the pinnacle of apparent success, the NCDH confronted a crisis. Within the broader Civil Rights movement, goal disputes emerged as black nationalists questioned integration and white activists urged greater concern with housing stabilization and housing for the poor. The movement also came up against insurmountable implementation obstacles. Confronted with a resulting atrophy of grass-roots activism, the Ford Foundation imposed a reorganization of the NCDH, transforming it from an "organization of organizations" into a free-standing professional advocacy organization with an elected board, a new executive director, and a new program of litigation, congressional lobbying, and the monitoring of federal housing and development agencies (Saltman 1978, pp. 78–112). At this point, the organization's vital grass-roots links withered and, frustrated by the inability to tackle private housing desegregation, lost its cutting edge. Allowing the transformed NCDH to survive, foundation funding expanded rapidly.

The Welfare Rights movement was spawned by local protest and the organizing efforts of George Wiley, a former staff member of CORE and the Ford Foundation–funded Citizen's Crusade Against Poverty. Social scientists Frances Piven and Richard Cloward also played crystallizing roles, building off their experience with Mobilization for Youth to propose a strategy of welfare reform based on promoting welfare claims, thereby bankrupting the welfare budget and forcing a comprehensive reorganization of the welfare system. In June 1966, they opened the Poverty Rights Action Project (later reorganized as the National Welfare Rights Organization, or NWRO) with small grants from the Norman, Stern, and Field Foundations and several white liberal donors. As Wiley quickly found, foundations and wealthy liberals were extremely reluctant to fund welfare rights organizing. Many favored a "jobs not dole" program and others feared a white backlash and political controversy. In one notable case, the New York affiliate organized demonstrations against Governor Nelson Rockefeller, who was simultaneously funding the NWRO out of the Rockefeller Brothers Fund. Nor did the low social status of the primary organizing target—black welfare mothers—endear patrons. Yet, by August 1967, Wiley had garnered enough support from national church organizations and a handful of New York foundations to christen the National Welfare Rights Organization with a budget of over $200,000, an office in Washington, D.C., and a paid staff. The initial strategy was to organize grass-roots groups, using information about welfare benefits and appeals for solidarity to mobilize welfare recipients.

For two years, this strategy produced rapid growth of chapters, escalating protests, expanding welfare benefits, and a growing NWRO budget. In 1969, however, welfare officials began restricting access to benefits, especially the special grants that WRO organizers had used to solicit support, and the protests and grass-roots expansion collapsed rapidly. The Nixon administration proposed a new Family Assistance Plan (FAP), totally revamping the welfare system. The NWRO staff then turned to new tactics, emphasizing congressional lobbying, court litigation, and formal negotiations with federal welfare officials (including accepting several federal grants to police the Work Incentive Program). Concentrating almost all of its resources on defeating the FAP proposals, the NWRO orchestrated a drawn-out two-year campaign that finally "ZAP(ped) FAP." But by this point the chapters had collapsed and the teetering organization had become the target of an attempted take-over by the white radicals of the National Caucus of Labor Committees. In addition, the simmering tensions between the white middle-class male organizers and the black welfare mothers exploded in a series of acrimonious disputes that, when Wiley finally proposed shifting to a new coalitional strategy with white liberals in the face of the impending Nixon 1972 electoral victory, the Executive Committee rebelled. Wiley resigned and, since he had been the major fund raiser, funding dissolved. Within a year, the NWRO was bankrupt, closing its offices (Kotz and Kotz 1977; West 1981). Although external funding had fueled the early organizing, changed political circumstances led to a termination. After the 1972 Nixon victory, the churches and foundations concluded that a new "jobs" approach was needed. Nor did the NWRO organizers solve the problem of sustained poor people's organizing. External support also encouraged staff dominance, stirring internal disputes. Still, for a brief period, the funding did help produce significant gains for welfare recipients.

The nuclear freeze movement of the early 1980s suggests a slightly different interpretation of the success problems of funded movements. Rooted in indigenous volunteer activism, the Nuclear Freeze Campaign organizers had originally planned to build a grass-roots base over a five- to seven-year period. But their ideas quickly met a favorable response from prominent liberal politicians and a sudden outpouring of support from churches and liberal foundations. The Nuclear freeze organizers then decided to telescope their campaign, moving immediately to a national letter-writing campaign focused on a congressional nuclear freeze resolution. The resolution failed, however; and, without alternative projects, the campaign collapsed overnight. The problem lay, not in the Machiavellian foundations but in the naivete of the organizers who thought they could bypass time-consuming organizing and win congressional debates on strictly technical grounds. The funders, of course, shared the same native optimism but their aims could not be classified as cooptative (Mack 1987).

These experiences support a far more complex *channeling interpretation* of philanthropic impact that I developed in interpreting the black Civil Rights movement (Jenkins and Eckert 1986). In this case, foundation funding did fit several of arguments put forward by the cooptation theorists. Foundation funding did expand in response to movement growth and militancy; it did go largely to professional organizations and moderates; it did increase the prominence of professional organizations; it was followed by some fall off in protests and indigenous organizing; and it did centralize movement activism around a single movement organization (the NAACP). Yet, this philanthrophy was also frequently constructive. Nor was movement transformation and decline due primarily to foundation support. This decline stemmed from forces at work within the movement as well as the broader political arena. Partial successes whetted the appetites of the moderates, tempering their militancy. Militant leaders advanced a disastrous new "black power" strategy that alienated former allies and provided no means for organizing the urban black underclass. The Nixon administration played on these trends to build a new conservative coalition that blocked movement victories. Still, the black movement did not actually die during the late 1970s but rather became more formally structured as national and professional movement organizations came to compose over half of all movement activity. Probably, this would have occurred without foundation support, but it was certainly accelerated by patronage. As protests proved ineffective against the diffuse problems of racial housing segregation and black poverty, the protests declined precipitously. Movement activity persisted at a lower, less volatile level and actually experienced another take-off in the late 1970s. Nor were goals completely transformed. Black power continued to represent the major demand, but it was now reframed in terms of voting rights and holding political office as opposed to redistributing economic power.

In this sense, foundation patronage *channeled* the movement into more organized, less militant directions. In this, it was powerfully reinforced by forces at work both within the movement and in the larger political arena. By funding the moderates, the foundations helped to institutionalize moderate social reforms, such as voting rights, affirmative action, and open housing laws. This also helped make the moderate leaders and organizations the dominant political representatives of the black community, helping to stifle the radicals (who were in trouble anyway) and allowing the foundations to play the role of political gatekeepers.

A similar picture emerges from the two other grass-roots challenges: the Peace movement and the Women's movement. In the Peace movement, mass insurgency developed in the mid-1960s, peaking in 1967–1969 and forcing a gradual and antiwar shift in Congress and the national security establishment. Foundation funding did not emerge until 1971, spurring the growth of professional advocacy and national movement organizations. Protests and grass-roots activism rapidly declined, especially after Kent State demonstrations and Vietnamization, leaving the new professional advocates to carry on the campaign. These new advocates managed to secure a range of moderate policy successes, trimming back military budgets and the use of military force.

Similarly, the Women's movement emerged indigenously in the late 1960s, stirring protests and raising new issues, such as sexism in advertising, restrictiveness of family roles, and abortion rights. Foundation funding arrived in 1973, centered largely on civil rights concerns and nurtured a new group of professional organizations that pursued the legally tractable aspects of these issues. In this case, grass-roots activism continued to spread through the early 1980s, directed especially at the pro-ERA campaign. Although more radical and militant varieties of feminism did not spread across the movement, this was due more to intra-movement tensions and institutional obstacles than foundation funding.

The public interest movements followed a similar script. In the environmental case, activism emerged along with funding in 1968–1969. Contrary to a simple cooptation thesis, protests and grass-roots activism grew in tandem with patronage, especially the upsurge of antinuke protests in 1977–1979. While professional activity remained prominent at about a fifth of movement activity, membership in the national organizations mushroomed and their activities continued to spur the movement. Of course, the environmental movement had already won several of its major victories, most notably the Environmental Protection Act of 1970 and its various extensions in the early 1970s. Again, the foundations were funding the institutionalization of basic legal victories through litigation, policy advocacy, and government monitoring. Similarly, grass-roots activism in the consumer rights movement was confined to the early 1970s with the movement almost entirely a professional campaign. In this case, unruliness was virtually nonexistent and the Nader organizations steadily dominated the scene, constituting 60 percent of the activism. The major impact was institutionalization of previous legislative victories and professionalization of the movement.

The more plausible way of making the social control argument is to focus on the steering effects of patronage on the social movement arena as a whole and thereby the general system of political representation. Foundation funding helped channel middle-class reformers into new issues, like environmentalism and consumerism, where direct challenges to corporate and government authority were less direct and militancy subdued. Similarly, patronage strengthened the moderates within the grass-roots challenges, reinforcing the weakness of the radicals. In this sense, the foundations could be seen as political gatekeepers, identifying the leaders and organizations that would eventually prevail as the legitimate representatives of new social interests.

Foundation patronage, then, could be interpreted as creating a neo-corporatist system of political representation in which elites exert increasing control over the representation of social interests (Handler 1978, pp. 222–33; Wilson 1983). By funding professional organizations, foundations (and government elites) strengthen the political advocacy of a limited number of centrally controlled organizations that frequently lack grass-roots base. Direct participation may wane as control becomes more centralized and alienated citizens (falsely) sense that their interests are being attended to. Political representation is increasingly structured around

a limited number of singular, compulsory, noncompetitive, hierarchically ordered, and functionally differentiated categories recognized and licensed (if not created) by the state (or institutional elites) and granted a deliberate representational monopoly within their respective categories in exchange for observing certain controls on their selection of leaders and articulation of demands and supports. (Schmitter 1979, p. 65)

Is this an accurate picture of the broader impact of movement philanthrophy? The tendencies are clearly present. Several analysts have documented the political vulnerabilities of professionalization. Patronage has frequently transformed advocacy organizations into service providers, undermining grass-roots participation (cf. Gittell 1980; Helfgot 1981). Nor are professional movements consistently effective shapers of policy. Congress is frequently skeptical of environmental organizations, viewing them as staff driven and lacking genuine member support (Tober 1984). There are also significant limits to legal advocacy as a method. Courts are better able to prohibit practices than issue new directives. Implementation is frequently contingent on restricting the scope of administrative discretion, eliminating the possibilities of subversion by government agencies (Handler 1978). The presence of professional advocates may make political leaders aware of a broader range of interests but does not ensure that their views will actually prevail in public policy.

Yet, the real choice may not be between grass-roots democracy and professionalization. It may be between professionalization and no advocacy whatsoever. If we judge by the experience of these recent movements, professionalization has frequently strengthened the grass-roots challenges, and movement decline was primarily due to strategic and political problems. Professionalized representation may not substitute for grass-roots participation, but it is better than a complete lack of representation. Clearly, it broadens the range of views formally represented in politics. There are some groups, such as prisoners and children, whose interests would otherwise go unrepresented. In many cases, professional movements have ensured that changes brought about by grass-roots social movements would be institutionalized. And, as in the civil rights case, professional advocates may be strong facilitators of grass-roots activism. In this sense, foundation patronage has helped to "represent the underrepresented," creating a more open and accessible political system.

NOTES

1. Other formulations were advanced by Roelofs (1983, 1986) and Arnove (1980).
2. Based on data for the years 1965, 1972, and 1980.
3. Also author's personal interviews.
4. The movement action counts were created by content coding the *New York Times Annual Index* summaries of new stories under the major topic headings (for example, "Negroes (or Blacks)—U.S.—General"). The date and type of movement actions were coded, as were the major issues and the organization of the movement actors. For a fuller discussion of the method, see Jenkins and Eckert (1986) and McAdam (1982).

REFERENCES

Adams, John H. 1974. "Responsible Militancy: The Anatomy of a Public Interest Law Firm." *Record of the Bar Association of the City of New York* 29:631–45.

Arnove, Robert. 1980. *Philanthropy and Cultural Imperialism*. Boston: G. K. Hall.

Bachrach, Peter. 1967. *The Theory of Democratic Elitism*. Boston: Little, Brown and Co.

Bailis, Lawrence. 1974. *Bread or Justice*. Lexington, Mass.: D. C. Heath.

Boris, Elizabeth. 1987. "Creation and Growth: A Survey of Private Foundations." In Teresa Odendahl, ed., *America's Wealthy and the Future of Foundations*. Washington, D.C.: Foundation Center, pp. 65–126.

Brownfield, Edward. 1969. *Financing Revolution*. Washington, D.C.: American Conservative Union.

Carden, Maureen. 1977. *Feminism in the Mid-1970s*. New York: Ford Foundation.

Carey, Sara. 1977. "Philanthropy and the Powerless." Commission on Private Philanthropy and Public Needs (Filer Commission). *Research Papers* 2:1109–64. Washington, D.C.: Government Printing Office.

Cleghorn, Reese. 1963. "The Angels Are White." *The New Republic* (August), pp. 12–14.

Garrow, David J. 1986. *Bearing the Cross*. New York: William Morrow.

Gale Research. 1982. "Public Affairs Organization" in *The Encyclopedia of Associations*. Detroit, Mich.: Gale Research.

Gittell, Marilyn. 1980. *Limits to Citizen Participation*. Beverly Hills: Sage Publications.

Goetz, Charles, and Gordon Brady. 1975. "Environmental Policy Formation and the Tax Treatment of Citizens Interest Groups." *Laws and Contemporary Problems* 39:213–16.

Haines, Herbert. 1988. *Black Radicals and the Civil Rights Mainstream, 1954–1970*. Knoxville: University of Tennessee Press.

Handler, Joel. 1978. *Social Movements and the Legal System*. New York: Academic Press.

Harrison, Gordon, and Sanford Jaffe. 1971. "Public Interest Law Firms: New Voices for New Constituencies." *American Bar Association Journal* 58:459–67.

Hart, Jeffrey. 1973. "Foundations and Public Controversy: A Negative View." In Fritz Heiman (ed.) *The Future of Foundations*. Englewood Cliffs, N.J.: Prentice-Hall, pp. 99–116.

Helfgot, Joseph. 1981. *Professional Reforming*. Lexington, Mass.: D. C. Heath.

Human Resources Corporation. 1975. *U.S. Foundations and Minority Group Interests*. San Antonio, Texas: Mexican American Cultural Center.

Jenkins, J. Craig. 1977. "The Radical Transformation of Organizational Goals." *Administrative Science Quarterly* 22:248–67.

———. 1985. *The Politics of Insurgency*. New York: Columbia University Press.

Jenkins, J. Craig, and Craig Eckert. 1986. "Channeling Black Insurgency." *American Sociological Review* 51:812–29.

Kanter, Rosabeth, Marcy Muringhan, Barry Stein, and May Wheatley. 1977. *Review of Grant-Making for Women's Issues in the 1970s*. Cambridge, Mass.: Social Analysis Associates.

Kotz, Nick, and Mary Kotz. 1977. *A Passion for Equality*. New York: W. W. Norton and Co.

Lomax, Louis. 1962. *The Negro Revolt*. New York: Harper and Row.

Lowi, Theodore. 1979. *The End of Liberalism*. New York: W. W. Norton and Co.

Mack, Kenneth. 1987. "An Analysis of the Nuclear Weapons Freeze Campaign." B.A. honors paper, Hampshire College, Amherst, Mass.

Maslow, Abraham. 1962. *Towards a Psychology of Being*. New York: Van Nostrand Reinhold Co.

McAdam, Doug. 1982. *Political Process and the Development of Black Insurgency*. Chicago: University of Chicago Press.

McCarthy, John, and Mayer Zald. "Resource Mobilization and Social Movements." *American Journal of Sociology* 82:1212–41.

———. 1973. *The Trend of Social Movements: Professionalization and Resource Mobilization*. Morristown, N.J.: General Learning.

———. 1975. "Organizational Intellectuals and the Criticism of Society." *Social Service Review* 49:344–62.

McLlaney, William H. 1980. *The Tax-Exempt Foundations*. Westport, Conn.: Arlington House.

Meier, August, and Elliott Rudwick. 1973. *CORE*. New York: Oxford University Press.

Metzger, Peter. 1979. *The Coercive Utopians: Their Hidden Agenda*. Denver: Public Service Company of Colorado.

Morris, Aldon. 1984. *The Origins of the Civil Rights Movement*. New York: Free Press.

Moynihan, Daniel P. 1969. *Maximum Feasible Misunderstanding*. New York: Free Press.

Nadel, Mark. 1971. *The Politics of Consumer Protection*. Indianapolis: Bobbs-Merrill.

National Council of La Raza. 1977. "Philanthropic Foundations of the U.S. and Their Responsiveness to the Special Needs, Problems, and Concerns of the Hispanic Community." Commission on Private Philanthropy and Public Needs (Filer Commission) *Research Papers* 2:1283–1304. Washington, D.C.: Government Printing Office.

Nielsen, Waldemar. 1985. *The Golden Donors*. New York: G. P. Dutton and Co.

———. 1972. *The Big Foundations*. New York: Columbia University Press.

Pateman, Carole. 1970. *Participation and Democratic Theory*. New York: Cambridge University Press.

Peterson Commission. 1971. *Foundations, Private Giving and Public Policy*. Chicago: University of Chicago Press.

Pitkin, Hanna. 1967. *The Concept of Representation*. Berkeley: University of California Press.

Piven, Frances Fox, and Richard Cloward. 1977. *Poor Peoples' Movements*. New York: Pantheon Books.

Roelofs, Joan. 1986. "Do Foundations Set the Agenda? From Social Protest to Social Service." Unpublished paper, Department of Political Science, Keene College, Keene, N.J.

———. 1983. "Foundation Influence on Supreme Court Decision-Making." *Telos* 62:59–87.

Saltman, Janet. 1978. *Opening Housing as a Social Movement*. New York: Praeger.

Schmitter, Phillipe. 1979. "Modes of Interest Intermediation and Models of Societal Change in Western Europe." In Phillipe Schmitter and G. Lehmbruch, eds., *Trends Towards Corporatist Intermediation*. Beverly Hills, Calif.: Sage. Pp. 117–46.

Sierra, Christine Marie. 1983. "The Political Transformation of a Minority Organization: The Council of La Raza, 1965–80."

Taylor, Charles Lewis, and David A. Jodice. 1983. *World Handbook of Political and Social Indicators*, 3rd Edition. New Haven, Conn.: Yale University Press.

Tilly, Charles. 1978. *From Mobilization to Revolution*. Reading, Mass.: Addison-Wesley.

Tober, James. 1984. "Wildlife and the Public Interest." Program on Non-Profit Organizations, Working Paper #80, Yale University.

Tully, Mary Jean. 1977. "Who's Funding the Women's Movement?" Commission on Private Philanthropy and Public Needs (Filer Commission). *Research Papers* 2:1383–84. Washington, D.C.: Government Printing Office.

Watters, Pat, and Reese Cleghorn. 1967. *Climbing Jacob's Ladder*. New York: Harcourt, Brace and World.

West, Guida. 1981. *The National Welfare Rights Movement*. New York: Praeger.

Wilson, John. 1983. "Corporatism and the Professionalization of Reform." *Journal of Political and Military Sociology* 11:53–68.

Wofford, Harris. 1980. *Of Kennedys and Kings*. New York: Farrar, Straus and Giroux.

Wolfe, Tom. 1970. *Radical Chic and Mau Mauing the Flack Catchers*. New York: Farrar, Straus and Giroux.

17

Giving for Children and Youth

ROBERT H. BREMNER

Public attitudes and government policy toward children in the United States are based on the assumption that, except in times of natural disaster or economic cataclysm, parents themselves can and should take care of their children with a minimum of assistance of interference from the state. Private philanthropy, less doctrinaire than government, recognizes that even in good times, some parents are unable to meet their children's needs, for one reason or another. They may be absent, dead, sick, immature, or merely unfortunate. The children themselves may suffer from such severe emotional, physical, or mental handicaps that the best and wisest parents could not cope with on their own. Moreover, regardless of the resources and ability of parents, philanthropy supports children's causes on the grounds that they benefit from association with their peers and adults other than their parents in organizations that offer wholesome outlets for their energies, broaden their interests, and foster their talents.

At present it is possible only to estimate how much of America's philanthropic giving is intended for the benefit of children and youth—in the neighborhood of 20 percent of the total annually contributed to philanthropy ($80 billion in 1985) by individuals, bequests, corporations, and foundations.[1] One hundred twenty national service and professional organizations, each with state branches and local chapters, play an important part in raising and dispensing funds for children. Through volunteers and paid workers the organizations offer a variety of services to young people and their families, act as advocates for children and protectors of their legal rights, and represent the concerns of the child-caring professions such as pediatrics, nursing, recreation, and social work. In 1985, six of the organizations (Boy Scouts, Girl Scouts, YMCA, YWCA, Boys Clubs of America, and Campfire Girls) enlisted the help of 2.5 million volunteers.[2]

Philanthropic foundations give a high priority to grants for child welfare and child development. Grants for children and youth rank highest in dollar-value and percentage of foundation giving among grants designed for special population groups.[3] Every household is bombarded with direct-mail solicitations for children's causes. Television and movie celebrities hold well-publicized "pitythons" to raise money for sufferers of children's diseases. Corporate philanthropy dotes on youth organizations like the Boy Scouts and Boys Clubs of America, and both foundations and corporations vie for sponsorship of the Special Olympics and similar events for handicapped and underprivileged children.

This chapter deals with the development of organized efforts for the care and education of orphans, handicapped, dependent, delinquent, neglected, and former slave children in the eighteenth and nineteenth centuries, the emergence and activities of service and advocacy organizations for children and young people in the late nineteenth and early twentieth centuries, and recent trends in philanthropy and public policy affecting children and youth.

Benjamin Franklin's account of his response to evangelist George Whitefield's call for an orphan home in Georgia testifies to the power an appeal for children can exert on the most unlikely donor. The children in question were relics of the poor debtors who had perished in the difficult struggle to found Georgia. Franklin had steeled himself not to contribute because he thought it would be cheaper to bring the orphans to Philadelphia where they could be bound out to service or otherwise provided for than to erect a costly institution for their use in a remote, poverty-stricken area. Against his will and better judgment Franklin found himself so moved by Whitefield's eloquence and arguments that he emptied his pockets of all he had in them, first coppers, then silver, and finally gold. A friend who shared Franklin's reservation about the project had taken the precaution of leaving his money at home. But now, carried away by Whitefield's message, he tried to borrow from a neighbor to add to the collection. The neighbor, less moved than most of Whitefield's listeners, refused on the grounds the would-be donor had departed his senses. On that single Sunday in April 1740 Whitefield collected nearly £200 for the orphanage.[4]

Whitefield was one of America's earliest and most successful fund raisers for children and youth. In addition to Bethesda Orphanage, which remained the principal object of his collections, he raised money and obtained books for Harvard, Dartmouth, Princeton, and the University of Pennsylvania. His success was the more remarkable in that the great age of orphanage founding, as of college establishments, lay almost a century in the future. By turning religious fervor from preoccupation with individual salvation to practical humanitarian objectives, he accustomed Americans to spasmodic concern for the welfare of their neighbors' children as well as their own.[5]

Orphanages were rare in colonial America. As Franklin indicated, the usual practice was to bind out or indenture orphans not adopted or cared for by relatives or friends at as early an age and upon the best terms public authorities could arrange. The same procedures were followed for children whose parents were too poor to support them and also, with some variations, for adult paupers. The plan, although subject to abuse by tight-fisted officials and unscrupulous masters, had the advantage of relieving the public of expense while providing children with shelter, food and clothing of a sort, and rudiments of training in a trade, husbandry, or housekeeping. By the latter part of the eighteenth century the larger towns on the eastern seaboard, where the poor were too numerous to be bound out to private contractors, maintained almshouses for the poor and helpless of all ages. In these institutions, despite efforts to segregate inmates by age and sex, children mingled with adult paupers and persons suffering from all sorts of mental and physical ills. Shortly after the turn of the nineteenth century, women in a number of communities organized orphan asylum societies to shelter children, especially girls, who would otherwise have been exposed to the unwholesome surroundings and debasing influences of the poor-house.[6]

A child's particular condition—sex, age, race, sickness, or handicap—may make it more or less appealing to donors. Factors of greater importance in determining charitable attitudes toward children are the esteem in which the parents are held and the extent to which other adults can relate to them because of religion, class, occupation, or national origin. During the first half of the nineteenth century the number of dependent children kept in poorhouses or supported by other public poor relief systems multiplied but would have been still larger had it not been for the establishment of more than seventy orphan asylums, most of which were founded in the two decades after 1830. Individual philanthropists, religious denominations, and benevolent and fraternal organizations associated with ethnic and occupational groups established them to assure that children would be raised in the faith and according to the traditions and values of their deceased or disabled parents. Many of these institutions, although privately supported and managed, received financial aid from government sources because the services they performed were deemed beneficial to society. Most observers believed the private orphanages were restrictive, and altogether they served only a fraction of the children who needed help.

Beginning in the 1820s, the city of Boston and private citizens and philanthropists in New York, Philadelphia, and other cities established institutions for juvenile delinquents that received both legal recognition and financial assistance from public authorities. The motivating force behind these organizations was not sympathy but prudence and concern for the public interest. Although the founders sought to spare youthful offenders from the harsh conditions and evil associations of the adult prison system, at the same time, they sought to protect their communities from the depredations and expenses they were likely to experience if the children's behavior went uncorrected. Believing that irresponsible or ineffective parents were at least partly responsible for delinquency, the managers of the reform schools accepted vagrant and neglected children as well as delinquents and subjected them all to a regimen of work, school, and moral instruction. The avowed purpose of confinement in the reformatories was not punishment but training in sound work habits, good behavior, and self-discipline. By mid-century the presumed success of the early reform schools had encouraged the founding of municipal and state institutions for the reformation or, at least, confinement of juvenile delinquents.[7]

The many faces of early and mid-nineteenth century humanitarians can be observed in provision for the education of deaf, blind, and mentally retarded children. The Connecticut Asylum for the Education and Instruction of Deaf and Dumb Persons, a private institution incorporated in 1817, received charity and tuition-paying students from the New England states and was supported in part by annual appropriations from New England state legislatures. A Kentucky institution founded in 1822 and supported, according to its charter, "by the donations and legacies of the charitable, by such aid the legislature may be pleased to afford and by the money to be received for the education of children whose parents, guardians or friends are of ability to pay" received children from the southern states. Congress provided land grants to both Connecticut and Kentucky asylums. The New England Asylum for the blind, founded in 1832 by Samuel Gridley Howe, quickly raised $20,000 from private sources in order to qualify for an appropriation of $10,000 from the state legislature. It obtained funds from fees, subscriptions, fairs, and exhibitions at which the pupils displayed their accomplishments. In 1839 the asylum's name was changed to Perkins Institution in recognition of gifts from Thomas H. Perkins.

Compassion seems to have played a part in arousing such wide support for these and other institutions for afflicted children. But the concern Thomas H. Gallaudet, principal founder of the Connecticut Asylum, and Samuel Gridley Howe felt for their charges went far beyond pity, and the school programs sought to provide students with the skills, self-respect, and self-confidence needed for independent living. As Howe declared in a successful effort to obtain state funds for the education of retarded children, extending a helping hand to the unfortunate was more than a kindly impulse: in the case of those "who can never of themselves step up upon the platform of humanity" it was an "imperative duty."[8]

One of the oldest and best-established forms of charity, giving for the education of children whose parents or guardians were not "of an ability to pay" continued to be practiced by religious groups operating charity schools in the larger cities and towns. In bequeathing large sums for the education of poor boys Stephen Girard (1750–1831) and John McDonough (1779–1850) set a pattern followed by many other childless rich men. In the early nineteenth century, public schools, where in existence, were free only to children whose parents took pauper oaths. In the middle third of the century concern for the education of children whose parents were not paupers but unable to pay school fees without stinting or sacrificing other crucial family needs was a factor in winning support for the common school as a civic obligation. By the time of the Civil War, northern and western states had instituted free public tax-supported elementary education for all normal children. Charitable societies in New York and other cities continued to maintain night schools for working children and industrial schools for children too ragged and irregular in habits for public schools. Except in a few states higher education remained an exclusive charge of philanthropy.

In the North, the coming of the Civil War transformed antislavery sympathies into zeal for educating those freed. Even before emancipation religious and benevolent societies began to send teachers and educational supplies and equipment to parts of the South under the control of Union forces. During Reconstruction a government agency, the Freedmen's Bureau, facilitated the efforts of northern religious and benevolent organizations whose missionary and educational efforts among the freed slaves involved them in rivalry with each other and hostility from white southerners. The earliest American philanthropic foundation, the Peabody Education Fund, founded in 1867 to foster sectional reconciliation, made grants to southern communities to assist them in establishing public schools. As southern states gradually developed systems of public education, northern missionary societies turned their efforts toward support of secondary schools and colleges to train black teachers and ministers.[9]

"What harms the child endangers the future—what helps the child protects the future."[10] Nearly all friends of children can subscribe to this proposition, but whether a particular program or course of action helps or harms a child often gives rise to sharp disagreement. In the latter half of the nineteenth century, individuals and agencies serving children differed on whether children whose parents were unable to care for them should be sent to institutions or "placed out" in family homes. The issue, as Franklin's objection to Whitefield's plan for an orphan house in Georgia indicates, was not new, but two factors made the problem urgent. One was concern about the hordes, at once pitiful and menacing, of neglected or abandoned, seemingly homeless, children who roamed the streets of New York and other large cities. The other was the conviction that society owed destitute children, especially orphans of Civil War soldiers, a better start in life than being shut up in a poorhouse with adult paupers. Advocates of institutional care cited discipline, protection from exploitation, and close supervision of religious and moral training as advantages of asylums. Their opponents countered that congregate living exposed children to many dangers, physical and moral, and that family life was a better participation for adult responsibilities than the routine of institutional existence. The debate, although acrimonious, was not clear-cut since supporters and administrators of children's asylums acknowledged, in principle, that sooner or later children should leave the institution for a foster home; on the other hand some advocates of placing out believed a brief period of institutionalization should precede a child's placement in a family. In addition to matters of policy the dispute involved competition for public recognition, public favor, and, literally, public support.

The leading critic of institutions and apostle of placing out was Charles Loring Brace, a liberal Protestant clergyman who founded the Children's Aid Society of New York in 1853 and remained its spokesman until his death in 1890. Under Brace's direction the society operated lodging houses for newsboys and working girls, industrial and foreign language schools, and many other services for poor children in New York City. It was best-known, however, for its "emigration" program, the shipment of trainloads of children from New York City to towns and villages in the Midwest where local committees helped place the children in private homes. During its first ten years of operation, the society assisted 7500 children emigrate from New York City; in the late 1860s and early 1870s, the yearly average of children sent west approached 1000.[11] Because of the large number of children involved, and also because of Brace's boundless confidence in the beneficent influence of family and rural life, the society made little effort to supervise the placement process, and follow-up visits by its agents to check the children's condition were perfunctory. The families receiving the children agreed to send them to school, take them to church, and treat them as members of the family. The children were not indentured, but they were expected to make themselves useful, and the possibility of using them as farm or domestic laborers was one of the inducements for taking them in. Brace, a believer in the gospel of work, seems not to have worried about exploitation. Henry W. Thurston, writing in the 1920s when attitudes toward child labor had shifted, said of the emigration plan: "It is the wolf of the old indenture philosophy of child labor in the sheepskin disguise of a so-called good or Christian family home."[12]

In seeking funds for the society Brace appealed to self-interest and altruism, prudence and pity. In the first report of the agency he declared, "there are no dangers to the value of property or to the permanency of our institutions, so great as those from the existence of such a class of vagabond, ignorant, ungoverned children," and warned, "Those who were too negligent or selfish to notice them as children, will be fully aware of them as men." Almost in the same breath Brace asked donors to "feel what it is for a boy or girl to be thrown out alone in the bleak city." The little ragged outcasts were all "children of our common Father"; each had an "immortal destiny"; and the fortunate to whom God had given homes, friends, and rich gifts of kindness must accept responsibility "for the condition of these little ones in the great city, poor, neglected and friendless."[13]

Contributions came to the Children's Aid Society from individual donors, old and young, rich and poor, in sums ranging from 50 cents to $100 or more, from churches in and outside of New York, and from collections in widely scattered Sunday schools. The society's first treasurer pointed out on several occasions that if 10,000 Sunday-school scholars could be induced to give 2 cents a week the Children's Aid Society would gain $10,000 a year. Brace often contrasted the cheapness ($12 to $15 dollars per child) of the emigration program with the high cost ($100 or more a year) of keeping children in asylums. He also cited the money the society saved the city, property owners, and taxpayers by sending potential criminals and troublemakers out of the city. Not surprisingly, he sought and gratefully accepted subventions from New York City and applied for appropriations from the state legislature and a share in the "Common School Fund" of New York state.[14]

Contemporaries praised the work of Brace and his organization for its objectives but found fault with some of its results. One usually blunt objection was voiced in 1880 by L. P. Alden, principal of the Michigan State School for Poor Children: "The great majority of even respectable well-to-do families are unfit to train up their own children, to say nothing of training up the children of others, many of whom are unattractive in appearance, have unpleasant habits and bad dispositions."[15] The Boston Children's Aid Society, founded in 1863, refined the rough-and-ready methods originally followed by Brace and his agents. Under the direction of Charles W. Birtwell, the Boston society gained wide recognition for care in placing children in homes suited to their capabilities and limitations. Beginning in 1883, children's home societies in a number of middle western and southern states concentrated their efforts on placing young children for adoption in childless homes. Children removed from almshouses as a result of laws passed after the Civil War might, as in Massachusetts, be boarded in private homes at public expense or, as in the Midwest, be transferred to county or state "homes" or "schools."

For a variety of reasons children's institutions continued to flourish long after Brace had proclaimed a poor home better than the best institution. Some served special classes of children such as soldiers' orphans or children of particular religious denominations. Brace's efforts had the ironic effect of strengthening the allegiance of Catholics and Jews to institutional care because they did not want children of their faiths to be placed in rural Protestant homes. The public policy of making per capita or lump-sum payments to private institutions for the care of children who would otherwise be public charges undoubtedly contributed to the survival of institutional care. As late as 1929, two decades after the White House Conference on the Care of Dependent Children of 1909 had gone on record as opposing the removal of children from their own homes solely because of the poverty of parents, and at least a decade after a number of states had passed mothers' pension laws to help dependent children in their own homes, 1500 children's institutions, with annual budgets totaling $60 to $70 million, still held 150,000 residents.[16]

Twentieth century philanthrophy has been marked by proliferation of organizations seeking to serve children and youth at all socioeconomic levels, not just the poor, and intended to supplement rather than supplant the influence of family, church, and school.[17] In 1900, one of the oldest and most widespread of the nation's youth service organizations was the Young Men's Christian Association. The original purpose of the YMCA, a British import introduced into the United States in the 1850s, was to protect the clerks, office workers, and aspiring businessmen of America's growing cities from the dangers of irreligion, intemperance, and immorality. By 1900, the 1500 local associations boasted 270,000 members, mostly young men from eighteen to twenty-nine years of age, but also including 30,000 teenaged boys. Until 1931, only members in good standing of evangelical Protestant churches could join the local associations as full members but associates and nonmembers participated in the association's programs and used their facilities. Facilities became increasingly important to the YMCA as the associations shifted their emphases from evangelism and conducting religious revivals in the 1850s and 1860s, to Bible study and character-and-career-strengthening programs in the 1870s, and to physical education and recreation in the 1880s and 1890s. Before 1890, many local associations had been involved in relief of distress in times of depression or disaster, but as communities developed more specialized welfare activities the YWCAs tended to concentrate on providing the services their members and clients seemed to need and want: gymnasiums, swimming pools, inexpensive housing, and summer camps.[18]

Like other service organizations, YMCAs met their expenses by contributions, dues, users' fees, and returns from operations such as cafeterias and dormitories. Contributions from businessmen sympathetic to the Y's aims and program were always an important source of support but, of course, seldom came without solicitation. "Begging" was therefore a necessary part of the work of secretaries of local associations. In the 1890s, Charles Sumner Ward, secretary of the Grand Rapids YMCA, conducted intensive, short-term campaigns at the start of each year in order, as he said, "to get the agony over quickly." Between 1905 and 1915, Ward and a colleague, Lyman L. Pierce, directed a series of fund drives to help local YMCAs raise money for new buildings. The techniques perfected in these campaigns—a specified goal to be reached in a limited period of time, careful planning, use of high-prestige leaders, large gifts to be matched by public donations, competition among teams of carefully selected volunteer solicitors, and powerful publicity—have been adopted by other agencies and associations and constitute one of YMCA's major contributions to philanthropy.[19]

Local associations of Christian women, patterned after but distinct from the YMCA, grew up in a number of American cities in the years after the Civil War. During the 1870s and 1880s, similar associations developed on college campuses. Both city and student associations formed national organizations that in 1906, united to form the National Board of the YWCA of the United States. Like the YMCA, the women's association established a broad range of secular services including residence and dining halls, employment offices, classes, clubs, gymnasiums, and swimming pools for working women and students.

To a much greater extent than the YMCA, the YWCA supported social action to improve working and living conditions. During the decade 1910–1920, the YWCA took a strong stand in favor of child labor legislation and maximum hour and minimum wage legislation for women. In the 1920s, denunciation of the YWCA's "social service labor program" and threats to withhold funds by employers' associations failed to shake YWCA's commitment to social legislation. After 1933, the YWCA backed New Deal relief, welfare and labor programs, and during and after World War II, gave vigorous support to efforts to end racial discrimination in employment and housing, to uphold civil liberties, and extend civil rights.[20]

The YMCA played an important role in bringing together the diverse groups seeking to organize an American version of the Boy Scouts, founded in England in 1908 by Lord Robert Baden-Powell. Among those contending for leadership of the American Boy Scout movement were Ernest Thompson Seton and Daniel Beard, both heading their own boys' organizations; William D. Boyce, a Chicago publisher impressed by the "good turn" philosophy of English Scouts; and William Randolph Hearst, newspaper publisher and unsuccessful Democratic candidate for president of the United States in 1908. In 1910, YMCA officials succeeded in amalgamating most of the contending forces into the Boy Scouts of America. In 1911, President Taft accepted the position of honorary president of the Boy Scout National Board. Early contributors to the BSA treasury included Mrs. Russell Sage, Andrew Carnegie, and John D. Rockefeller, Jr.

Under the aggressive leadership of James E. West, chief scout executive from 1911 to 1948, BSA developed a strong central organization with many staff positions filled by former YMCA workers. The Boy Scout program, stressing both individual achievement and community service, quickly proved its appeal to American boys, their parents, and business and political leaders. One factor in BSA's growth was effective use of volunteers as troop leader and in local councils. By 1919, Boy Scouts outnumbered YMCA boy members by more than two to one. Shortly before West's retirement, Boy Scout membership exceeded 1 million, YMCA had not quite half a million members under seventeen, and Boys Clubs of America (which included younger and poorer boys than those ordinarily found in BSA or YMCA) counted about 300,000 members.[21]

Camp Fire Girls and Girl Scouts of the United States were both founded in 1912, the former by Luther and Charlotte Gulick (he was the former president of the Playground Association of America), the latter by Juliette Low, a friend of Lord Baden-Powell. West and other Boy Scout leaders, regarding scouting as a male prescriptive and prerogative, were less than enthusiastic about a similar organization for girls. The Gulicks had no intention of copying the Boy Scouts but instead sought to fashion a program that would foster girls' womanly and domestic interests. Mrs. Low had no qualms about modeling her troops after the Boy Scouts. At the outset, with support from Jane Addams; Mary Richmond, a prominent social worker; and Grace H. Dodge, national president of the YWCA, Camp Fire Girls seemed to have an advantage over the Girl Scouts, but both organizations were firmly established by the time the United States entered World War I and each won plaudits for participation in the nation's war effort.[22]

In later years, Camp Fire Girls and Girl Scouts seemed more alike than different, at least to outsiders. In any given community one or the other might be more active, visible, and popular than the other. Deciding which to join seems to have been as much a matter of chance or circumstance as choice. Each tried to prepare girls for their roles as women; both rendered useful services. After World War II, the Girl Scouts' Clothes for Friendship campaign gathered quantities of clothing for destitute children in Europe and Asia. During the 1960s, in the Giant Step Program financed by the U.S. Children's Bureau, Camp Fire Girls made a determined effort to extend services to the hitherto neglected girls of inner cities.[23]

About the time Ward and Pierce were perfecting the YMCA "drive," Jacob Riis, whose writings and illustrated lectures about children of the poor had stirred the nation's conscience, called attention to a European fund-raising device destined to have a great impact on American philanthropy. In 1904, Riis received a letter from his native Denmark whose envelope bore a special Christmas stamp. Riis learned that proceeds from sale of the stamp were used to build a hospital for tubercular children. Six of Riis's brothers had died of tuberculosis and he knew of its ravages in the slums of American cities. In 1907, he wrote an article about Danish Christmas seals and suggested adoption of the idea in the United States. "Why should we not have a Christmas stamp," Riis wrote, "printed by a tuberculosis committee, and sold not for the purpose of building a hospital—let each state or town build its own—but for the purpose of rousing up and educating the people on this most important matter. . . . It is because they do not know a few amazing simple things that people die of tuberculosis."[24]

Christmas seals proved to be both popular and, from a fund-raising point of view, profitable. In little more than fifty years sale of the stamps produced $500 million dollars for the eradication of tuberculosis. Founded in 1904, the National Tuberculosis Association (later renamed the American Lung Association), which conducted and reaped the benefits of the annual Christmas seal campaign, was one of the earliest, and in many respects the prototype, of America's voluntary health associations devoted to the conquest of specific diseases. As Riis had hoped, the association made the seal the symbol of its message that tuberculosis was infectious, preventable, and curable. Funds from the Christmas seal campaign enabled the National Tuberculosis Association to pioneer in educational, nursing, diagnostic, and clinical services, many of which, having demonstrated their value in combating and preventing tuberculosis, were taken over by tax-supported agencies. The association has supposed undergraduate and postgraduate medical education, medical research, and industrial hygiene programs related to tuberculosis.[25]

One of the most highly regarded of the American health organizations, the National Society for Crippled Children (now the National Easter Seal Society for Crippled Children and Adults) began in 1919 as a father's memorial to his son and grew rapidly in the 1920s with the help of Rotary Clubs. The society's highly successful Easter Seal campaign, begun in 1934, financed a broad range of physical and rehabilitation programs for the handicapped. The organization's strategy, however, was not only to supplement existing community services but to push for better funding and establishment of more public provisions for the crippled. With the cooperation of the United States Children's Bureau, the society worked for inclusion in the Social Security Act of provision for federal grants in aid to states for development and expansion of services for crippled children. The major achievements cited by the society included increasing public awareness of the needs and abilities of the crippled, creation of state and federal legislation and funding on their behalf, and promotion of education for the handicapped.[26]

It is not easy to draw a sharp distinction between service and advocacy organizations.[27] Like the YWCA and National Easter Seal Society, most service agencies feel an obligation to represent their causes and the interests of their clients in the public arena. The American Lung Association, for example, is concerned not only with helping victims and fostering research on emphysema and asthma but in preventing illness by seeking to influence public attitudes and policy on smoking and air pollution. Conversely, circumstances and the desire to hold or attract members may involve advocacy organizations in a variety of service functions.

Continuity and change in an advocacy organization are well illustrated in the record of the National Child Labor Committee, founded in 1904 and still active in behalf of children and youth more than eighty years later. The founders included Felix Adler of the Ethical Culture movement; Homer Folks, a social worker prominent in every child-related cause of the early twentieth century; settlement house workers like Jane Addams, Robert Hunter, Lillian Wald, and Graham Taylor; Robert W. DeForest, president of the Charity Organization Society of New York; business people; financiers; lawyers; and a former president of the United States, Grover Cleveland. The object of the committee was not to deprive children of the experience, responsibility, and rewards of work but to safeguard them, and the national stake in their well-being, from the harmful effects of premature and exploitative labor. Despite the endorsement of John D. Rockefeller, Andrew Carnegie, and J. P. Morgan, the committee's financial resources were modest and derived mainly from gifts and contributions of members: around 4,600 in 1910; 11,000 in 1927; and 17,000 in 1947. Its strength lay in the tenacity and longevity of its leaders and staff members and the capabilities and dedication they brought to their work.[28]

One of the committee's first acts was to draft a model child labor bill based on the laws of New York, Massachusetts, and Illinois, providing for a minimum age of fourteen in manufacturing jobs, sixteen in mining, an eight-hour workday, no night work, and documentary proof of age. While lobbying for adoption of the standards of the model law, the committee investigated and (with the help of Lewis Hine's documentary photographs) reported on child labor in anthracite coal mines, glass factories, textile mills, canneries, the street trades, and night messenger service. The committee did not support the first federal child labor bill introduced in Congress in 1906 but enthusiastically backed President Theodore Roosevelt's proposal for an investigation of the condition of women and child laborers and for a model child labor law in the District of Columbia. The committee's first, and in many ways its most important, achievement was its successful campaign for the establishment of the United States Children's Bureau, an agency charged with investigating and reporting "upon all matters pertaining to the welfare of children and child life among all classes of our people."[29]

By 1913, the majority of NCLC leaders had become convinced that effective control of child labor could not be obtained by state action alone but also required national standards enforced by the federal government. For the next decade the committee worked for adoption of federal child labor laws. After the second of these measures had been declared unconstitutional by the Supreme Court in 1922, NCLC began a fifteen-year drive for ratification of a constitutional amendment granting Congress the right to regulate child labor. Before ratification by the required number of states had been obtained, Congress adopted (1938) the Fair Labor Standards Act which, with its child-labor provisions, was upheld by the Supreme Court in 1941. New legislation was at a standstill during the war years; but in 1949, the child-labor provisions of the Fair Labor Standards Act were strengthened and extended. The committee continued its watchdog activities regarding the administration of child labor at both state and federal levels.[30]

As NCLC entered its second half-century, the organization began to devote an increasing share of its attention to youth unemployment. In the 1950s, NCLC promoted work-study, cooperative education projects and service programs to help youths locate and keep satisfactory jobs. Since the 1930s, except for the war years, those working with youths had regarded finding jobs for young people legally entitled to work as challenging a problem as protecting children from overwork. The shift was also related to and, in part, intended to deflect criticism of child labor laws as depriving children of work experience, contributing to juvenile delinquency, and constraining children's rights. NCLC continued to emphasize that child labor laws were barriers against undesirable employment and had not deprived children fortunate enough to find jobs of the opportunity to work. The real problems were employers' attitudes and a changing economy that curtailed job opportunities formerly available to the unskilled and inexperienced. "Young people need work experience," declared Gertrude Folks Zimand, NCLC's general secretary in 1955, "and they also need assistance in preparing for and finding opportunities for suitable work." She urged that "the same determination, fervor and cooperation that were exhibited by agencies and individuals who worked to eliminate the early abuses of child labor . . . be brought to bear on the development of useful work experience programs."[31]

In 1974, according to an official of the United States Children's Bureau, there were fewer than 25,000 full orphans in the United States; 20,000 of them had been legally adopted or lived with relatives; 4,000 were in foster homes; and only about 1,000 were in institutions. The decline in number of orphans was largely the result of medical advances against premature death of parents. The turn away from institutional care came about because of the expansion and liberalization of the social insurance and public assistance programs inaugurated by the Social Security Act of 1935. In the postwar years, amendments to the original act extended the protection of old age, survivors, unemployment, and disability insurance to the great majority of American workers and significantly enlarged the number of mothers and children assisted by the federal and state Aid to Families of Dependent Children (AFDC) program. By the 1970s, some of the old orphan homes had closed, a few had been converted to senior citizens centers, and many had reshaped their programs to meet the needs of "orphans of the living." These were the approximately 500,000 children whose parents were unable to look after them, because of gaps in the social security system or personal inadequacy, or whose own emotional or behavioral problems required more care than parents alone could provide.[32]

One residential institution for children that survived and prospered in the era of AFDC and foster homes was Boys Town, founded in 1917 in Omaha, Nebraska, by Father Edward J. Flanagan as a refuge for homeless boys. During the 1940s, Boys Town began to send out letters at Christmas and Easter soliciting small contributions to "bring happiness to homeless and unwanted boys." Contributors, responding to the love, kindness, and good sense Boys Town represented, gave so willingly that in 1970 (the first year in which the figures became available to the public) 34 million requests for help produced $17.7 million in donations. A statement of the institution's financial condition, required by the Tax Reform Act of 1969, revealed that as a result of success in fund raising and money management Boys Town had an endowment of more than $200 million (larger than all but the very richest colleges and universities and four times as large as Notre Dame's), an annual return on investment of $6 million, and operating expenses of $9 million, one-third of which went for fund raising.

As a result of these disclosures the institution's board of directors sought professional counsel in devising programs to put Boys Town's wealth to more productive use. By 1986 the *Boys Town Quarterly Newsletter* contained a tear-out coupon requesting not money but daily prayers for the 430 boys and girls living on the main campus. Boys Town also operated an inner-city high school for troubled youth in Omaha; a hospital for children with hearing, speech, and language disorders; and provided technical assistance to help fifty-four other residential institutions for children in different parts of the country develop "Boys Town quality care."[33] The latter program suggests a revival of interest in the establishment and use of institutional, or at least group-home, care for children who are beyond the control of their parents or who do not function well in ordinary foster homes.

Greatly improved financial condition also brought change in the program of an influential New York City service agency, the Association for the Aid of Crippled Children. Organized in 1900 as an auxiliary of a Children's Aid Society class for crippled children, the association became an independent organization in 1908. It promoted education of crippled children in the public schools, and provided social and medical services to them and their families. Receipt of a large bequest in 1944 enabled the association to inaugurate research, education, and demonstration programs in child development as well as to continue its service to crippled children. During the 1950s, the association turned over its service functions to other agencies; investment income from the bequest continued to grow, and in 1972 the association reorganized and renamed itself the Foundation for Child Development. The foundation supports medical research, policy studies, advocacy projects, and service demonstrations for children in general. It has become well known for sponsorship of research on the well-being of children in New York City and for a national survey in which a sample of children aged seven to eleven were interviewed and given a chance to speak out about their lives and upbringing.[34]

Not affluence but scarcity is the normal state of affairs in the treasuries of nonprofit organizations. Patterns of support vary from agency to agency and according to function: child welfare, health, or youth services. Overall the largest source of revenue from nonprofit human service organizations in 1982 was government (38.4 percent); fees, dues, and charges made up the second largest revenue source (29.6 percent); and private giving, including contributions from individuals, United Way allocations, foundation grants, and corporate gifts, the third (21.3 percent). The important role government plays in funding voluntary agencies reflects the public-private partnership that has long characterized the American social welfare system: the federal government and, to a lesser extent, state and local governments fund many of the social services delivered by voluntary agencies. Put another way, voluntary nonprofit agencies deliver about as many government-funded human services as does government itself.[35]

Cuts in federal expenditures for social welfare effected during the Reagan administration worked hardships not only on poor children but also on the charitable agencies the administration praised as vital expressions of the American spirit. Between 1981 and 1985, federal budget cuts cost nonprofit human service organizations an estimated $30 billion dollars. Some organizations closed; others survived by putting greater reliance on fees and charges, thus making their service less charitable and less available to the poor. By 1986, increases in private giving had compensated for only about one-fourth of the lost federal revenue.[36]

While one organization for children, the Foundation for Child Development, undertook a study of the damage done to children's services by the administration and Congress in 1981 and 1982, another, the Children's Defense Fund, lobbyed and rallied public opinion against further cuts in children's programs. The Children's Defense Fund had been organized in 1973 as part of the Washington Research Project, a group of lawyers, researchers, and public policy analysts that monitored implementation of education and civil rights legislation by federal, state, and local agencies.

The Nixon administration's downgrading and dismemberment of the United States Children's Bureau in 1969 deprived the nation and those responsible for making public policy of much official information "pertaining to the welfare of children and child life among all classes of our people," and muffled a government voice that since 1912 had been outspoken in behalf of justice for all children. No subsequent administration has seen fit to restore the Children's Bureau to its former place as advocate for children's interests, but in the 1970s a number of private organizations contended for recognition as advocates or surrogate lobbyists for children. Most of them, unlike the Children's Bureau, addressed special problems or represented particular categories of children. The National Child Labor Committee continued to work for legislation affecting job training and work-study programs and took a leading part in monitoring federal programs for the education of migrant children. The Child Welfare League of America concentrated on legislation protecting and improving standards for services to neglected and dependent children. The National Association for Retarded Citizens, a parents' organization founded in 1950, proved highly effective in winning consideration and appropriations to help the retarded at all levels of government. Innumerable advocacy organizations championed causes whose significance might otherwise have gone unnoticed. Thus, Children before Dogs, founded in 1971, challenged the values of a pet-oriented society in which "many abused children go to bed hungry while at the same time pet food companies are spending millions of dollars on advertising." Its program stressed the need for enforcement of pet control laws, educating the public to the danger of diseases transmitted from dogs to children, and opposed the use of children in Earth Day cleanups.[37]

The Children's Defense Fund, led by Marian Wright Edelman, has undertaken to protect and to secure adequate appropriations for existing public programs for children, to point out flaws and inequities in AFDC and other federal and state programs affecting children and families, to inform families of their rights in dealing with bureaucracies, and to show what needs to be done to meet the needs of poor, minority, and handicapped children. In arousing public opinion on issues affecting children, lobbying for, and monitoring public programs, CDF revives memories of the National Child Labor Committee at the peak of its influence in the early years of its history. Of all the advocacy groups that emerged in the 1970s, CDF most closely resemblies the Children's Bureau in identifying problems and researching and publishing information on topics of urgent concern to children and families: *Children Out of School in America* (1974), *Children in Adult Jails* (1976), *Children Without Homes* (1978), *Children Without Health Care* (1981), and studies of children who need but do not receive mental health care (1982) and of the educational rights of handicapped children.[38]

In 1979, the United Nations International Year of the Child, experts estimated that 250 million of the world's children were hungry, sick, and homeless. In many countries, where there was a desperate need for educated citizens and leaders, young children worked instead of going to school. At least eighty American voluntary organizations, many associated with churches and others secular, sent food, clothing, medicine, and school supplies to children overseas. Despite the number of agencies and the broad support they obtain from individuals and groups of all kinds, voluntary efforts in foreign aid are small in comparison to those of government, and much of the food and freight costs for goods distributed by the voluntary agencies are supplied or paid for by the government. Voluntary efforts are important, however, both to donors and recipients, because they seek to deliver emergency aid and reconstruction assistance, not just to allies or friends but wherever it is needed and to whoever needs it. Voluntary agencies not only give direct relief but point the way to self-help. In this respect the United Nations Children's Fund (UNICEF), Save the Children Federation, Foster Parents Plan International, and similar organizations recapitulate the methods and teachings American philanthropy has developed for the relief and prevention of misery among children in our own country over more than 250 years.[39]

The hardships children face, whether abroad or at home, are rarely of their own making and seldom entirely or mainly the fault of their parents. Other than natural disasters, the cause is usually that in pursuing some political-economic goals nations fail to consider the interests of children or assume that they will be looked out for by their parents. Today, we need to remind ourselves that contemporary family life differs in many respects from the ideal official image. In place of, or at least in addition to, our traditional fear of government interference in family matters, we must recognize that almost every government budget or policy decision has some impact on children. The most important thing philanthropy can do for children is to see that data on their condition and needs are held in the foreground of the national consciousness and receive the attention prudence no less than affection dictates.[40]

NOTES

1. The National Center for Charitable Statistics, Washington, D.C., is developing a classification system for philanthropic giving that will make possible a more precise determination of the amount of giving intended for children and youth. The figure of 20 percent is based on percentages of total distribution of contributions for various categories (that is, religion, education, health and hospitals, and social service) in American Association of Fund

Raising Counsel (AAFRC), *Giving U.S.A. 1985* (New York, 1986), p. 7; see also The Foundation Center, *Grants for Children and Youth* (New York: 1986), and United Way of America, *Allocation Profiles* (Alexandria, Va.: 1985), pp. 3–4. I am grateful to Martha E. Taylor, Research and Evaluation Director, The United Way of Franklin County for making available information on national and local United Way allocations.

2. Marian L. Peterson, comp. *National Directory of Children and Youth Services '86–87* (Denver and Longmont, Colo.: 1985), 553–58; AAFRC, *Giving USA, 1985*, p. 93.

3. Patricia Read and Lorenz Renz, eds., *Foundation Directory, 10th Edition* (New York: 1985), p. xxviii.

4. *The Autobiography of Benjamin Franklin*, Leonard W. Larrabee, et al., eds., (New Haven, Conn.: 1964), pp. 176–78. The first orphan home in the present boundaries of the United States was established in 1729 at Ursuline Convent, New Orleans; the first orphanage in Georgia had been founded in 1738 by German settlers at Ebenezer.

5. Whitefield's contributions to American philanthropy are discussed in Robert H. Bremner, *American Philanthropy* (Chicago: 1988), pp. 22–23.

6. Robert H. Bremner, et al., eds., *Children and Youth in America, A Documentary History*, 3 vols. (Cambridge, Mass.: 1970–1974), I:262, 277–81. Hereafter cited as *Children and Youth*.

7. Homer Folks, *Care of Destitute, Neglected, and Delinquent Children* (New York: 1900, 30–41, 111–26; *Children and Youth*, I:671–72.

8. *Children and Youth*, I:786.

9. Robert H. Bremner, *The Public Good* (New York: 1980), pp. xvi–xvii, 129–31, 186–89.

10. The Children's Aid Society, *Ninetieth Annual Report for the Year 1942* (New York: 1943), p. 7.

11. Henry W. Thurston, *The Dependent Child* (New York: 1930), p. 121. Donald Dale Jackson, "It Took Trains to Put Street Kids on the Right Tracks out of the Slums," *Smithsonian* 17 (August 1986), pp. 98–99, says that by 1910 the Children's Aid Society had sent 105,000 children to the West.

12. Thurston, ibid., p. 136.

13. The Children's Aid Society, *First Annual Report, 1854* (New York: 1854), 13–14.

14. The Children's Aid Society, *Ninth Annual Report, 1862* (New York: 1862), pp. 21–22.

15. Quoted in *Children and Youth* II:294.

16. R. R. Reeder, "The Place of Children's Institutions," *Survey* 61 (January 15, 1929), p. 483. On the subsidy system see Folks, pp. 69–73; on the White House Conference of 1909 and the Mothers' Aid movement, see *Children and Youth* II:357–97.

17. Only a few of the agencies are discussed in this chapter. For information on others see Peter Romanofsky, ed., *Social Service Organizations*, 2 vols. (Westport, Conn.: 1978); developments in the movement can be followed in the articles on Youth Service Organizations in *The Social Work Yearbook, 1929–60* (New York: 1930–1960) and *Social Work Encyclopedia, 15th, 16th, and 17th Issues* (New York: 1965–1977).

18. Mayer Zald, *Organizational Change, The Political Economy of the YMCA* (Chicago: 1970), pp. 28–48.

19. Scott M. Cutlip, *Fundraising in the United States, Its Role in America's Philanthropy* (New Brunswick, N.J.: 1965), pp. 38–45.

20. Ann W. Nichols, "Young Women's Christian Association of the N.S.A. National Board of the (YWCA-USA)" in Romanofsky, *Social Service Organizations*, II:764–72; Elsie D. Harper, *The Past Is Prelude, Fifty Years of Social Action in the YWCA* (New York: 1963), pp. 5–27, 40–42, 50–65.

21. C. Howard Hopkins, *History of the Y.M.C.A. in North America* (New York: 1951), pp. 468–69; David I. Macleod, *Building Character in the American Boy: The Boy Scouts, YMCA, and Their Forerunners, 1870–1920* (Madison, Wisc.: 1983), pp. 146–67, 300; "Boy Scouts of America (BSA)" in Romanofsky, *Social Service Organization*, I:181–86.

22. Macleod, *Building Character*, pp. 50–51; Campfire Girls, Inc. (CFG)," and "Girl Scouts

of the United States of America (GSUSA)" in Romanofsky, *Social Service Organizations*, I:200–207, 327–31.

23. F. Emerson Andrews, *Philanthropic Giving* (New York: 1950), p. 84; Catherine P. Papell, "Youth Service Organization," *Social Work Encyclopedia, 16th Issue* (New York: 1971), II:1554.

24. Jacob A. Riis, "The Christmas Stamp," *The Outlook* 36 (1907), p. 513.

25. Richard H. Shrylock, *National Tuberculosis Association, 1904–1954: A Study of the Voluntary Health Movement in the United States* (New York: 1957), pp. 127–35; Selskar M. Gunn and Philip S. Platt, *Voluntary Health Agencies, An Interpretive Study* (New York: 1945), p. 173; on the development of the Christmas Seal campaign see also Cutlip, Fundraising in the United States, pp. 53–58, and Richard Carter, *The Gentle Legions* (New York: 1961), pp. 75–90.

26. Carter, ibid., pp. 189–94; Gunn and Platt, ibid., p. 175; "National Easter Seal Society for Crippled Children and Adults (NESSCCA)" in Romanofsky, *Social Service Organizations*, II:527–31; *Children and Youth*, II:1213, 1218–19, 1221–22.

27. In 1983, the American Red Cross and Girl Scouts of America joined a number of other predominantly service organizations in protesting that proposed regulations excluding advocacy organizations from the Combined Federal Campaign would have severely restricted their political activities. Council on Foundations, *Newsletter* 2 (November 8, 1983), p. 2.

28. Walter I. Trattner, *Crusade for the Children* (Chicago: 1970), pp. 45–67, and "Working for Youth for Eighty Years," *New Generation* 64 (Winter 1984), p. 5.

29. *Children and Youth*, II:774.

30. Ibid., III:342–48; Gertrude Folks Zimand, *Child Labor after Ten Years of Federal Regulation* (New York: 1948), p. 4; and *Trends in Employment of Young Workers* (New York: 1949), pp. 9–11.

31. Gertrude Folks Zimand, *Child Labor vs. Work Experience* (New York: 1955), pp. 9–10.

32. Linda Amster, "Orphanages Vanishing for Lack of Orphans," *New York Times* (December 26, 1974), Section 1, p. 67; John Matsushima, "Child Welfare: Institutions for Children," *Social Work Encyclopedia, Seventeenth Issue* (New York: 1977), p. 146; James K. Whittaker, *"Residential Treatment for Children,"* 1983–84 Supplement to the Encyclopedia of Social Work, Seventeenth Edition (New York: 1983), p. 135.

33. "Father Flanagan is Dead in Berlin," *New York Times* (May 15, 1948), p. 15; *Newsweek* 79 (April 10, 1972), p. 55; *Time* 102 (April 2, 1972), pp. 17–18, and 104 (August 5, 1974), p. 178; Marge Peterson, "Boys Town—A 'Miracle of the Heart,' " *Home and Away* 7 (July–August 1986), pp. 26–28.

34. "Foundation for Child Development (FCD)" in Romanofsky, Jane Dustan, "Marking Seventy-Five Years: Foundation for Child Development Looks Ahead," *Foundation News* 16 (January–February 1975), pp. 30–35; Trude W. Lash and Heidi Sigal, *State of the Child: New York City* (New York: 1976) and *State of the Child: New York City*, vol. 2 (New York: 1980); Nicholas Zill, Heidi Sigal, and Orville G. Brim, Jr., "Development of Childhood Social Indicators," in Edward F. Zigler, Sharon Lynn Kagan, and Edgar Klugman, eds., *Children, Families, and Government, Perspectives on American Social Policy* (Cambridge, London, and New York: 1983), pp. 198–203, 207.

35. Lester M. Salamon, James C. Musselwhite, Jr., and Carol J. DeVita, "Partners in Public Service: Government and the Non-profit Sector in the Welfare State," in Independent Sector, *Philanthropy, Voluntary Action and the Public Good* (Washington, D.C.: 1986), pp. 8–10, 22.

36. Lester M. Salamon, "The Results Are Coming In," *Foundation News* 25 (July–August 1984), pp. 16–23; and "Federal Budget: Deeper Cuts Ahead," *Foundation News* 26 (March–April 1985), pp. 48–54; Council on Foundations *Newsletter* 5 (October 6, 1986), pp. 1–3.

37. *Children and Youth*, II:viii and III:522; Gilbert Y. Steiner, *The Children's Cause* (Washington, D.C.: 1976), pp. 143–75; National Child Labor Committee, *The First 80 Years, A Chronology* (New York: 1984); Robert Segal, "National Association for Retarded Citizens (NARC)," in Romanofsky, *Social Service Organizations*, II:436–43; Annie M. Brewer, ed.,

Youth-Serving Organizations Directory (Detroit: 1980), pp. 164–65; Peterson, *National Directory of Children and Youth Services, 86–87,* p. 554. Both Brewer and Peterson provide information on numerous organizations not discussed in this chapter.

38. Steiner, *The Children's Cause,* pp. 172–75; Catherine J. Ross, "Advocacy Movements in the Century of the Child," in Zigler et al., *Children, Families, and Government,* p. 175.

39. Curtis J. Sitomer, "Global Perspectives on Children in Need," *Foundation News* 20 (May–June, 1979), pp. 29–35; James P. Grant, *The State of the World's Children 1984* (New York: 1984), p. 1, 69–70; *New York Times* (July 14, 1982), pp. 7, 22. For Foster Parents Plan International and Save the Children Federation see Romanofsky, Social Service Organizations, I:317–19 and II:657–62.

40. Marian Wright Edelman, "Children Still Get Hungry in U.S.," *Foundation News* 21 (May–June 1980), pp. 37–39.

18

Private Foundations and Black Education and Intellectual Talent Development

JOHN H. STANFIELD II

During the hundred-year history of the American private foundation sector, numerous efforts have been made by several foundations to grapple with a pressing public issue. One of the most important occurred in the late nineteenth and early twentieth centuries in the institutionalization of black education in the South, particularly in rural areas. Ever since, foundations have been a major source of aid in the development of black educational institutions and intelligensia. Indeed, between 1880 and 1960, the foundation sector was the major funding force behind public and private black education and the development of black intellectual talent.[1]

The historical configuration of this intervention has been dependent upon four factors that shaped the general history of foundation giving: (1) localism and regionalism in the national political culture; (2) foundation giving as a stimulant for government action; (3) the historical sensitivity of black issues in state public policy circles and institutions; and (4) the role of individuals and private foundation philanthropy. This chapter will relate these factors to foundation contributions to the development of black education and intellectual talent during three historical periods exemplified by case studies: pre-World War I; between the world wars; and the post-World War II decades.

LOCALISM AND REGIONALISM IN NATIONAL POLITICAL CULTURE AND PRIVATE FOUNDATION PHILANTHROPY

Historically, American citizens have preferred a weak central government and extensive local and regional control over public and social policy formulation. This preference in areas such as social welfare, health, education, and race relations was particularly apparent before World War I.

Except during major war years, the American federal government rarely intervened extensively into policy areas outside those assigned federal jurisdiction in the U.S. Constitution. When the central government has gained extensive intervention rights, these rights have tended to persist as controversies and often have been eventually reversed or subverted. Also, even during times of state intervention, such as the New Deal or the Great Society, intervention was far from complete. A number of policies remain the domain of local and regional administrative policy circles and institutions.

Private foundations have played increasingly crucial roles in reinforcing the tradition of localism and regionalism. Particularly before the 1960s, even national and international foundations tended to subsidize projects that were local (municipal or state) or regional (especially in the South). Only a few foundation-funding efforts, such as agricultural reform, were directed toward stimulating federal intervention.

Even if most foundations subsidize local and regional projects, the foundation sector has not been divorced from the central government. From the late nineteenth century to the present, the governing boards of foundations with regional, national, or international foci have served as informal meeting grounds for local and regional leaders and central government officials. Often, private foundation capital has enabled central government officials informally to advocate policy stances that may be contrary to or of no interest to official stands. For example, before World War I, numerous prominent officials such as U.S. Supreme Court justices, presidents, and ambassadors sat on the boards of foundations involved in southern affairs (for example, the John F. Slater Fund and the Peabody Education Fund). Since the New Deal, the informal, interlocked character of the foundation sector and the central government has become more explicit as foundation executives became government officials and vice versa; for example, Dean Rusk, president of the Rockefeller Foundation became Secretary of State, and McGeorge Bundy, National Security Advisor, became president of the Ford Foundation.

PRIVATE FOUNDATIONS AND GOVERNMENT ACTION

From the late nineteenth century to the present, the stimulation of local and federal government action has been a primary strategy of private foundation funding. Given the relatively limited capital of even the largest foundations, this has proven to be an ingenious donations principle. Agricultural reform, public housing, scientific research, social welfare, and public health measures now financed by local and federal government agencies owe at least part of their origins to foundation-sponsored demonstration projects. As shall be seen, this giving principle was instrumental in forging the historical contributions foundations made to the development of black education and intellectual talent.

RACE RELATIONS AS A SENSITIVE PUBLIC ISSUE

Foundations have at times paved the way for significant social changes well before society at large was ready or even equipped for them, and not a few of such vanguard efforts by foundations were quite controversial. Yet, when it has come to issues related to the black experience and race relations in general, philanthropists and their foundation administrators, for the most part, have dragged their feet or promoted programs that accommodate prevailing race relations conventions. The foundation sector has followed rather than led society. During the pre-Civil Rights era, all but one of the few foundations that officially reported an interest in race relations areas (the Julius Rosenwald Fund)[2] respected racial caste traditions in the South and North. At most, foundations involved in race relations work pursued racial betterment for blacks in a publicly sanctioned, racially segregated society. During the Civil Rights era of the 1950s and 1960s, the advent of a few foundations concerned with racial desegregation, and staff changeovers in older foundations, led to funding priorities that were in line ideologically with dramatic changes in race relations. But even then, and indeed now in the 1980s, most foundation efforts in race relations have been symbolic attempts to "work on and for victims" rather than on transforming society at large.[3]

The reluctance of philanthropists and officers of large foundations to be more bold in issues related to the black experience and race relations is an extension of the historical discomfort of discussing or dealing realistically with issues of racial inequality. When not ignored, diluted, or talked around in popular culture and public policy circles, racial inequality tends to be addressed in a moralistic or individualistic manner or discussed evasively: that is, calling racism something else, such as *urban problems*.

Thus, throughout the history of foundation involvement in black affairs, foundation-supported projects, programs, and institutions have tended to encourage "culture of poverty" and other victim-oriented paradigms. Few foundations have attempted to tamper with the political economy of institutional arrangements of racism. When foundation officials have attempted to use grants to promote significant institutional and political changes in race relations, they encountered great public or congressional resistance or criticism. Among the most obvious cases were the McCarthyite attack on the Julius Rosenwald Fund and other foundations; the Ford Foundation's controversial school decentralization experiment in New York City in the late 1960s, and the racial politics behind the passage of the 1969 Tax Reform Act.[4]

The preceding discussion helps us to understand why major, older foundations supported Jim Crow public policies for so long and began to gradually change their funding policies during the 1960s; within social and political limits, post-World War II foundations were able to gradually develop pro-desegregation funding policies; and there has been so little effort in the foundation sector to innovate positive institutional and system changes that affect blacks and other racial minorities and race relations in general.

Table 18.1 Foundation Contributions in 1937 and 1940, by Field

Grant Field Foundation	1937	1940
Aesthetics	$ 835,038.75	$ 792,821.56
Agriculture and Forestry	90,585.44	253,198.94
Aviation	600,000.00	50,450.00
Child Welfare	783,307.01	769,040.12
City and Regional Plan- ning and Housing	141,670.44	102,293.83
Economics	1,353,386.09	749,406.34
Education	9,170,317.93	11,690,605.79
Engineering	546,334.28	63,770.85
Government and Public Administration	1,710,598.08	1,062,916.68
Heroism	185,564.00	168,110.00
Humanities	805,520.02	614,557.19
International Relations	897,026.23	688,848.65
Medicine and Public Health	13,495,898.38	12,273,590.19
Physical and Biological Sciences	2,253,298.24	3,783,642.81
Publications	15,562.31	142,137.12
Public Service	1,000.00	270.00
Race Relations	108,081.95	29,923.21
Religion	274,664.35	1,224,044.22
Social Sciences	969,067.43	1,528,510.23
Social Welfare	4,695,879.63	4,395,897.92
Unclassified	144,532.25	—
Total	$38,477,932.81	$40,390,035.74

SOURCE: Data from: *American Foundations and Their fields* vols. 4 and 5 (New York: Raymond Rich Associates, 1939, 1942).

In comparison to other major public concerns such as education, health, and the arts, relatively few foundations and relatively few dollars have been devoted to race relations.[5] Tables 18.1 and 18.2 are tabulated from the 1939 and 1942 volumes of *American Foundations and Their Fields,* which offer data on foundations assets and grant appropriations, respectively, from 1937 and 1940. Table 18.1 documents the minute official reporting of race relations projects by foundations as contrasted with other major public policy areas. However, a number of foundations probably appropriated funds for race relations projects but reported such donations under other broad categories. Table 18.2 indicates that for both years, 1937 and 1940, the same few foundations reported on official interest in race relations (out of, respectively, 121 and 162 foundations). And, among these, race relations was a minute fraction of the total grant making. Directories during the 1950s and later decades continue to document the continued scarcity of officially reported race relations grant making in foundation circles.

Another indication of the continued discomfort with race relations issues was a transformation of labels. What was reported as *race relations* prior to World War II, became known as *intercultural relations* in foun-

Table 18.2   Foundations Involved in Race Relations, Official
Reporting (top figure shows race relations grants: lower figure
shows total grants)

Foundation	1937	1940
Carnegie Corporation	$24,750.00	$0
	(3,695,534.03)	0
Phelps-Stokes Fund	3,665.00	2,995.73
	(32,375.06)	(27,045.83)
Rockefeller Foundation	44,700.90	0
	(8,996,016.19)	0
Julius Rosenwald Fund	12,021.97	20,927.58
	(731,300.26)	(637,293.77)
Russell Sage Foundation	12,250.00	6,000.00
	(307,305.00)	(203,600.00)
Spelman Fund of N.Y.	10,694.08	0
	(1,103,172.24)	0

SOURCE: Data from *American Foundations and Their Fields* vols. 4 and 5 (New York:
Raymond Rich Associates, 1939, 1942).

dation directories. Respected contemporary observers of the foundation
sector have discussed the continued reluctance of most foundations to can-
didly address the problems and needs of racial minorities and other powerless populations (for example, see the U.S. Human Resources Corporation's report, "U.S. Foundations and Minority Group Interests"[6]).

Private foundations have, at times, become more involved in a public
policy area than government when officials are reluctant to intervene because of political delicacy. After Reconstruction, conservative southerners
and their northern allies created and institutionalized a passive, if not hostile, central government policy when it came to blacks. Therefore, issues
such as black education, welfare, health, and civil rights became the burden of foundations and other nonprofit organizations with the central government's informal blessings and participation. Although foundations contributions to black institutional and community development was small, it
was all the significant funding blacks had, given their meager sources of
indigenous philanthropic capital and their exclusion from the sources of
white investment capital that made possible the development of foundations in white Anglo-Saxon and Jewish communities. Thus, for many decades, the foundation sector was an *auxiliary patrimonial* state over so-
called Negro problem, holding sway over black affairs, especially in education, health care, and civil rights organization, through the maintainence
of highly personalized ties with favorite individual and institutional beneficiaries.

INDIVIDUAL AND PRIVATE FOUNDATION PHILANTHROPY

Because private foundations historically have largely been accountable to
the public, their policies and organizational behavior have been influenced

more sharply by the personalities of "corporate owners and managers" than other institutions involved in public affairs. The mix among the influence of donor families, governing board members, and staff members varies from foundation to foundation. Therefore, in order to understand trends in the organizational behavior and funding policies of foundations, a researcher must first study the donors, governing board members, and administrators. Thus, in the ensuing discussion, much emphasis is placed on the persons who contributed to the patterns of foundation giving to black education and intellectual development in three major historical periods.

Era 1: Private Foundations and the Funding of Local and State-Supported Black Education, 1867–1920

The few scholars who have studied the private foundation sponsorship of black education between 1867 and 1920 have focused on the roles of Rockefeller Foundation and the Phelps-Stokes Fund. Thus, most attention has been directed toward exploring the philanthropic contributions made to the Tuskegee and Hampton Institutes.[7]

The academic preoccupation with philanthropy and private black education during the post-Civil War–pre-1920s era distorts greatly what foundations interested in black education actually were doing during that period of time. When the total picture is examined through foundation records, institutional histories, and pertinent biographies, it is apparent that foundation interest in private black education between 1867 and 1920 was short lived.[8] As in the case of missionary associations, the major foundation thrust toward private black education was dominant in the late nineteenth century and began to taper off sharply by 1905. To the extent that foundations continued to fund private black education from 1905 on, it was more selective, with Tuskegee and Hampton becoming the darlings of the emerging foundation sector. The movement of foundations away from concentration on extensive financing of private black institutions was accelerated by the publication of Thomas Jesse Jones' 1917 *Negro Education: A Study of the Private and Higher Schools for Colored People in the United States,* which assisted foundations in their efforts to streamline black education funding priorities.

During the first decades of the twentieth century, philanthropists and foundation administrators shifted dramatically from private black education to black education financed and managed by individual southern states. This interest can be traced to three concerns. First, within the regional political framework of the turn-of-the century South, educational efforts by blacks with no or marginal white supervision was considered to be a grave threat to white hegemony.[9] If blacks were to be educated, and if foundations were to be safely involved in such endeavors, local and state-controlled black education was the safest route to follow. Second, due to the relative scarcity of foundation capital, cost-sharing with local states stretched the dollars foundations earmarked for black education. Third, encouraging local and state interest in black education was part and parcel of growing foundation efforts to use their funds to stimulate government intervention.

Whatever the reasons, foundation interest in black public education in the South became one of the most well-orchestrated efforts in the history of the foundation sector. It involved a small group of philanthropists, foundation administrators, and southern educational administrators who pooled and distributed venture capital and quietly lobbied local and state officials about the virtues of publicly supported black education. Given the conservative public political attitudes about race relations in the South, the efforts and accomplishments of these people in their historical context was extraordinary.

The principal architects and agencies of the foundation effort to stimulate local state support of black education were John D. Rockefeller and The General Education Board, James Dillard and the Anna Jeanes Fund, the Peabody Education Fund, and Julius Rosenwald and his namesake fund. Rockefeller was a member of a group of northern philanthropists and Progressive Southerners interested in expanding universal educational policies into the South; particularly as it affected the education of blacks. They organized influencial conferences on universal and black education such as the 1890 Mohonk Conference on the Negro Problem, and the Capton Spring Conferences. They were also behind the organization of the Southern Education Board, the immediate predecessor to the General Education Board's (GEB) work in the South.

Although Rockefeller was most interested in the black dimension of southern education, racial etiquette prevented the GEB from making "Negro education" a prominent initial goal. While the GEB began to legitimate its activities in other areas of education, its officials assured southern opinion leaders that their planned efforts in black education would not compromise regional race relations' traditions and patterns.[10]

The General Education Board was organized in 1902. Its officers were in the midst of planning their black education funding priorities when the Anna Jeanes Fund was established in 1907. Anna Jeanes was a Philadelphia Quaker philanthropist who donated $1 million for the organization of a foundation dedicated to rural black education. James Hardy Dillard, a Tulane University dean, became its first president. While developing the Jeanes Fund's funding priorities, a proposal submitted by Jackson Davis, a young Virginia county school administrator, came to his attention. He requested funds for subsidizing a rotating black teacher program that attempted to meet the needs of poorly financed and understaffed black schools in his county. Dillard not only funded Davis's project, but made it the focal point of the Anna Jeanes Fund. For decades, Jeanes teachers would be crucial personnel in black school systems throughout the South.[11]

Davis's mentor and friend, Hollis Frissell, president of Virginia's Hampton Institute, was a strong advocate of local and state-supported black education.[12] Shortly after Davis made a strong and lasting impression on

Dillard and his foundation, Frissell persuaded the Peabody Education Fund to stimulate local and state interest in black education, using Virginia as a model. He suggested the employment of a foundation-paid official—a state agent for Negro schools—in the Department of Education, who would encourage and oversee local and state-supported black schools. In 1910 Frissell recommended the appointment of Jackson Davis as the first state agent.[13]

By 1913, the Virginia state Negro schools agent program was working out so well that the GEB, taking over from the then defunct Peabody Education Fund, funded six other state agents for Negro schools. By 1915, most of the southern states had such agents and Jackson Davis was hired by the General Education Board as general field agent to work with the state agents of Negro schools in all the southern states."[14] For three decades, under Davis's leadership, the agent for Negro Schools Project would be the most significant contribution the GEB made to black education.

Although the GEB provided funds for the salaries and expenses of (white) state agents for Negro schools and the Anna Jeanes Fund was a major source of black teachers in the expansion of state-supported black schools, Julius Rosenwald and his foundation also played a most important role in this movement. Rosenwald, the president of Sears and Roebuck, first became interested in black education in 1910 through reading the autobiography of Booker T. Washington, principal of the Tuskegee Institute, and the biography of William Baldwin, Jr., a major Washington supporter. By 1912, Rosenwald was a major financial backer of Tuskegee Institute and was appointed as a trustee for life.[15] In 1914, Rosenwald gave Washington several thousand dollars to allocate to communities around Tuskegee interested in building publicly supported schoolhouses for blacks. The community interest, Rosenwald stipulated, had to be demonstrated through white and black citizens raising matching funds before his funds were released. This was the beginning of the famous massive Rosenwald school program, which operated out of Tuskegee for several years and then out of Nashville after Rosenwald organized his foundation in 1917. By the time the Rosenwald schoolhouse program closed down in 1935, he and his foundation contributed to building nearly 5000 schoolhouses for blacks throughout the rural South.[16]

For many years, Rosenwald schoolhouse design and building funds were a critical resource of state agents for Negro schools. With such funds, state agents were able to stimulate local and state interest in black education that did not offend prevailing racial mores. The Rosenwald funds also enabled each state agent to hire black assistant agents, who were crucial observers and evaluators of the status of publicly supported black schools.

The coordination of the foundation sector effort to promote local state black schools was possible through an extensive overlap of foundation governing board membership. The GEB served as the big foundation hub. This is seen most clearly in the appointment of James Dillard and Julius Rosenwald to the GEB governing board, along with Anson Stokes of the

Phelps-Stokes Fund. Also, for many years, the GEB would appropriate grants to the smaller foundations to maintain the intricate private foundation effort toward a well-institutionalized, albeit segregated, black school system in the South.

Era 2: The Julius Rosenwald Fund's Black Education and Intellectual Talent Development Programs

In 1928, Julius Rosenwald reorganized his eleven-year old family foundation into a more-corporate donation institution. Like its predecessor, the newly organized Rosenwald Fund would focus on issues related to blacks, particularly in health, urban and rural development, education, the development of intellectual talent, racial betterment associations, and social science research.[17]

Between 1928 and 1930, Rosenwald picked three men who would be the fund's chief decision makers during its twenty-year life-span: Edwin Embree, Charles S. Johnson, and Will W. Alexander. Rosenwald recruited Embree from the Rockefeller Foundation as the fund's first and only president; Johnson, a sociologist, from the National Urban League and Fisk University as the primary adviser on blacks and race relations; and Alexander, executive director of the Atlanta-based Commission on Interracial Cooperation, as the chief adviser on southern problems and issues.[18] Embree, Johnson, and Alexander formulated funding priorities that would gradually move away from Julius Rosenwald's approach to black and southern education. This shift would become particularly apparent after Rosenwald's death in 1932.

By the time Rosenwald died, the men he put in charge of his foundation grew weary of his approach. It would be phased out by the mid-1930s.[19] For one thing, the school program was becoming too expensive and impractical. More important, it was out of step with the times. Demographically, the period between world wars was a peculiar time in white-black relations that has yet to be fully appreciated by social scientists with a historical perspective. On the one hand, the great black migration North dramatically transformed the place of blacks in society and began to transform northern cities through the creation or expansion of urban black communities. With the urbanization of the black population came the development of new, more sophisticated forms of black leadership and intellectual life implanted in an emerging network of middle-class oriented nonprofit and business organizations. On the other hand, two-thirds of the black population still resided in the South, especially outside large cities, through the 1940s.

Between the wars, most foundations interested in the condition of blacks at all continued to support rural black education projects in the South and increasingly in Africa. But the chief decision makers of the Julius Rosenwald Fund went the other way. Embree, Johnson, and Alexander were future-oriented men who spent much time speculating where society was going.[20] As the years wore on, as they became more openly critical of Jim Crow and attempted to influence the direction of white-black relations. This was apparent in the evolution of their funding priorities in black education and the development of intellectual talent.

During the fund's planning period, Embree, Johnson, and Alexander established two funding priorities that reflected their respective interest in expanding the urban black middle class and black liberal arts education: the Fellowship Program for Negroes and the founding of Dillard University. The fellowship program's purpose was to give promising blacks opportunities to attend graduate school in the arts and sciences or in other ways attain additional advanced training. From 1930 to 1948, the fellowship committee awarded hundreds of fellowships to blacks directly or through the other agencies. Virtually every renowned black in the arts, humanities, and sciences who came of intellectual age during the 1930s and 1940s at least applied for a Rosenwald fellowship some time in his or her career. Not a few black scholars who became prominent in the 1950s and 1960s, were Julius Rosenwald fellows during their younger years.

During the 1940s, the Rosenwald fellowship program served as a form of creative philanthropy to stimulate the racial integration of white universities. Edwin Embree and his staff used lists of black Rosenwald fellows to quietly lobby white university administrators to offer regular and summer school faculty positions to black scholars.[21] The most noted case was the Rosenwald Fund's sponsorship of the career of anthropologist Allison Davis, their star fellow in the social sciences. Embree tracked Davis's career from Harvard graduate school through Dillard University and to the University of Chicago in the early 1940s. The Rosenwald Fund paid Davis's initial Chicago faculty salary.[22] But there were numerous, less publicized efforts of the Rosenwald Fund to promote the integration of their black fellows into white higher education. In so doing, the Rosenwald Fund deviated from the norm in race relations philanthropy that tended to adhere to status quo race relation traditions. Indeed, this creative effort was two decades ahead of what would be called affirmative action in higher education.

The major example of the Rosenwald Fund's interest in black liberal arts colleges was its pivotal role in founding Dillard University, named after James Hardy Dillard. Embree and his staff decided a liberal arts college was needed to serve the black population of the Deep South. They decided on New Orleans not only because of its location but also because it was the home of Edgar Stern. Stern, a brother-in-law of Julius Rosenwald and a trustee of the Rosenwald Fund, was a wealthy civic leader in New Orleans. He would become the most instrumental fund raiser and negotiator for the establishment of Dillard University and would serve as chairman of its governing board for many years.[23]

Rosenwald administrators had to wage two crucial political battles waged in the attempt to build a black liberal arts college in New Orleans: consolidating the resources of two black church-controlled colleges and siting a campus that would not offend the white community.[24] The first president was Will Alexander, Rosenwald Fund vice-president and executive director of the Commission on Interracial Cooperation. With Alexander as president and Stern as chairman of the board, Dillard University also drew generous GEB assistance. It also became renowned for the best managed black teaching hospital in the country, under the able hand of Rosenwald Fellow Albert Dent.[25]

The Rosenwald concept of Dillard as a regional black liberal arts college was a predecessor to a major effort in foundation funding of black education during the late 1930s and the 1940s. This was the regional center plan endorsed by Jackson Davis and others. An attempt to develop several magnet black colleges, the plan was a foundation response to Supreme Court decisions that began to pressure southern states to provide for the undergraduate and graduate educational needs of black residents. Rather than encouraging the integration of white universities and colleges with undergraduate programs, the GEB's Davis and other foundation officers advocated the development of regional black colleges to meet the growing educational demands of blacks in southern states. It was a stalling tactic that would eventually be discredited by the 1954 *Brown* decision and subsequent civil rights legislation.[26]

Era 3: The Post-World War II Decades

Since World War II, foundation intervention in black education has declined sharply. First, during the 1940s, the older, large foundations interested in race relations either shifted or curtailed their sponsorship of programs on black education issues and problems. The Julius Rosenwald Fund, the most progressive of them, was phased out in 1948. The absolute decline of foundation activity in the field of black education was paralleled by the retirement or death of "Negro experts" in the old foundation sector, who were the major proponents of black education and intellectual talent development, such as Davis, Embree, and Thomas Jesse Jones of the Phelps-Stokes Fund.

Second, the pre-World War II private foundation sponsored black schooling sector had adhered to Jim Crow norms and laws. In the 1950s and early 1960s, as the Civil Rights movement gained momentum in the legal system and through political protest, traditional foundation approaches to the education of blacks became both illegal and sociopolitically unacceptable. Indeed, there is much fascinating evidence that, given the depth of "Jim Crow thinking" about black education in the older foundations, the 1950s was a time of internal confusion, discord, and intellectual paralysis in their boardrooms over the implications of a racially desegregated society. Therefore, it is not surprising to find that the major foundations that would take the lead in racial desegregation issues emerging in the 1940s and early 1950s were newer ones, organized during the war years or in the early 1950s, such as the Field Foundation and the Ford Foundation.

Third, and perhaps most important, the emergence of significant central government intervention in education and race relations issues during the 1960s made extensive foundation sponsorship of black educational institutions less necessary. With the federal government becoming the major patron of black colleges and universities, the foundation role shifted general from extensive funding to more specific, program-project funding.

The change in foundation giving to black colleges and universities during the post-1960s was also due to transformations in how officers of major foundations designed and implemented grant programs and structured relationships with the beneficiaries. Before World War II, most major foundations appropriated general funds to particular black colleges and universities for years, if not decades, and developed highly personalized (patrimonial) ties with the beneficiaries.[27] Due to tax law reforms and changes in managerial style, major post-World War II foundations have tended to develop more circumscribed time limits on grant programs targeted to black colleges and universities. Relations with the beneficiaries are short lived and less patrimonial.[28] Still, foundations played crucial roles in the survival of black colleges and universities, particularly in subsidizing attempts to provide remedial curricula and high-tech training programs and to develop human resources.

Although in terms of subsidization and sociopolitical hegemony, foundation intervention in black education and intellectual development has declined during the last forty years, foundations have continued to play vital roles. During the 1950s, while the United States was beginning to grapple with the realities of a desegregating society, the Ford and Field Foundations financed experiments in desegregation and public relations campaigns designed to educate citizens about the advantages of racial desegregation. Foundation-financed nonprofit organizations such as the National Association against Discrimination in Housing, the Southern Regional Council, and the Southern Education Reporting Service played instrumental roles in educating the public and in research on the virtues of a racially desegregated society.

Perhaps the major contribution of post-World War II foundations in black affairs has been the development of intellectual talent. One of the most pressing problems in this period has been the scarce supply of intellectual talent for a mature industrial–high tech society. During the last forty years, government agencies, universities, foundations, and nonacademic research institutions have spent vast sums of money developing intellectual talent. For blacks, the issue of intellectual development has been intertwined with the intricate and sometimes contradictory matters of racial desegregation in white institutional sectors (for example, academia, the corporate world, media, law, medicine) and the maintenance of black institutional sectors. No matter the sector, the social and political dynamics of racism and structural poverty have mitigated against the development of an ample supply of talented blacks to take advantage of the employment and educational opportunities that have emerged through desegregation processes and policies.

Private foundation efforts to address the shortage of black intellectual talent can be divided into three periods, each with a distinct focus. The first period ranged from the early 1950s through the mid-1960s, during which private foundations concentrated on strengthening black colleges and developing a few career training programs for blacks. Working primarily through the United Negro College Fund, foundation officers believed that although racial desegregation was occurring, black colleges would remain the major sources of black academic talent for some time to come.[29] This gradualistic assumption resulted in foundation officers choosing to maintain the racial status quo—oriented black education programs rather than programs that would have pushed aggressively toward racial desegregation. The latter programming strategy would have meant funding projects to stimulate significant desegregation of administrations, faculties, and student bodies of white academic institutions and key decision making areas in government and private industry.[30]

During the mid-1960s, a confluence of forces began to slowly change foundation approaches to black issues. First were the external societal conditions: black protest movements, urban race riots, and liberal government policies. Second was the emergence of prominent foundation spokesmen, notably McGeorge Bundy, who boldly stood against racism and insisted that foundations fight against it.[31]

Most foundations continued to ignore race relations issues or gave only a minuscule percentage of their appropriations for race relations projects. But a few, particularly the Ford Foundation and Carnegie Corporation, launched major race-related programs. This was especially easy for Ford Foundation officers, since several Ford race-related programs were in operation when the race riots broke out.[32] The appointment of McGeorge Bundy as president of the Ford Foundation in the mid-1960s was a major factor behind Ford's leadership in race relations matters thereafter.

Perhaps the most important Ford Foundation contribution to black intellectual development from the mid-1960s through the mid-1970s, was their predoctoral fellowship program for blacks and other racial minority groups. At first, this program, which allowed talented blacks to attend prominent white graduate schools, was meant to meet the demand for teachers in black colleges.[33] But after a few years, it became apparent that the program had another effect, which became its major function—meeting the needs of a growing number of white universities that, due to the development of affirmative action policies, were searching for black administrators and faculty members.

The Ford Foundation's predoctoral program for minorities was crucial in providing financial assistance for the first significant wave of black scholars and academic administrators in white universities. The program was phased out in the late 1970s and then resumed in the early 1980s due to increasing concern over the declining number of blacks entering and completing graduate training.

During this second postwar period, there were also a few foundation attempts to desegregate white professional educational and occupational sectors: particularly law, medicine, and journalism. For example, the Josiah Macy, Jr., Foundation financed a small program between 1966 and 1971 to assist black students through white medical schools and residency programs.[34]

During the third and contemporary period, government agencies and foundations have increased emphasis on postdoctoral programs. This is a response to the shrinkage of the academic marketplace and the growth of specialization in academic disciplines. Given the high professional demands blacks encounter, whether employed in predominantly white or black institutions, two foundations have launched large-scale racial minority postdoctoral programs: the Rockefeller Foundation Research Program for Minority-Group Scholars and the Ford Foundation National Research Council's Minority Postdoctoral Fellowship Program. These programs are just beginning to be evaluated.

CONCLUSION

In the beginning of this chapter, we noted that with respect to race relations and black education private foundations historically tended to accommodate to rather than change the racial status quo. This was even the case for the Progressive Julius Rosenwald Fund; only during its last years did its officers begin to boldly attack Jim Crow public policies. They were several years behind more militant organizations such as the National Association for the Advancement of Colored People. That the historical norm has been for most foundations to ignore racial problems and issues altogether belies the conventional rhetoric that foundations are in the vanguard of social issues, that they are not averse to taking risks, and that their horizons are longer than other social institutions, which must respond to current pressures and demanding constituencies.

As argued earlier, the prevalent foundations tradition of ignoring racial matters or of making only symbolic gestures is an extension of the cultural discomfort most Americans feel about addressing racism, particularly institutional and political change-oriented solutions. Like most Americans, many foundation executives close their eyes, wishing that racism will magically disappear through symbolic efforts or pretend that racism does not really exist.

The history of foundation giving cannot be clearly divorced from the life histories of the givers: donors, foundation governing boards, and administrators. Therefore, to understand why foundation support for black education took the form it did, we must take into account the philosophies of race relations as well as the organizational resources of philanthropists and philanthropoids during various time periods. Certainly, more life-history analysis of the people in private foundations is sorely needed, so that we may leave a deeper understanding than impressionistic or abstract statistical studies have thus far provided of the role of the foundation in such complex fields as the development of black education and intellectual talent.

NOTES

1. Carter G. Woodson, *The Miseducation of the Negro* (Washington, D.C.: Associated Press, 1933); Dwight Oliver Wendell Holmes, *The Evolution of the Negro College,* (New York: Teachers College, Columbia University, 1934); Louis P. Harlem, *Separate and Unequal Public School Campaigns and Racism in the Southeast Seaboard States, 1901–1915* (Chapel Hill: University of North Carolina Press, 1958); Kenneth J. King, Pan-Africanism and Education (London: Oxford University Press, 1971); John H. Stanfield, "Philanthropic Regional Consciousness and Institution-Building in the American South: The Formative Years, 1867–1920," in Jack Salzman, ed. *Philanthropy and American Society: Selected Papers* (New York: Center for American Culture Studies, Columbia University, 1987).

2. John H. Stanfield, "Dollars for the Silent South," in Merle Black and John Shelton Reed, ed., *Perspectives on the American South* (New York: Gordon and Beach, 1984); John H. Stanfield, *Philanthropy and Jim Crow in American Social Science* (Westport, Conn.: Greenwood Press, 1985).

3. Human Resources Corporation, *U.S. Foundations and Minority Group Interests,* in U.S. Department of the Treasury, Commission on Private Philanthropy and Public Needs (Washington, D.C.: Government Printing Office, 1977), pp. 1206–11; Waldemar A. Nielsen, *The Big Foundations* (New York: Columbia University Press, 1972).

4. Human Resources Corporation, *U.S. Foundations and Minority Group Interests,* ibid., pp. 1201–1204; Thomas C. Reeves, *Foundations under Fire* (Ithaca, N.Y.: Cornell University Press, 1970), p. 98.

5. Nielsen, *Big Foundations;* Human Resources Corporation, ibid., pp. 1165–1282.

6. Human Resources Corporation, ibid., pp. 1165–1287.

7. Stanfield, "Philanthropic Regional Consciousness."

8. Ibid.

9. King, *Pan-Africanism and Education;* Stanfield, "Philanthropic Regional Consciousness."

10. King, ibid.

11. Arthur D. Wright, *The Negro Rural School Fund, Inc.—Anna T. Jeanes Foundation, 1907–1933* (Washington, D.C.: Negro Rural School Fund, 1933); Lance G. E. Jones, *The Jeanes Teacher in the United States 1908–1933* (Chapel Hill: University of North Carolina Press, 1937).

12. Jones, ibid., p. 16.

13. S. L. Smith, *Builders of Goodwill: The Story of the State Agents of Negro Education in the South 1910–1950* (Nashville: Tennessee Book Company, 1950), p. 7.

14. Ibid., p. 9.

15. Stanfield, "Dollars for the Silent South."

16. Smith, *Builders of Goodwill,* pp. 15, 29, 44, 66, 86, 88, 92, 119, 132.

17. Edwin R. Embree and Julia Waxman, *Investment in People: The Story of the Julius Rosenwald Fund* (New York: Harper and Brothers, 1949).

18. Ibid.

19. Ibid.

20. Ibid.

21. Edwin R. Embree Correspondence Files, Julius Rosenwald Fund Archives, Fisk University, Nashville, Tenn.

22. Ibid.

23. Edgar Stern Correspondence Files, Rosenwald Archives.

24. Dillard University Files, Rosenwald Archives.

25. Ibid.

26. Jackson Davis Diary, General Education Board Archives, Rockefeller Archive Center, Pocantico Hills, N.Y.

27. Wilma Dykeman and James Stokely, *Seeds of Southern Change: The Life of Will Alexander.* (Chicago: University of Chicago Press, 1962); Raymond Fosdick, *Adventure in Giving: The Story of the General Education Board* (New York: Harper and Row, 1962); King, *Pan-Africanism and Education;* Stanfield, *Philanthropy and Jim Crow.*

28. This observation is most applicable to the larger national and international foundations.

29. For example, *The Ford Foundation Annual Report,* 1964, p. 9.

30. Nielsen, *The Big Foundations.*

31. McGeorge Bundy, "The President's Review," *The Ford Foundation Annual Report,* 1967, pp. 2–6.

32. *The Ford Foundation Annual Reports,* 1960–1967.

33. Interview with Dr. Samuel Nabrit (December 1982).

34. John H. Stanfield, "The Josiah Macy, Jr. Foundation Postbaccalaureate Fellowship Program for Minorities. (Unpublished manuscript, 1985).

Index